Regression Analysis by Example
Third Edition

Regression Analysis By Example

Third Edition

SAMPRIT CHATTERJEE
New York University

ALI S. HADI
Cornell University

BERTRAM PRICE
Price Associates, Inc.

A Wiley-Interscience Publication
JOHN WILEY & SONS, INC.
New York • Chichester • Weinheim • Brisbane • Singapore • Toronto

For ordering and customer service, call 1-800-CALL-WILEY.

Library of Congress Cataloging in Publication Data is available.

ISBN 0-471-31946-5

Printed in the United States of America

10 9 8 7

Dedicated to:

Allegra, Martha, and Rima – S. C.

My mother and the memory of my father – A. S. H.

Ann and Anne – B. P.

It's a gift to be simple . . .

Old Shaker hymn

True knowledge is knowledge of why things are
as they are, and not merely what they are.

Isaiah Berlin

Contents

Preface

Regression analysis has become one of the most widely used statistical tools for analyzing multifactor data. It is appealing because it provides a conceptually simple method for investigating functional relationships among variables. The standard approach in regression analysis is to take data, fit a model, and then evaluate the fit using statistics such as t, F, and R^2. Our approach is much broader. We view regression analysis as a set of data analytic techniques that examine the interrelationships among a given set of variables. The emphasis is not on formal statistical tests and probability calculations. We argue for an informal analysis directed towards uncovering patterns in the data.

We utilize most standard and some not so standard summary statistics on the basis of their intuitive appeal. We rely heavily on graphical representations of the data, and employ many variations of plots of regression residuals. We are not overly concerned with precise probability evaluations. Graphical methods for exploring residuals can suggest model deficiencies or point to troublesome observations. Upon further investigation into their origin, the troublesome observations often turn out to be more informative than the well-behaved observations. We notice often that more information is obtained from a quick examination of a plot of residuals than from a formal test of statistical significance of some limited null-hypothesis. In short, the presentation in the chapters of this book is guided by the principles and concepts of exploratory data analysis.

Our presentation of the various concepts and techniques of regression analysis relies on carefully developed examples. In each example, we have isolated one or two techniques and discussed them in some detail. The data were chosen to highlight the techniques being presented. Although when analyzing a given set of data it is usually necessary to employ many techniques, we have tried to choose the various data sets so that it would not be necessary to discuss the same technique more than once. Our hope is that after working through the book, the reader will be ready and able to analyze his/her data methodically, thoroughly, and confidently.

The emphasis in this book is on the analysis of data rather than on formulas, tests of hypotheses, or confidence intervals. Therefore no attempt has been made to derive the techniques. Techniques are described, the required assumptions are given, and finally, the success of the technique in the particular example is assessed. Although derivations of the techniques are not included, we have tried to refer the reader in each case to sources in which such discussion is available. Our hope is that some of these sources will be followed up by the reader who wants a more thorough grounding in theory.

We have taken for granted the availability of a computer and a statistical package. Recently there has been a qualitative change in the analysis of linear models, from model fitting to model building, from overall tests to clinical examinations of data, from macroscopic to the microscopic analysis. To do this kind of analysis a computer is essential and we have assumed its availability. Almost all of the analyses we use are now available in software packages.

The material presented is intended for anyone who is involved in analyzing data. The book should be helpful to those who have some knowledge of the basic concepts of statistics. In the university, it could be used as a text for a course on regression analysis for students whose specialization is not statistics, but, who nevertheless, use regression analysis quite extensively in their work. For students whose major emphasis is statistics, and who take a course on regression analysis from a book at the level of Rao (1973), Seber (1977), or Sen and Srivastava (1990), this book can be used to balance and complement the theoretical aspects of the subject with practical applications. Outside the university, this book can be profitably used by those people whose present approach to analyzing multifactor data consists of looking at standard computer output $(t, F, R^2$, standard errors, etc.), but who want to go beyond these summaries for a more thorough analysis.

The book has a Web site: http://www.ilr.cornell.edu/~hadi/RABE. This Web site contains, among other things, all the data sets that are included in this book and more. These and other data sets can be found in the book's Web site.

This is not just another edition of the book; it is a major rewriting and reorganization of the previous edition. Major changes have been made in the third edition of the book. The book now has twelve chapters instead of the earlier nine. Every major topic has its own chapter. New material has been added which reflects the advances in the field. Some of the examples have been replaced by new ones and additional new examples have been included.

The graphics have also been updated to reflect the vast strides made in this area. Attention is now drawn to major changes made in this edition. The first chapter is new and is an introduction to the subject of regression analysis. The material is descriptive and it draws attention to the richness of possible applications. Chapter 2 has been extensively rewritten and it treats simple regression. Concepts of covariance and correlation are discussed for completeness even though we expect most readers to be familiar with the concepts. Chapter 3 introduces multiple regression. Regression diagnostics is given particular attention in the new Chapter 4, reflecting the new importance of this topic. Transformation of variables in regression analysis is now treated in a separate chapter (Chapter 6). Weighted least squares is also given its own chapter (Chapter 7) to reflect its importance.

Although new material is added to Chapters 8–11 the general arrangement remains the same as in the earlier editions. Chapter 12 on logistic regression is new. In the earlier editions logistic regression was given a cursory treatment in the chapter on weighted least squares. Logistic regression is now being widely used in statistical analysis. To reflect this we have devoted a chapter to it. Some comments are made in that chapter on discriminant analysis, an alternative tool for the classification problem. In response to repeated requests by teachers who have used the book for exercises at the end of each chapter we provide them. We hope they prove challenging. More exercises are available on the book's Web site and they will be periodically updated.

We have attempted to write a book for a group of readers with diverse backgrounds. We have also tried to put emphasis on the art of data analysis rather than on the development of statistical theory.

We are fortunate to have had assistance and encouragement from several friends, colleagues, and associates. Some of our colleagues at New York University and Cornell University have used portions of the material in their courses and have shared with us their comments and comments of their students. The students in our classes on regression analysis have all contributed by asking penetrating questions and demanding meaningful and understandable answers. Our special thanks go to Nedret Billor (Cukurova University, Turkey), for the invaluable feedback she provided on an earlier draft of this edition, and to Sahar El-Sheneity (Cornell University), for her very careful reading of the manuscript.

SAMPRIT CHATTERJEE
ALI S. HADI

Brooksville, Maine
Ithaca, New York
September 1999

1

Introduction

1.1 WHAT IS REGRESSION ANALYSIS?

Regression analysis is a conceptually simple method for investigating functional relationships among variables. A real estate appraiser may wish to relate the sale price of a home from selected physical characteristics of the building and taxes (local, school, county) paid on the building. We may wish to examine whether cigarette consumption is related to various socioeconomic and demographic variables such as age, education, income, and price of cigarette. The relationship is expressed in the form of an equation or a model connecting the *response* or *dependent* variable and one or more *explanatory* or *predictor* variables. In the cigarette consumption example, the response variable is cigarette consumption (measured by the number of packs of cigarette sold in a given state on a per capita basis during a given year) and the explanatory or predictor variables are the various socioeconomic and demographic variables. In the real estate appraisal example, the response variable is the price of a home and the explanatory or predictor variables are the characteristics of the building and taxes paid on the building.

We denote the response variable by Y and the set of predictor variables by $X_1, X_2, \ldots, X_p$, where p denotes the number of predictor variables. The true relationship between Y and $X_1, X_2, \ldots, X_p$ can be approximated by the regression model

$$Y = f(X_1, X_2, \ldots, X_p) + \varepsilon, \tag{1.1}$$

where ε is assumed to be a random error representing the discrepancy in the approximation. It accounts for the failure of the model to fit the data exactly.

The function $f(X_1, X_2, \ldots, X_p)$ describes the relationship between Y and X_1, $X_2, \ldots, X_p$. An example is the linear regression model

$$Y = \beta_0 + \beta_1 X_1 + \beta_2 X_2 + \cdots + \beta_p X_p + \varepsilon, \tag{1.2}$$

where $\beta_0, \beta_1, \ldots, \beta_p$, called the *regression parameters* or *coefficients*, are unknown constants to be determined (estimated) from the data. We follow the commonly used notational convention of denoting unknown parameters by Greek letters.

The predictor or explanatory variables are also called by other names such as *independent* variables, *covariates*, *regressors*, *factors*, and *carriers*. The name independent variable, though commonly used, is the least preferred, because in practice the predictor variables are rarely independent of each other.

1.2   PUBLICLY AVAILABLE DATA SETS

Regression analysis has numerous areas of applications. A partial list would include economics, finance, business, law, meteorology, medicine, biology, chemistry, engineering, physics, education, sports, history, sociology, and psychology. A few examples of such applications are given in Section 1.3. Regression analysis is learned most effectively by analyzing data that are of direct interest to the reader. We invite the readers to think about questions (in their own areas of work, research, or interest) that can be addressed using regression analysis. Readers should collect the relevant data and then apply the regression analysis techniques presented in this book to their own data. To help the reader locate real-life data, this section provides some sources and links to a wealth of data sets that are available for public use.

A number of datasets are available in books and on the Internet. The book by Hand et al. (1994) contains data sets from many fields. These data sets are small in size and are suitable for use as exercises. The book by Chatterjee, Handcock, and Simonoff (1995) provides numerous data sets from diverse fields. The data are included in a diskette that comes with the book and can also be found in the World Wide Web site.[1]

Data sets are also available on the Internet at many other sites. Some of the Web sites given below allow the direct copying and pasting into the statistical package of choice, while others require downloading the data file and then importing them into a statistical package. Some of these sites also contain further links to yet other data sets or statistics-related Web sites.

The Data and Story Library (DASL, pronounced "dazzle") is one of the most interesting sites that contains a number of data sets accompanied by the "story" or background associated with each data set. DASL is an online

---

[1] http://www.stern.nyu.edu/~ jsimonof/Casebook.

library[2] of data files and stories that illustrate the use of basic statistical methods. The data sets cover a wide variety of topics. DASL comes with a powerful search engine to locate the story or data file of interest.

Another Web site, which also contains data sets arranged by the method used in the analysis, is the Electronic Dataset Service.[3] The site also contains many links to other data sources on the Internet.

Finally, this book has a Web site: http://www.ilr.cornell.edu/~hadi/RABE. This site contains, among other things, all the data sets that are included in this book and more. These and other data sets can be found in the book's Web site.

1.3 SELECTED APPLICATIONS OF REGRESSION ANALYSIS

Regression analysis is one of the most widely used statistical tools because it provides simple methods for establishing a functional relationship among variables. It has extensive applications in many subject areas. The cigarette consumption and the real estate appraisal, mentioned above, are but two examples. In this section, we give a few additional examples demonstrating the wide applicability of regression analysis in real-life situations. Some of the data sets described here will be used later in the book to illustrate regression techniques or in the exercises at the end of various chapters.

1.3.1 Agricultural Sciences

The Dairy Herd Improvement Cooperative (DHI) in Upstate New York collects and analyzes data on milk production. One question of interest here is how to develop a suitable model to predict current milk production from a set of measured variables. The response variable (current milk production in pounds) and the predictor variables are given in Table 1.1. Samples are taken once a month during milking. The period that a cow gives milk is called lactation. Number of lactations is the number of times a cow has calved or given milk. The recommended management practice is to have the cow produce milk for about 305 days and then allow a 60-day rest period before beginning the next lactation. The data set, consisting of 199 observations, was compiled from the DHI milk production records. The Milk Production data can be also be found in the book's Web site.

---

[2]DASL's Web site is: http://lib.stat.cmu.edu/DASL/.
[3]http://www-unix.oit.umass.edu/~statdata/.

Table 1.1   Variables for the Milk Production Data.

Variable	Definition
Current	Current month milk production in pounds
Previous	Previous month milk production in pounds
Fat	Percent of fat in milk
Protein	Percent of protein in milk
Days	Number of days since present lactation
Lactation	Number of lactations
I79	Indicator variable (0 if $Days \leq 79$ and 1 if $Days > 79$)

Table 1.2   Variables for the Right-To-Work Laws Data.

Variable	Definition
COL	Cost of living for a four-person family
PD	Population density (person per square mile)
URate	State unionization rate in 1978
Pop	Population in 1975
Taxes	Property taxes in 1972
Income	Per capita income in 1974
RTWL	Indicator variable (1 if there is right-to-work laws in the state and 0 otherwise)

1.3.2   Industrial and Labor Relations

In 1947, the United States Congress passed the Taft-Hartley Amendments to the Wagner Act. The original Wagner Act had permitted the unions to use a *Closed Shop Contract*[4] unless prohibited by state law. The Taft-Hartley Amendments made the use of Closed Shop Contract illegal and gave individual states the right to prohibit union shops[5] as well. These right-to-work laws have caused a wave of concern throughout the labor movement. A question of interest here is: What are the effects of these laws on the cost of living for a four-person family living on an intermediate budget in the United States? To answer this question a data set consisting of 38 geographic locations has been assembled from various sources. The variables used are defined in Table 1.2. The Right-To-Work Laws data are given in Table 1.3 and can also be found in the book's Web site.

---

[4]Under a Closed Shop Contract provision, all employees must be union members at the time of hire and must remain members as a condition of employment.
[5]Under a Union Shop clause, employees are not required to be union members at the time of hire, but must become a member within two months, thus allowing the employer complete discretion in hiring decisions.

Table 1.3   The Right-To-Work Laws Data.

City	COL	PD	URate	Pop	Taxes	Income	RTWL
Atlanta	169	414	13.6	1790128	5128	2961	1
Austin	143	239	11	396891	4303	1711	1
Bakersfield	339	43	23.7	349874	4166	2122	0
Baltimore	173	951	21	2147850	5001	4654	0
Baton Rouge	99	255	16	411725	3965	1620	1
Boston	363	1257	24.4	3914071	4928	5634	0
Buffalo	253	834	39.2	1326848	4471	7213	0
Champaign-Urbana	117	162	31.5	162304	4813	5535	0
Cedar Rapids	294	229	18.2	164145	4839	7224	1
Chicago	291	1886	31.5	7015251	5408	6113	0
Cincinnati	170	643	29.5	1381196	4637	4806	0
Cleveland	239	1295	29.5	1966725	5138	6432	0
Dallas	174	302	11	2527224	4923	2363	1
Dayton	183	489	29.5	835708	4787	5606	0
Denver	227	304	15.2	1413318	5386	5982	0
Detriot	255	1130	34.6	4424382	5246	6275	0
Green Bay	249	323	27.8	169467	4289	8214	0
Hartford	326	696	21.9	1062565	5134	6235	0
Houston	194	337	11	2286247	5084	1278	1
Indianapolis	251	371	29.3	1138753	4837	5699	0
Kansas City	201	386	30	1290110	5052	4868	0
Lancaster, PA	124	362	34.2	342797	4377	5205	0
Los Angeles	340	1717	23.7	6986898	5281	1349	0
Milwaukee	328	968	27.8	1409363	5176	7635	0
Minneapolis, St. Paul	265	433	24.4	2010841	5206	8392	0
Nashville	120	183	17.7	748493	4454	3578	1
New York	323	6908	39.2	9561089	5260	4862	0
Orlando	117	230	11.7	582664	4613	782	1
Philadelphia	182	1353	34.2	4807001	4877	5144	0
Pittsburgh	169	762	34.2	2322224	4677	5987	0
Portland	267	201	23.1	228417	4123	7511	0
St. Louis	184	480	30	2366542	4721	4809	0
San Diego	256	372	23.7	1584583	4837	1458	0
San Francisco	381	1266	23.7	3140306	5940	3015	0
Seattle	195	333	33.1	1406746	5416	4424	0
Washington	205	1073	21	3021801	6404	4224	0
Wichita	206	157	12.8	384920	4796	4620	1
Raleigh-Durham	126	302	6.5	468512	4614	3393	1

Table 1.4   Variables for the Egyptian Skulls Data.

Variable	Definition
Year	Approximate Year of Skull Formation (negative = B.C.; positive = A.D.)
MB	Maximum Breadth of Skull
BH	Basibregmatic Height of Skull
BL	Basialveolar Length of Skull
NH	Nasal Height of Skull

1.3.3   History

A question of historical interest is how to estimate the age of historical objects based on some age-related characteristics of the objects. For example, the variables in Table 1.4 can be used to estimate the age of Egyptian skulls. Here the response variable is Year and the other four variables are possible predictors. The original source of the data is Thomson and Randall-Maciver (1905), but they can be found in Hand et al. (1994), pp. 299–301. An analysis of the data can be found in Manly (1986). The Egyptian Skulls data can be found in the book's Web site.

1.3.4   Government

Information about domestic immigration (the movement of people from one state or area of a country to another) is important to state and local governments. It is of interest to build a model that predicts domestic immigration or to answer the question of why do people leave one place to go to another? There are many factors that influence domestic immigration, such as weather conditions, crime, tax, and unemployment rates. A data set for the 48 contiguous states has been created. Alaska and Hawaii are excluded from the analysis because the environments of these states are significantly different from the other 48, and their locations present certain barriers to immigration. The response variable here is net domestic immigration, which represents the net movement of people into and out of a state over the period 1990–1994 divided by the population of the state. Eleven predictor variables thought to influence domestic immigration are defined in Table 1.5. The data are given in Tables 1.6 and 1.7, and can also be found in the book's Web site.

1.3.5   Environmental Sciences

In a 1976 study exploring the relationship between water quality and land use, Haith (1976) obtained the measurements (shown in Table 1.8) on 20 river basins in New York State. A question of interest here is how the land use around a river basin contributes to the water pollution as measured by

Table 1.5   Variables for the Study of Domestic Immigration.

Variable	Definition
State	State name
NDIR	Net domestic immigration rate over the period 1990–1994
Unemp	Unemployment rate in the civilian labor force in 1994
Wage	Average hourly earnings of production workers in manufacturing in 1994
Crime	Violent crime rate per 100,000 people in 1993
Income	Median household income in 1994
Metrop	Percentage of state population living in metropolitan areas in 1992
Poor	Percentage of population who fall below the poverty level in 1994
Taxes	Total state and local taxes per capita in 1993
Educ	Percentage of population 25 years or older who have a high school degree or higher in 1990
BusFail	The number of business failures divided by the population of the state in 1993
Temp	Average of the 12 monthly average temperatures (in degrees Fahrenheit) for the state in 1993
Region	Region in which the state is located (northeast, south, midwest, west)

the mean nitrogen concentration (mg/liter). The data are shown in Table 1.9 and can be found in the book's Web site.

1.4   STEPS IN REGRESSION ANALYSIS

Regression analysis includes the following steps:

- Statement of the problem

- Selection of potentially relevant variables

- Data collection

- Model specification

- Choice of fitting method

- Model fitting

- Model validation and criticism

- Using the chosen model(s) for the solution of the posed problem.

These steps are examined below.

Table 1.6   First Six Variables of the Domestic Immigration Data.

State	NDIR	Unemp	Wage	Crime	Income	Metrop
Alabama	17.47	6.0	10.75	780	27196	67.4
Arizona	49.60	6.4	11.17	715	31293	84.7
Arkansas	23.62	5.3	9.65	593	25565	44.7
California	−37.21	8.6	12.44	1078	35331	96.7
Colorado	53.17	4.2	12.27	567	37833	81.8
Connecticut	−38.41	5.6	13.53	456	41097	95.7
Delaware	22.43	4.9	13.90	686	35873	82.7
Florida	39.73	6.6	9.97	1206	29294	93.0
Georgia	39.24	5.2	10.35	723	31467	67.7
Idaho	71.41	5.6	11.88	282	31536	30.0
Illinois	−20.87	5.7	12.26	960	35081	84.0
Indiana	9.04	4.9	13.56	489	27858	71.6
Iowa	0.00	3.7	12.47	326	33079	43.8
Kansas	−1.25	5.3	12.14	469	28322	54.6
Kentucky	13.44	5.4	11.82	463	26595	48.5
Louisiana	−13.94	8.0	13.13	1062	25676	75.0
Maine	−9.770	7.4	11.68	126	30316	35.7
Maryland	−1.55	5.1	13.15	998	39198	92.8
Massachusetts	−30.46	6.0	12.59	805	40500	96.2
Michigan	−13.19	5.9	16.13	792	35284	82.7
Minnesota	9.46	4.0	12.60	327	33644	69.3
Mississippi	5.33	6.6	9.40	434	25400	34.6
Missouri	6.97	4.9	11.78	744	30190	68.3
Montana	41.50	5.1	12.50	178	27631	24.0
Nebraska	−0.62	2.9	10.94	339	31794	50.6
Nevada	128.52	6.2	11.83	875	35871	84.8
New Hampshire	−8.72	4.6	11.73	138	35245	59.4
New Jersey	−24.90	6.8	13.38	627	42280	100.0
New Mexico	29.05	6.3	10.14	930	26905	56.0
New York	−45.46	6.9	12.19	1074	31899	91.7
North Carolina	29.46	4.4	10.19	679	30114	66.3
North Dakota	−26.47	3.9	10.19	82	28278	41.6
Ohio	−3.27	5.5	14.38	504	31855	81.3
Oklahoma	7.37	5.8	11.41	635	26991	60.1
Oregon	49.63	5.4	12.31	503	31456	70.0
Pennsylvania	−4.30	6.2	12.49	418	32066	84.8
Rhode Island	−35.32	7.1	10.35	402	31928	93.6
South Carolina	11.88	6.3	9.99	1023	29846	69.8
South Dakota	13.71	3.3	9.19	208	29733	32.6
Tennessee	32.11	4.8	10.51	766	28639	67.7
Texas	13.00	6.4	11.14	762	30775	83.9
Utah	31.25	3.7	11.26	301	35716	77.5
Vermont	3.94	4.7	11.54	114	35802	27.0
Virginia	6.94	4.9	11.25	372	37647	77.5
Washington	44.66	6.4	14.42	515	33533	83.0
West Virginia	10.75	8.9	12.60	208	23564	41.8
Wisconsin	11.73	4.7	12.41	264	35388	68.1
Wyoming	11.95	5.3	11.81	286	33140	29.7

Table 1.7   Last Six Variables of the Domestic Immigration Data.

State	Poor	Taxes	Educ	BusFail	Temp	Region
Alabama	16.4	1553	66.9	0.20	62.77	South
Arizona	15.9	2122	78.7	0.51	61.09	West
Arkansas	15.3	1590	66.3	0.08	59.57	South
California	17.9	2396	76.2	0.63	59.25	West
Colorado	9.0	2092	84.4	0.42	43.43	West
Connecticut	10.8	3334	79.2	0.33	48.63	Northeast
Delaware	8.3	2336	77.5	0.19	54.58	South
Florida	14.9	2048	74.4	0.36	70.64	South
Georgia	14.0	1999	70.9	0.33	63.54	South
Idaho	12.0	1916	79.7	0.31	42.35	West
Illinois	12.4	2332	76.2	0.18	50.98	Midwest
Indiana	13.7	1919	75.6	0.19	50.88	Midwest
Iowa	10.7	2200	80.1	0.18	45.83	Midwest
Kansas	14.9	2126	81.3	0.42	52.03	Midwest
Kentucky	18.5	1816	64.6	0.22	55.36	South
Louisiana	25.7	1685	68.3	0.15	65.91	South
Maine	9.4	2281	78.8	0.31	40.23	Northeast
Maryland	10.7	2565	78.4	0.31	54.04	South
Massachusetts	9.7	2664	80.0	0.45	47.35	Northeast
Michigan	14.1	2371	76.8	0.27	43.68	Midwest
Minnesota	11.7	2673	82.4	0.20	39.30	Midwest
Mississippi	19.9	1535	64.3	0.12	63.18	South
Missouri	15.6	1721	73.9	0.23	53.41	Midwest
Montana	11.5	1853	81.0	0.20	40.40	West
Nebraska	8.8	2128	81.8	0.25	46.01	Midwest
Nevada	11.1	2289	78.8	0.39	48.23	West
New Hampshire	7.7	2305	82.2	0.54	43.53	Northeast
New Jersey	9.2	3051	76.7	0.36	52.72	Northeast
New Mexico	21.1	2131	75.1	0.27	53.37	Midwest
New York	17.0	3655	74.8	0.38	44.85	Northeast
North Carolina	14.2	1975	70.0	0.17	59.36	South
North Dakota	10.4	1986	76.7	0.23	38.53	Midwest
Ohio	14.1	2059	75.7	0.19	50.87	Midwest
Oklahoma	16.7	1777	74.6	0.44	58.36	South
Oregon	11.8	2169	81.5	0.31	46.55	West
Pennsylvania	12.5	2260	74.7	0.26	49.01	Northeast
Rhode Island	10.3	2405	72.0	0.35	49.99	Northeast
South Carolina	13.8	1736	68.3	0.11	62.53	South
South Dakota	14.5	1668	77.1	0.24	42.89	Midwest
Tennessee	14.6	1684	67.1	0.23	57.75	South
Texas	19.1	1932	72.1	0.39	64.40	South
Utah	8.0	1806	85.1	0.18	46.32	West
Vermont	7.6	2379	80.8	0.30	42.46	Northeast
Virginia	10.7	2073	75.2	0.27	55.55	South
Washington	11.7	2433	83.8	0.38	46.93	Midwest
West Virginia	18.6	1752	66.0	0.17	52.25	South
Wisconsin	9.0	2524	78.6	0.24	42.20	Midwest
Wyoming	9.3	2295	83.0	0.19	43.68	West

Table 1.8 Variables for Study of Water Pollution in New York Rivers.

Variable	Definition
Y	Mean nitrogen concentration (mg/liter) based on samples taken at regular intervals during the spring, summer, and fall months
X_1	Agriculture: percentage of land area currently in agricultural use
X_2	Forest: percentage of forest land
X_3	Residential: percentage of land area in residential use
X_4	Commercial/Industrial: percentage of land area in either commercial or industrial use

Table 1.9 The New York Rivers Data.

Row	River	Y	X_1	X_2	X_3	X_4
1	Olean	1.10	26	63	1.2	0.29
2	Cassadaga	1.01	29	57	0.7	0.09
3	Oatka	1.90	54	26	1.8	0.58
4	Neversink	1.00	2	84	1.9	1.98
5	Hackensack	1.99	3	27	29.4	3.11
6	Wappinger	1.42	19	61	3.4	0.56
7	Fishkill	2.04	16	60	5.6	1.11
8	Honeoye	1.65	40	43	1.3	0.24
9	Susquehanna	1.01	28	62	1.1	0.15
10	Chenango	1.21	26	60	0.9	0.23
11	Tioughnioga	1.33	26	53	0.9	0.18
12	West Canada	0.75	15	75	0.7	0.16
13	East Canada	0.73	6	84	0.5	0.12
14	Saranac	0.80	3	81	0.8	0.35
15	Ausable	0.76	2	89	0.7	0.35
16	Black	0.87	6	82	0.5	0.15
17	Schoharie	0.80	22	70	0.9	0.22
18	Raquette	0.87	4	75	0.4	0.18
19	Oswegatchie	0.66	21	56	0.5	0.13
20	Cohocton	1.25	40	49	1.1	0.13

1.4.1 Statement of the Problem

Regression analysis usually starts with a formulation of the problem. This includes the determination of the question(s) to be addressed by the analysis. The problem statement is the first and perhaps the most important step in regression analysis. It is important because an ill-defined problem or a misformulated question can lead to wasted effort. It can lead to the selection of irrelevant set of variables or to a wrong choice of the statistical method of analysis. A question that is not carefully formulated can also lead to the wrong choice of a model. Suppose we wish to determine whether or not an employer is discriminating against a given group of employees, say women. Data on salary, qualifications, and sex are available from the company's record to address the issue of discrimination. There are several definitions of employment discrimination in the literature. For example, discrimination occurs when on the average (a) women are paid less than equally qualified men, or (b) women are more qualified than equally paid men. To answer the question: "On the average, are women paid less than equally qualified men?" we choose salary as a response variable, and qualification and sex as predictor variables. But to answer the question: "On the average, are women more qualified than equally paid men?" we choose qualification as a response variable and salary and sex as predictor variables, that is, the roles of variables have been switched.

1.4.2 Selection of Potentially Relevant Variables

The next step after the statement of the problem is to select a set of variables that are thought by the experts in the area of study to explain or predict the response variable. The response variable is denoted by Y and the explanatory or predictor variables are denoted by $X_1, X_2, \ldots, X_p$, where p denotes the number of predictor variables. An example of a response variable is the price of a single family house in a given geographical area. A possible relevant set of predictor variables in this case is: area of the lot, area of the house, age of the house, number of bedrooms, number of bathrooms, type of neighborhood, style of the house, amount of real estate taxes, etc.

1.4.3 Data Collection

The next step after the selection of potentially relevant variables is to collect the data from the environment under study to be used in the analysis. Sometimes the data are collected in a controlled setting so that factors that are not of primary interest can be held constant. More often the data are collected under nonexperimental conditions where very little can be controlled by the investigator. In either case, the collected data consist of observations on n units. Each of these n observations consists of measurements for each of the potentially relevant variables. The data are usually recorded as in Table 1.10. A column in Table 1.10 represents a variable, whereas a row represents an

Table 1.10   Notation for the Data Used in Regression Analysis.

Observation Number	Response Y	X_1	X_2	...	X_p
1	y_1	x_{11}	x_{12}	...	x_{1p}
2	y_2	x_{21}	x_{22}	...	x_{2p}
3	y_3	x_{31}	x_{32}	...	x_{3p}
⋮	⋮	⋮	⋮	⋮	⋮
n	y_n	x_{n1}	x_{n2}	...	x_{np}

observation, which is a set of $p+1$ values for a single unit (e.g., a house); one value for the response variable and one value for each of the p predictors. The notation x_{ij} refers to the ith value of the jth variable. The first subscript refers to observation number and the second refers to variable number.

Each of the variables in Table 1.10 can be classified as either *quantitative* or *qualitative*. Examples of quantitative variables are the house price, number of bedrooms, age, and taxes. Examples of qualitative variables are neighborhood type (e.g., good or bad neighborhood) and house style (e.g., ranch, colonial, etc.). In this book we deal mainly with the cases where the response variable is quantitative. A technique used in cases where the response variable is *binary*[6] is called *logistic regression*. This is introduced in Chapter 12. In regression analysis, the predictor variables can be either quantitative and/or qualitative. For the purpose of computations, however, the qualitative variables, if any, have to be coded into a set of *indicator* or *dummy* variables as discussed in Chapter 5.

If all predictor variables are qualitative, the techniques used in the analysis of the data are called the *analysis of variance* techniques. Although the analysis of variance techniques can be introduced and explained as methods in their own right[7], it is shown in Chapter 5 that they are special cases of regression analysis. If some of the predictor variables are quantitative while others are qualitative, regression analysis in these cases is called the *analysis of covariance*.

1.4.4   Model Specification

The form of the model that is thought to relate the response variable to the set of predictor variables can be specified initially by the experts in the area of study based on their knowledge or their objective and/or subjective

---

[6] A variable that can take only one of two possible values such as yes or no, 1 or 0, and success or failure, is called a binary variable

[7] See, for example, the books by Scheffé (1959), Iversen (1976), Wildt and Ahtola (1978), Krishnaiah (1980), Iversen and Norpoth (1987), Lindman (1992), and Christensen (1996)

judgments. The hypothesized model can then be either confirmed or refuted by the analysis of the collected data. Note that the model need to be specified only in form, but it can still depend on unknown parameters. We need to select the form of the function $f(X_1, X_2, \ldots, X_p)$ in (1.1). This function can be classified into two types: *linear* and *nonlinear*. An example of a linear function is

$$Y = \beta_0 + \beta_1 X_1 + \varepsilon \tag{1.3}$$

while a nonlinear function is

$$Y = \beta_0 + e^{\beta_1 X_1} + \varepsilon. \tag{1.4}$$

Note that the term *linear* (*nonlinear*) here does not describe the relationship between Y and $X_1, X_2, \ldots, X_p$. It is related to the fact that the regression parameters enter the equation linearly (nonlinearly). Each of the following models are linear

$$
\begin{aligned}
Y &= \beta_0 + \beta_1 X + \beta_2 X^2 + \varepsilon, \\
Y &= \beta_0 + \beta_1 \log X + \varepsilon,
\end{aligned}
$$

because in each case the parameters enter linearly although the relationship between Y and X is nonlinear. This can be seen if the two models are re-expressed, respectively, as follows:

$$
\begin{aligned}
Y &= \beta_0 + \beta_1 X_1 + \beta_2 X_2 + \varepsilon, \\
Y &= \beta_0 + \beta_1 X_1 + \varepsilon,
\end{aligned}
$$

where in the first equation we have $X_1 = X$ and $X_2 = X^2$ and in the second equation we have $X_1 = \log X$. The variables here are *re-expressed* or *transformed*. Transformation is dealt with in Chapter 6. All nonlinear functions that can be transformed into linear functions are called *linearizable* functions. Accordingly, the class of linear models is actually wider than it might appear at first sight because it includes all linearizable functions. Note, however, that not all nonlinear functions are linearizable. For example, it is not possible to linearize the nonlinear function in (1.4). Some authors refer to nonlinear functions that are not linearizable as *intrinsically* nonlinear functions.

A regression equation containing only one predictor variable is called a *simple regression equation*. An equation containing more than one predictor variable is called a *multiple regression equation*. An example of simple regression would be an analysis in which the time to repair a machine is studied in relation to the number of components to be repaired. Here we have one response variable (time to repair the machine) and one predictor variable (number of components to be repaired). An example of a very complex multiple regression situation would be an attempt to explain the age-adjusted mortality rates prevailing in different geographic regions (response variable) by a large number of environmental and socioeconomic factors (predictor variables). Both types of problems are treated in this book. These two particular examples are studied, one in Chapter 2, the other in Chapter 11.

In certain applications the response variable can actually be a set of variables, $Y_1, Y_2, \ldots, Y_q$, say, which are thought to be related to the same set of predictor variables, $X_1, X_2, \ldots, X_p$. For example, Bartlett, Stewart, and Abrahamowicz (1998) present a data set on 148 healthy people. Eleven variables are measured; six variables represent different types of measured sensory thresholds (e.g., vibration, hand and foot temperatures) and five a priori selected baseline covariates (e.g., age, sex, height, and weight) that may have systematic effects on some or all of the six sensory thresholds. Here we have six response variables and five predictor variables. This data set, which we refer to as the QST (*quantitative sensory testing*) data, is not listed here due to its size (148 observations) but it can be found in the book's Web site. For further description of the data and objectives of the study, see Bartlett, Stewart, and Abrahamowicz (1998).

When we deal only with one response variable, regression analysis is called *univariate* regression and in cases where we have two or more response variables, the regression is called *multivariate* regression. Simple and multiple regressions should not be confused with univariate versus multivariate regressions. The distinction between simple and multiple regressions is determined by the number of predictor variables (simple means one predictor variable and multiple means two or more predictor variables), whereas the distinction between univariate and multivariate regressions is determined by the number of response variables (univariate means one response variable and multivariate means two or more response variables). In this book we consider only univariate regression (both simple and multiple, linear and nonlinear). Multivariate regression is treated in books on multivariate analysis such as Rencher (1995), Johnson and Wichern (1992), and Johnson (1998). In this book the term regression will be used to mean univariate regression.

The various classifications of regression analysis we discussed above are shown in Table 1.11.

1.4.5 Method of Fitting

After the model has been defined and the data have been collected, the next task is to estimate the parameters of the model based on the collected data. This is also referred to as *parameter estimation* or *model fitting*. The most commonly used method of estimation is called the *least squares* method. Under certain assumptions (to be discussed in detail in this book), least squares method produce estimators with desirable properties. In this book we will deal mainly with least squares method and its variants (e.g., weighted least squares). In some instances (e.g., when one or more of the assumptions does not hold) other estimation methods may be superior to least squares. The other estimation methods that we consider in this book are the *maximum likelihood* method, the *ridge method*, and the *principal components* method.

Table 1.11 Various Classifications of Regression Analysis.

Type of Regression	Conditions
Univariate	Only one quantitative response variable
Multivariate	Two or more quantitative response variables
Simple	Only one predictor variable
Multiple	Two or more predictor variables
Linear	All parameters enter the equation linearly, possibly after transformation of the data
Nonlinear	The relationship between the response and some of the predictors is nonlinear or some of the parameters appear nonlinearly, but no transformation is possible to make the parameters appear linearly
Analysis of Variance	All predictors are qualitative variables
Analysis of Covariance	Some predictors are quantitative variables and others are qualitative variables
Logistic	The response variable is qualitative

1.4.6 Model Fitting

The next step in the analysis is to estimate the regression parameters or to fit the model to the collected data using the chosen estimation method (e.g., least squares). The estimates of the regression parameters $\beta_0, \beta_1, \ldots, \beta_p$ in (1.1) are denoted by $\hat{\beta}_0, \hat{\beta}_1, \ldots, \hat{\beta}_p$. The estimated regression equation then becomes

$$\hat{Y} = \hat{\beta}_0 + \hat{\beta}_1 X_1 + \hat{\beta}_2 X_2 + \cdots + \hat{\beta}_p X_p. \tag{1.5}$$

A *hat* on top of a parameter denotes an estimate of the parameter. The value $\hat{Y}$ (pronounced as *Y-hat*) is called the *fitted* value. Using (1.5), we can compute n fitted values, one for each of the n observations in our data. For example, the ith fitted value $\hat{y}_i$ is

$$\hat{y}_i = \hat{\beta}_0 + \hat{\beta}_1 x_{i1} + \hat{\beta}_2 x_{i2} + \cdots + \hat{\beta}_p x_{ip}, \quad i = 1, 2, \ldots, n, \tag{1.6}$$

where $x_{i1}, x_{i2}, \ldots, x_{ip}$ are the values of the p predictor variables for the ith observation.

Note that (1.5) can be used to predict the response variable for any values of the predictor variables not observed in our data. In this case, the obtained $\hat{Y}$ is called the *predicted* value. The difference between fitted and predicted values is that the fitted value refers to the case where the values used for the predictor variables correspond to one of the n observations in our data, but the predicted values are obtained for any set of values of the predictor variables. It is generally not recommended to predict the response variable for a set of values of the predictor variables far outside the range of our data. In cases where the values of the predictor variables represent future values of the predictors, the predicted value is referred to as the *forecasted* value.

1.4.7   Model Criticism and Selection

The validity of a statistical method, such as regression analysis, depends on certain assumptions. Assumptions are usually made about the data and the model. The accuracy of the analysis and of the conclusions derived from an analysis depends crucially on the validity of these assumptions. Before using (1.5) for any purpose, we first need to determine whether the specified assumptions hold. We need to address the following questions:

1. What are the required assumptions?

2. For each of these assumptions, how do we determine whether or not the assumption is valid?

3. What can be done in cases where one or more of the assumptions does not hold?

The standard regression assumptions will be specified and the above questions will be addressed in great detail in various parts of this book. We emphasize here that validation of the assumptions must be made *before* any conclusions are drawn from the analysis. Regression analysis is viewed here as a *cyclical* process, a process in which the outputs are used to diagnose, validate, criticize, and possibly modify the inputs. The process has to be repeated until a satisfactory output has been obtained. A satisfactory output is an estimated model that satisfies the assumptions and fits the data reasonably well. This cyclic process is illustrated schematically in Figure 1.1.

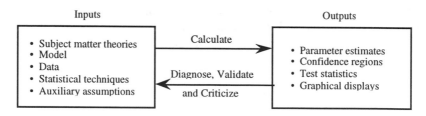

Fig. 1.1   A schematic illustration of the cyclic nature of the regression process.

1.4.8   Objectives of Regression Analysis

The explicit determination of the regression equation is the most important product of the analysis. It is a summary of the relationship between Y (the response variable) and the set of predictor variables $X_1, X_2, \ldots, X_p$. The equation may be used for several purposes. It may be used to evaluate the importance of individual predictors, to analyze the effects of policy that involves changing values of the predictor variables, or to forecast values of the response variable for a given set of predictors. Although the regression equation is the final product, there are many important by-products. We view regression analysis as a set of data analytic techniques that are used to help understand the interrelationships among variables in a certain environment. The task of regression analysis is to learn as much as possible about the environment reflected by the data. We emphasize that what is uncovered along the way to the formulation of the equation may often be as valuable and informative as the final equation.

1.5   SCOPE AND ORGANIZATION OF THE BOOK

This book can be used by all who analyze data. A knowledge of matrix algebra is not necessary. We have seen excellent regression analysis done by people who have no knowledge of matrix theory. A knowledge of matrix algebra is certainly very helpful in understanding the theory. We have provided appendices which use matrix algebra for readers who are familiar with that topic. Matrix algebra permits expression of regression results much more compactly and is essential for the mathematical derivation of the results.

Lack of knowledge of matrix algebra should not deter anyone from using this book and doing regression analysis. For readers who are not familiar with matrix algebra but who wish to benefit from the material in the appendices, we recommend reading the relatively short book by Hadi (1996), *Matrix Algebra As a Tool*. We believe that the majority, if not all, of our readers can read it entirely on their own or with minimal assistance.

There are no formal derivations in the text and readers interested in mathematical derivations are referred to a number of books that contain formal derivations of the regression formulas. Formulas are presented, but only for purposes of reference. It is assumed throughout the book that the necessary summary statistics will be computer generated from an existing regression package.[8]

The book is organized as follows: It begins with the simple linear regression model in Chapter 2. The simple regression model is then extended to the multiple regression model in Chapter 3. In both chapters, the model is formulated, assumptions are specified, and the key theoretical results are stated and illustrated by examples. For simplicity of presentation and for ped-

---

[8]Many commercial statistical packages include regression analysis routines. We assume that these programs have been thoroughly tested and produce numerically accurate answers. For the most part the assumption is a safe one, but for some data sets, different programs have given dramatically different results.

agogical reasons, the analysis and conclusions in Chapters 2 and 3 are made under the presumption that the standard regression assumptions are valid. Chapter 4 addresses the issue of assumptions validation and the detection and correction of model violations.

Each of the remaining chapters deals with a special regression problem. Chapter 5 deals with the case where some or all of the predictor variables are qualitative. Chapter 6 deals with data transformation. Chapter 7 presents situations where a variant of the least squares method is needed. This method is called the *weighted least squares* method. Chapter 8 discusses the problem that arises when the observations are correlated. This problem is known as the *autocorrelation* problem. Chapters 9 and 10 present methods for the detection and correction of an important problem called *collinearity*. Collinearity occurs when the predictor variables are highly correlated.

Chapter 11 presents variable selection methods – computer methods for selecting the best and most parsimonious model(s). Before applying any of the variable selection methods, we assume in this chapter that questions of assumptions validation and model violations have already been addressed and settled satisfactorily.

The book concludes with Chapter 12 on logistic regression, which is used when the response variable is binary. Logistic regression is studied because it is an important tool with many practical applications. It is in the last chapter of the book because all the previous chapters deal with the case where the response variable is quantitative.

We recommend that the chapters be covered in the same sequence they are presented, although Chapters 5 to 12 can be covered in any order after Chapter 4, as long as Chapter 9 is covered before Chapter 10 and Chapter 7 is covered before Chapter 12.

EXERCISES

1.1 Classify each of the following variables as either quantitative or qualitative. If a variable is qualitative, state the possible categories.

    (a) Geographical region   (b) Number of children in a family

    (c) Price of a house   (d) Race

    (e) Temperature   (f) Fuel consumption

    (g) Employment rate   (h) Political party preference

1.2 Give two examples in any area of interest to you (other than those presented in Chapter 1) where regression analysis can be used as a data analytic tool to answer some questions of interest. For each example:

    (a) What is the question of interest?

    (b) Identify the response and the predictor variables.

    (c) Classify each of the variables as either quantitative or qualitative.

(d) Which type of regression (see Table 1.11) can be used to analyze the data?

(e) Give a possible form of the model and identify its parameters.

1.3 In each of the following sets of variables, identify which of the variables can be regarded as a response variable and which can be used as predictors? (Explain)

(a) Number of cylinders and gasoline consumption of cars.

(b) SAT scores, grade point average, and college admission.

(c) Supply and demand of certain goods.

(d) Company's assets, return on a stock, and net sales.

(e) The distance of a race, the time to run the race, and the weather conditions at the time of running.

(f) The weight of a person, whether or not the person is a smoker, and whether or not the person has a lung cancer.

(g) The height and weight of a child, his/her parents' height and weight, and the sex and age of the child.

1.4 For each of the sets of variables in Exercise 1.3:

(a) Classify each variable as either quantitative or qualitative.

(b) Which type of regression (see Table 1.11) can be used in the analysis of the data?

2

Simple Linear Regression

2.1 INTRODUCTION

We start with the simple case of studying the relationship between a response variable Y and a predictor variable X_1. Since we have only one predictor variable, we shall drop the subscript in X_1 and use X for simplicity. We discuss covariance and correlation coefficient as measures of the direction and strength of the linear relationship between the two variables. Simple linear regression model is then formulated and the key theoretical results are given without mathematical derivations, but illustrated by numerical examples. Readers interested in mathematical derivations are referred to the bibliographic notes at the end of the chapter, where books that contain a formal development of regression analysis are listed.

2.2 COVARIANCE AND CORRELATION COEFFICIENT

Suppose we have observations on n units consisting of a dependent or response variable Y and an explanatory variable X. The observations are usually recorded as in Table 2.1. We wish to measure both the *direction* and the *strength* of the relationship between Y and X. Two related measures, known as the *covariance* and the *correlation coefficient*, are developed below.

Table 2.1 Notation for the Data Used in Simple Regression and Correlation.

Observation Number	Response Y	Predictor X
1	y_1	x_1
2	y_2	x_2
$\vdots$	$\vdots$	$\vdots$
n	y_n	x_n

On the scatter plot of Y versus X, let us draw a vertical line at $\bar{x}$ and a horizontal line at $\bar{y}$, as shown in Figure 2.1, where

$$\bar{y} = \frac{\sum_{i=1}^{n} y_i}{n} \quad \text{and} \quad \bar{x} = \frac{\sum_{i=1}^{n} x_i}{n}, \tag{2.1}$$

are the sample mean of Y and X, respectively. The two lines divide the graph into four quadrants. For each point i in the graph, compute the following quantities:

- $y_i - \bar{y}$, the deviation of each observation y_i from the mean of the response variable,

- $x_i - \bar{x}$, the deviation of each observation x_i from the mean of the predictor variable, and

- the product of the above two quantities, $(y_i - \bar{y})(x_i - \bar{x})$.

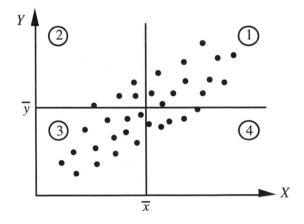

Fig. 2.1 A graphical illustration of the correlation coefficient.

It is clear from the graph that the quantity $(y_i - \bar{y})$ is positive for every point in the first and second quadrants, and is negative for every point in the third and fourth quadrants. Similarly, the quantity $(x_i - \bar{x})$ is positive for every point in the first and fourth quadrants, and is negative for every point in the second and third quadrants. These facts are summarized in Table 2.2.

Table 2.2   Algebraic Signs of the Quantities $(y_i - \bar{y})$ and $(x_i - \bar{x})$.

Quadrant	$y_i - \bar{y}$	$x_i - \bar{x}$	$(y_i - \bar{y})(x_i - \bar{x})$
1	+	+	+
2	+	−	−
3	−	−	+
4	−	+	−

If the linear relationship between Y and X is positive (as X increases Y also increases), then there are more points in the first and third quadrants than in the second and fourth quadrants. In this case, the sum of the last column in Table 2.2 is likely to be positive because there are more positive than negative quantities. Conversely, if the relationship between Y and X is negative (as X increases Y decreases), then there are more points in the second and fourth quadrants than in the first and third quadrants. Hence the sum of the last column in Table 2.2 is likely to be negative. Therefore, the sign of the quantity

$$Cov(Y, X) = \frac{\sum_{i=1}^{n}(y_i - \bar{y})(x_i - \bar{x})}{n - 1}, \tag{2.2}$$

which is known as the *covariance* between Y and X, indicates the direction of the linear relationship between Y and X. If $Cov(Y, X) > 0$, then there is a positive relationship between Y and X, but if $Cov(Y, X) < 0$, then the relationship is negative. Unfortunately, $Cov(Y, X)$ does not tell us much about the strength of such a relationship because it is affected by changes in the units of measurement. For example, we would get two different values for the $Cov(Y, X)$ if we report Y and/or X in terms of thousands of dollars instead of dollars. To avoid this disadvantage of the covariance, we *standardize* the data before computing the covariance. To standardize the Y data, we first subtract the mean from each observation then divide by the standard deviation, that is, we compute

$$z_i = \frac{y_i - \bar{y}}{s_y}, \tag{2.3}$$

where

$$s_y = \sqrt{\frac{\sum_{i=1}^{n}(y_i - \bar{y})^2}{n - 1}}, \tag{2.4}$$

is the sample *standard deviation* of Y. It can be shown that the standardized variable Z in (2.3) has mean zero and standard deviation one. We standardize X in a similar way by subtracting the mean $\bar{x}$ from each observation x_i then divide by the standard deviation s_x. The covariance between the standardized X and Y data is known as the *correlation coefficient* between Y and X and is given by

$$Cor(Y, X) = \frac{1}{n-1} \sum_{i=1}^{n} \left(\frac{y_i - \bar{y}}{s_y} \right) \left(\frac{x_i - \bar{x}}{s_x} \right). \tag{2.5}$$

Equivalent formulas for the correlation coefficient are

$$Cor(Y, X) \quad = \quad \frac{Cov(Y, X)}{s_y s_x} \tag{2.6}$$

$$= \quad \frac{\sum (y_i - \bar{y})(x_i - \bar{x})}{\sqrt{\sum (y_i - \bar{y})^2 \sum (x_i - \bar{x})^2}}. \tag{2.7}$$

Thus, $Cor(Y, X)$ can be interpreted either as the covariance between the standardized variables or the ratio of the covariance to the standard deviations of the two variables. From (2.5), it can be seen that the correlation coefficient is symmetric, that is, $Cor(Y, X) = Cor(X, Y)$.

Unlike $Cov(Y, X)$, $Cor(Y, X)$ is scale invariant, that is, it does not change if we change the units of measurements. Furthermore, $Cor(Y, X)$ satisfies

$$-1 \leq Cor(Y, X) \leq 1. \tag{2.8}$$

These properties make the $Cor(Y, X)$ a useful quantity for measuring both the direction and the strength of the relationship between Y and X. The magnitude of $Cor(Y, X)$ measures the strength of the linear relationship between Y and X. The closer $Cor(Y, X)$ is to 1 or -1, the stronger is the relationship between Y and X. The sign of $Cor(Y, X)$ indicates the direction of the relationship between Y and X. That is, $Cor(Y, X) > 0$ implies that Y and X are positively related. Conversely, $Cor(Y, X) < 0$, implies that Y and X are negatively related.

Note, however, that $Cor(Y, X) = 0$ does not necessarily mean that Y and X are not related. It only implies that they are not linearly related because the correlation coefficient measures only *linear* relationships. In other words, the $Cor(Y, X)$ can still be zero when Y and X are nonlinearly related. For example, Y and X in Table 2.3 have the perfect nonlinear relationship $Y = 50 - X^2$ (graphed in Figure 2.2), yet $Cor(Y, X) = 0$.

Furthermore, like many other summary statistics, the $Cor(Y, X)$ can be substantially influenced by one or few outliers in the data. To emphasize this point, Anscombe (1973) has constructed four data sets, known as Anscombe's quartet, each with a distinct pattern, but each having the same set of summary statistics (e.g., the same value of the correlation coefficient). The data and graphs are reproduced in Table 2.4 and Figure 2.3. The data can be found

Table 2.3 A Data Set With a Perfect Nonlinear Relationship Between Y and X, Yet $Cor(X, Y) = 0$.

Y	X	Y	X	Y	X
1	−7	46	−2	41	3
14	−6	49	−1	34	4
25	−5	50	0	25	5
34	−4	49	1	14	6
41	−3	46	2	1	7

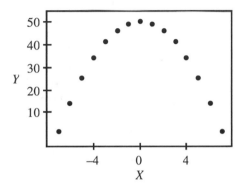

Fig. 2.2 A scatter plot of Y versus X in Table 2.3.

Table 2.4 Anscombe's Quartet: Four Data Sets Having Same Values of Summary Statistics.

Y_1	X_1	Y_2	X_2	Y_3	X_3	Y_4	X_4
8.04	10	9.14	10	7.46	10	6.58	8
6.95	8	8.14	8	6.77	8	5.76	8
7.58	13	8.74	13	12.74	13	7.71	8
8.81	9	8.77	9	7.11	9	8.84	8
8.33	11	9.26	11	7.81	11	8.47	8
9.96	14	8.10	14	8.84	14	7.04	8
7.24	6	6.13	6	6.08	6	5.25	8
4.26	4	3.10	4	5.39	4	12.50	19
10.84	12	9.13	12	8.15	12	5.56	8
4.82	7	7.26	7	6.42	7	7.91	8
5.68	5	4.74	5	5.73	5	6.89	8

Source: Anscombe (1973).

in the book's Web site.[1] An analysis based exclusively on an examination of summary statistics, such as the correlation coefficient, would have been unable to detect the differences in patterns.

[1] http://www.ilr.cornell.edu/~hadi/RABE

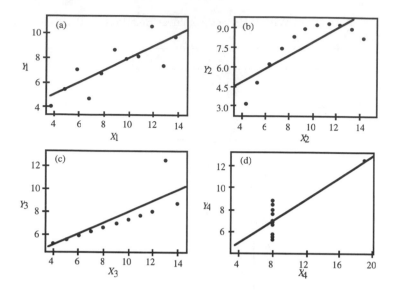

Fig. 2.3 Scatter plots of the data in Table 2.4 with the fitted lines.

An examination of Figure 2.3 shows that only the first set, whose plot is given in (a), can be described by a linear model. The plot in (b) shows the second data set is distinctly nonlinear and would be better fitted by a quadratic function. The plot in (c) shows that the third data set has one point that distorts the slope and the intercept of the fitted line. The plot in (d) shows that the fourth data set is unsuitable for linear fitting, the fitted line being determined essentially by one extreme observation. Therefore, it is important to examine the scatter plot of Y versus X before interpreting the numerical value of $Cor(Y, X)$.

2.3 EXAMPLE: COMPUTER REPAIR DATA

As an illustrative example, consider a case of a company that markets and repairs small computers. To study the relationship between the length of a service call and the number of electronic components in the computer that must be repaired or replaced, a sample of records on service calls was taken. The data consist of the length of service calls in minutes (the response variable) and the number of components repaired (the predictor variable). The data are presented in Table 2.5. The Computer Repair data can also be found in the book's Web site. We use this data set throughout this chapter as an illustrative example. The quantities needed to compute $\bar{y}$, $\bar{x}$, $Cov(Y, X)$, and

Table 2.5 Length of Service Calls (in Minutes) and Number of Units Repaired.

Row	Minutes	Units	Row	Minutes	Units
1	23	1	8	97	6
2	29	2	9	109	7
3	49	3	10	119	8
4	64	4	11	149	9
5	74	4	12	145	9
6	87	5	13	154	10
7	96	6	14	166	10

$Cor(Y, X)$ are shown in Table 2.6. We have

$$\bar{y} = \frac{\sum_{i=1}^{n} y_i}{n} = \frac{1361}{14} = 97.21 \text{ and } \bar{x} = \frac{\sum_{i=1}^{n} x_i}{n} = \frac{84}{14} = 6,$$

$$Cov(Y, X) = \frac{\sum_{i=1}^{n} (y_i - \bar{y})(x_i - \bar{x})}{n - 1} = \frac{1768}{13} = 136,$$

and

$$Cor(Y, X) = \frac{\sum (y_i - \bar{y})(x_i - \bar{x})}{\sqrt{\sum (y_i - \bar{y})^2 \sum (x_i - \bar{x})^2}} = \frac{1768}{\sqrt{27768.36 \times 114}} = 0.996.$$

Before drawing conclusions from this value of $Cor(Y, X)$, we should examine the corresponding scatter plot of Y versus X. This plot is given in Figure 2.4. The high value of $Cor(Y, X) = 0.996$ is consistent with the strong linear relationship between Y and X exhibited in Figure 2.4. We therefore conclude that there is a strong positive relationship between repair time and units repaired.

Although $Cor(Y, X)$ is a useful quantity for measuring the direction and the strength of linear relationships, it cannot be used for prediction purposes, that is, we cannot use $Cor(Y, X)$ to predict the value of one variable given the value of the other. Furthermore, $Cor(Y, X)$ measures only pairwise relationships. Regression analysis, however, can be used to relate one or more response variable to one or more predictor variables. It can also be used in prediction. Regression analysis is an attractive extension to correlation analysis because it postulates a model that can be used not only to measure the direction and the strength of a relationship between the response and predictor variables, but also to numerically describe that relationship. We discuss simple linear regression models in the rest of this chapter. Chapter 3 is devoted to multiple regression models.

Table 2.6   Quantities Needed for the Computation of the Correlation Coefficient Between the Length of Service Calls, Y, and Number of Units Repaired, X.

i	y_i	x_i	$(y_i - \bar{y})$	$(x_i - \bar{x})$	$(y_i - \bar{y})^2$	$(x_i - \bar{x})^2$	$(y_i - \bar{y})(x_i - \bar{x})$
1	23	1	−74.21	−5	5507.76	25	371.07
2	29	2	−68.21	−4	4653.19	16	272.86
3	49	3	−48.21	−3	2324.62	9	144.64
4	64	4	−33.21	−2	1103.19	4	66.43
5	74	4	−23.21	−2	538.90	4	46.43
6	87	5	−10.21	−1	104.33	1	10.21
7	96	6	−1.21	0	1.47	0	0.00
8	97	6	−0.21	0	0.05	0	0.00
9	109	7	11.79	1	138.90	1	11.79
10	119	8	21.79	2	474.62	4	43.57
11	149	9	51.79	3	2681.76	9	155.36
12	145	9	47.79	3	2283.47	9	143.36
13	154	10	56.79	4	3224.62	16	227.14
14	166	10	68.79	4	4731.47	16	275.14
Total	1361	84	0.00	0	27768.36	114	1768.00

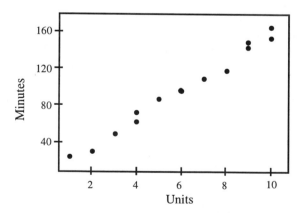

Fig. 2.4   Computer Repair data: Scatter plot of Minutes versus Units.

2.4   THE SIMPLE LINEAR REGRESSION MODEL

The relationship between a response variable Y and a predictor variable X is postulated as a linear[2] model

$$Y = \beta_0 + \beta_1 X + \varepsilon, \tag{2.9}$$

where β_0 and β_1, are constants called the *model regression coefficients* or *parameters*, and ε is a random disturbance or error. It is assumed that in the range of the observations studied, the linear equation (2.9) provides an acceptable approximation to the true relation between Y and X. In other words, Y is approximately a linear function of X, and ε measures the discrepancy in that approximation. In particular ε contains no systematic information for determining Y that is not already captured in X. The coefficient β_1, called the *slope*, may be interpreted as the change in Y for unit change in X. The coefficient β_0, called the *constant* coefficient or *intercept*, is the predicted value of Y when $X = 0$.

According to (2.9), each observation in Table 2.1 can be written as

$$y_i = \beta_0 + \beta_1 x_i + \varepsilon_i, \quad i = 1, 2, \ldots, n, \tag{2.10}$$

where y_i represents the ith value of the response variable Y, x_i represents the ith value of the predictor variable X, and ε_i represents the error in the approximation of y_i.

Regression analysis differs in an important way from correlation analysis. The correlation coefficient is symmetric in the sense that $Cor(Y, X)$ is the same as $Cor(X, Y)$. The variables X and Y are of equal importance. In regression analysis the response variable Y is of primary importance. The importance of the predictor X lies on its ability to account for the variability of the response variable Y and not in itself per se. Hence Y is of primary importance.

Returning to the Computer Repair Data example, suppose that the company wants to forecast the number of service engineers that will be required over the next few years. A linear model,

$$\text{Minutes} = \beta_0 + \beta_1 \cdot \text{Units} + \varepsilon, \tag{2.11}$$

is assumed to represent the relationship between the length of service calls and the number of electronic components in the computer that must be repaired or replaced. To validate this assumption, we examine the graph of

---

[2]The adjective *linear* has a dual role here. It may be taken to describe the fact that the relationship between Y and X is linear. More generally, the word *linear* refers to the fact that the regression parameters, β_0 and β_1, enter (2.9) in a linear fashion. Thus, for example, $Y = \beta_0 + \beta_1 X^2 + \varepsilon$ is also a linear model even though the relationship between Y and X is quadratic.

the response variable versus the explanatory variable. This graph, shown in Figure 2.4, suggests that the straight line relationship in (2.11) is a reasonable assumption.

2.5 PARAMETER ESTIMATION

Based on the available data, we wish to estimate the parameters β_0 and β_1. This is equivalent to finding the straight line that gives the *best fit* (representation) of the points in the scatter plot of the response versus the predictor variable (see Figure 2.4). We estimate the parameters using the popular *least squares method*, which gives the line that minimizes the sum of squares of the *vertical distances*[3] from each point to the line. The vertical distances represent the errors in the response variable. These errors can be obtained by rewriting (2.10) as

$$\varepsilon_i = y_i - \beta_0 - \beta_1 x_i, \quad i = 1, 2, \ldots, n. \tag{2.12}$$

The sum of squares of these distances can then be written as

$$S(\beta_0, \beta_1) = \sum_{i=1}^{n} \varepsilon_i^2 = \sum_{i=1}^{n} (y_i - \beta_0 - \beta_1 x_i)^2.$$

The values of $\hat{\beta}_0$ and $\hat{\beta}_1$ that minimize $S(\beta_0, \beta_1)$ are given by

$$\hat{\beta}_1 = \frac{\sum (y_i - \bar{y})(x_i - \bar{x})}{\sum (x_i - \bar{x})^2} \tag{2.13}$$

and

$$\hat{\beta}_0 = \bar{y} - \hat{\beta}_1 \bar{x}. \tag{2.14}$$

Note that we give the formula for $\hat{\beta}_1$ before the formula for $\hat{\beta}_0$ because $\hat{\beta}_0$ uses $\hat{\beta}_1$. The estimates $\hat{\beta}_0$ and $\hat{\beta}_1$ are called the least squares estimates of β_0 and β_1 because they are the solution to the *least squares method*, a procedure that gives the intercept, $\hat{\beta}_0$, and the slope, $\hat{\beta}_1$, of the line that has the smallest possible sum of squares of the vertical distances from each point to the line. This line is called the *least squares regression line*. The least squares regression line is given by

$$\hat{Y} = \hat{\beta}_0 + \hat{\beta}_1 X. \tag{2.15}$$

Note that a least squares line always exists because we can always find a line that gives the minimum sum of squares of the vertical distances. In fact, as

---

[3]An alternative to the vertical distance is the *perpendicular* (shortest) distance from each point to the line. The resultant line is called the *orthogonal regression* line. This will be discussed in Chapter 10.

we shall see later, in some cases a least squares line may not be unique. These cases are not common in practice.

For each observation in our data we can compute

$$\hat{y}_i = \hat{\beta}_0 + \hat{\beta}_1 x_i, \quad i = 1, 2, \ldots, n. \tag{2.16}$$

These are called the *fitted* values. Thus, the ith fitted value, $\hat{y}_i$, is the point on the least squares regression line (2.15) corresponding to x_i. The vertical distance corresponding to the ith observation is

$$c_i = y_i - \hat{y}_i. \tag{2.17}$$

These vertical distances are called the *ordinary[4] least squares residuals*.

Using the Computer Repair data and the quantities in Table 2.6, we have

$$\hat{\beta}_1 = \frac{\sum(y_i - \bar{y})(x_i - \bar{x})}{\sum(x_i - \bar{x})^2} = \frac{1768}{114} = 15.509,$$

and

$$\hat{\beta}_0 = \bar{y} - \hat{\beta}_1 \bar{x} = 97.21 - 15.509 \times 6 = 4.162.$$

Then the equation of the least squares regression line is

$$\text{Minutes} = 4.162 + 15.509 \cdot \text{Units}. \tag{2.18}$$

This least squares line is shown together with the scatter plot of Minutes versus Units in Figure 2.5. The fitted values in (2.16) and the residuals in (2.17) are shown in Table 2.7.

The coefficients in (2.18) can be interpreted in physical terms. The constant term represents the setup or startup time for each repair and is approximately 4 minutes. The coefficient of Units represents the increase in the length of a service call for each additional component that has to be repaired. From the data given, we estimate that it takes about 16 minutes (15.509) for each additional component that has to be repaired. For example, the length of a service call in which four components had to be repaired is obtained by substituting Units = 4 in the equation of the regression line (2.18) and obtaining $\hat{y} = 4.162 + 15.509 \times 4 = 66.20$. Since Units = 4, corresponds to two observations in our data set (observations 4 and 5), the value 66.198 is the fitted value for both observations 4 and 5, as can be seen from Table 2.7. Note, however, that since observations 4 and 5 have different values for the response variable Minutes, they have different residuals.

We should note here that by comparing (2.2), (2.7), and (2.13), an alternative formula for $\hat{\beta}_1$ can be expressed as

$$\hat{\beta}_1 = \frac{Cov(Y, X)}{Var(X)} = Cor(Y, X) \frac{s_y}{s_x}, \tag{2.19}$$

---

[4]To be distinguished from other types of residuals to be presented later.

Table 2.7   The Fitted Values, $\hat{y}_i$, and the Ordinary Least Squares Residuals, e_i, for the Computer Repair Data.

i	Units	$\hat{y}_i$	e_i	i	Units	$\hat{y}_i$	e_i
1	1	19.67	3.33	8	6	97.21	−0.21
2	2	35.18	−6.18	9	7	112.72	−3.72
3	3	50.69	−1.69	10	8	128.23	−9.23
4	4	66.20	−2.20	11	9	143.74	5.26
5	4	66.20	7.80	12	9	143.74	1.26
6	5	81.71	5.29	13	10	159.25	−5.25
7	6	97.21	−1.21	14	10	159.25	6.75

from which it can be seen that $\hat{\beta}_1$, $Cov(Y, X)$, and $Cor(Y, X)$ have the same sign. This makes intuitive sense because positive (negative) slope means positive (negative) correlation.

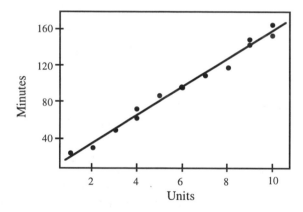

Fig. 2.5   Plot of Minutes versus Units with the fitted least squares regression line.

So far in our analysis we have made only one assumption, namely, that Y and X are linearly related. This assumption is referred to as the *linearity* assumption. This is merely an assumption or a hypothesis about the relationship between the response and predictor variables. An early step in the analysis should always be the validation of this assumption. We wish to determine if the data at hand support the assumption that Y and X are linearly related. An informal way to check this assumption is to examine the scatter plot of the response versus the predictor variable, preferably drawn with the least squares line superimposed on the graph (see Figure 2.5). If we observe a nonlinear pattern, we will have to take corrective action. For example, we may *re-express* or *transform* the data before we continue the analysis. *Data transformation* is discussed in Chapter 6.

If the scatter of points resemble a straight line, then we conclude that the linearity assumption is reasonable and continue with our analysis. The least squares estimators have several desirable properties when some additional assumptions hold. The required assumptions are stated in Chapter 4. The validity of these assumptions must be checked before meaningful conclusions can be reached from the analysis. Chapter 4 also presents methods for the validation of these assumptions. Using the properties of least squares estimators, one can develop statistical inference procedures (e.g., confidence interval estimation, tests of hypothesis, and goodness-of-fit tests). These are presented in Sections 2.6 to 2.9.

2.6 TESTS OF HYPOTHESES

As stated earlier, the usefulness of X as a predictor of Y can be measured informally by examining the correlation coefficient and the corresponding scatter plot of Y versus X. A more formal way of measuring the usefulness of X as a predictor of Y is to conduct a test of hypothesis about the regression parameter β_1. Note that the hypothesis $\beta_1 = 0$ means that there is no linear relationship between Y and X. A test of this hypothesis requires the following assumption. For every fixed value of X, the ε's are assumed to be independent random quantities normally distributed with mean zero and a common variance σ^2. With these assumptions, the quantities, $\hat{\beta}_0$ and $\hat{\beta}_1$ are unbiased[5] estimates of β_0 and β_1, respectively. Their variances are

$$Var(\hat{\beta}_0) = \sigma^2 \left[\frac{1}{n} + \frac{\bar{x}^2}{\sum(x_i - \bar{x})^2} \right],$$ (2.20)

and

$$Var(\hat{\beta}_1) = \frac{\sigma^2}{\sum(x_i - \bar{x})^2}.$$ (2.21)

Furthermore, the *sampling distributions* of the least squares estimates $\hat{\beta}_0$ and $\hat{\beta}_1$ are normal with means β_0 and β_1 and variance as given in (2.20) and (2.21), respectively.

The variances of $\hat{\beta}_0$ and $\hat{\beta}_1$ depend on the unknown parameter σ^2. So, we need to estimate σ^2 from the data. An unbiased estimate of σ^2 is given by

$$\hat{\sigma}^2 = \frac{\sum e_i^2}{n-2} = \frac{\sum(y_i - \hat{y}_i)^2}{n-2} = \frac{SSE}{n-2},$$ (2.22)

where SSE is the sum of squares of the residuals (errors). The number $n-2$ in the denominator of (2.22) is called the *degrees of freedom* (*df*). It is equal to the number of observations minus the number of estimated regression coefficients.

---

[5]An estimate $\hat{\theta}$ is said to be an unbiased estimate of a parameter θ if the expected value of $\hat{\theta}$ is equal to θ.

Replacing σ^2 in (2.20) and (2.21) by $\hat{\sigma}^2$ in (2.22), we get unbiased estimates of the variances of $\hat{\beta}_0$ and $\hat{\beta}_1$. An *estimate* of the *standard deviation* is called the *standard error* (*s.e.*) of the estimate. Thus, the standard errors of $\hat{\beta}_0$ and $\hat{\beta}_1$ are

$$s.e.(\hat{\beta}_0) = \hat{\sigma}\sqrt{\frac{1}{n} + \frac{\bar{x}^2}{\sum(x_i - \bar{x})^2}} \qquad (2.23)$$

and

$$s.e.(\hat{\beta}_1) = \frac{\hat{\sigma}}{\sqrt{\sum(x_i - \bar{x})^2}} , \qquad (2.24)$$

respectively, where $\hat{\sigma}$ is the square root of $\hat{\sigma}^2$ in (2.22). The standard errors of $\hat{\beta}_1$ is a measure of how precisely the slope has been estimated. The smaller the standard error the more precise the estimator.

With the sampling distributions of $\hat{\beta}_0$ and $\hat{\beta}_1$, we are now in position to perform statistical analysis concerning the usefulness of X as a predictor of Y. Under the normality assumption, an appropriate test statistic for testing the null hypothesis $H_0 : \beta_1 = 0$ against the alternative $H_1 : \beta_1 \neq 0$ is the t-test,

$$t_1 = \frac{\hat{\beta}_1}{s.e.(\hat{\beta}_1)} . \qquad (2.25)$$

The statistic t_1 is distributed as a Student's t with $(n-2)$ degrees of freedom. The test is carried out by comparing this observed value with the appropriate critical value obtained from the t-table given in the Appendix to this book (see Table A.2), which is $t_{(n-2,\alpha/2)}$, where α is a specified significance level. Note that we divide α by 2 because we have a two-sided alternative hypothesis. Accordingly, H_0 is to be rejected at the significance level α if

$$|t_1| \geq t_{(n-2,\alpha/2)}, \qquad (2.26)$$

where $|t_1|$ denotes the absolute value of t_1. A criterion equivalent to that in (2.26) is to compare the p-value for the t-test with α and reject H_0 if

$$p(|t_1|) \leq \alpha, \qquad (2.27)$$

where $p(|t_1|)$, called the *p-value*, is the probability that a random variable having a Student t distribution with $(n-2)$ is greater than $|t_1|$ (the absolute value of the observed value of the t-test). Figure 2.6 is a graph of the density function of a t-distribution. The p-value is the sum of the two shaded areas under the curve. The p-value is usually computed and supplied as part of the regression output by statistical packages. Note that the rejection of $H_0 : \beta_1 = 0$ would mean that β_1 is likely to be different from 0, and hence the predictor variable X is a statistically significant predictor of the response variable Y.

To complete the picture of hypotheses testing regarding regression parameters, we give here tests for three other hypotheses that may arise in practice.

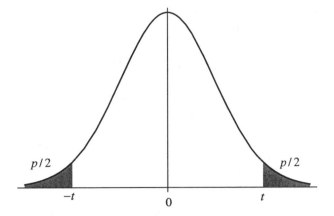

Fig. 2.6 A graph of the probability density function of a *t*-distribution. The *p*-value for the *t*-test is the shaded areas under the curve.

Testing H_0: $\beta_1 = \beta_1^0$

The above *t*-test can be generalized to test the more general hypothesis H_0 : $\beta_1 = \beta_1^0$, where β_1^0 is a constant chosen by the investigator, against the two-sided alternative $H_1 : \beta_1 \neq \beta_1^0$. The appropriate test statistic in this case is the *t*-test,

$$t_1 = \frac{\hat{\beta}_1 - \beta_1^0}{s.e.(\hat{\beta}_1)} \, . \tag{2.28}$$

Note that when $\beta_1^0 = 0$, the *t*-test in (2.28) reduces to the *t*-test in (2.25). The statistic t_1 in (2.28) is also distributed as a Student's *t* with $(n-2)$ degrees of freedom. Thus, $H_0 : \beta_1 = \beta_1^0$ is rejected if (2.26) holds (or, equivalently, if (2.27) holds).

For illustration, using the Computer Repair data, let us suppose that the management expected the increase in service time for each additional unit to be repaired to be 12 minutes. Do the data support this conjecture? The answer may be obtained by testing $H_0 : \beta_1 = 12$ against $H_1 : \beta_1 \neq 12$. The appropriate statistic is

$$t_1 = \frac{\hat{\beta}_1 - 12}{s.e.(\hat{\beta}_1)} = \frac{15.509 - 12}{0.505} = 6.948,$$

with 12 degrees of freedom. The critical value for this test is $t_{(n-2,\alpha/2)} = t_{(12,0.025)} = 2.18$. Since $t_1 = 6.948 > 2.18$, the result is highly significant, leading to the rejection of the null hypothesis. The management's estimate of the increase in time for each additional component to be repaired is not supported by the data. Their estimate is too low.

Table 2.8   A Standard Regression Output. The Equation Number of the Corresponding Formulas are Given in Parentheses.

Variable	Coefficient (Formula)	s.e. (Formula)	t-test (Formula)	p-value
Constant	$\hat{\beta}_0$ (2.14)	$s.e.(\hat{\beta}_0)$ (2.23)	t_0 (2.30)	p_0
X	$\hat{\beta}_1$ (2.13)	$s.e.(\hat{\beta}_1)$ (2.24)	t_1 (2.25)	p_1

Table 2.9   Regression Output for the Computer Repair Data.

Variable	Coefficient	s.e.	t-test	p-value
Constant	4.162	3.355	1.24	0.2385
Units	15.509	0.505	30.71	< 0.0001

Testing $H_0: \beta_0 = \beta_0^0$

The need for testing hypotheses regarding the regression parameter β_0 may also arise in practice. More specifically, suppose we wish to test $H_0 : \beta_0 = \beta_0^0$ against the alternative $H_1 : \beta_0 \neq \beta_0^0$, where β_0^0 is a constant chosen by the investigator. The appropriate test in this case is given by

$$t_0 = \frac{\hat{\beta}_0 - \beta_0^0}{s.e.(\hat{\beta}_0)} . \qquad (2.29)$$

If we set $\beta_0^0 = 0$, a special case of this test is obtained as

$$t_0 = \frac{\hat{\beta}_0}{s.e.(\hat{\beta}_0)} , \qquad (2.30)$$

which tests $H_0 : \beta_0 = 0$ against the alternative $H_1 : \beta_0 \neq 0$.

The least squares estimates of the regression coefficients, their standard errors, the t-tests for testing that the corresponding coefficient is zero, and the p-values are usually given as part of the regression output by statistical packages. These values are usually displayed in a table such as the one in Table 2.8. This table is known as the *coefficients table*. To facilitate the connection between a value in the table and the formula used to obtain it, the equation number of the formula is given in parentheses.

As an illustrative example, Table 2.9 shows a part of the regression output for the Computer Repair data in Table 2.5. Thus, for example, $\hat{\beta}_1 = 15.509$, the $s.e.(\hat{\beta}_1) = 0.505$, and hence $t_1 = 15.509/0.505 = 30.71$. The critical value for this test using $\alpha = 0.05$, for example, is $t_{(12,0.025)} = 2.18$. The $t_1 = 30.71$ is much larger than its critical value 2.18. Consequently, according to (2.26), $H_0 : \beta_1 = 0$ is rejected, which means that the predictor variable

Units is a statistically significant predictor of the response variable Minutes. This conclusion can also be reached using (2.27) by observing that the p-value ($p_1 < 0.0001$) is much less than $\alpha = 0.05$ indicating very high significance.

A Test Using Correlation Coefficient

As mentioned above, a test of $H_0 : \beta_1 = 0$ against $H_1 : \beta_1 \neq 0$ can be thought of as a test for determining whether the response and the predictor variables are linearly related. We used the t-test in (2.25) to test this hypothesis. An alternative test, which involves the correlation coefficient between Y and X, can be developed. Suppose that the population correlation coefficient between Y and X is denoted by ρ. If $\rho \neq 0$, then Y and X are linearly related. An appropriate test for testing $H_0 : \rho = 0$ against $H_1 : \rho \neq 0$ is given by

$$t_1 = \frac{Cor(Y, X)\sqrt{n - 2}}{\sqrt{1 - [Cor(Y, X)]^2}}, \tag{2.31}$$

where $Cor(Y, X)$ is the sample correlation coefficient between Y and X, defined in (2.6), which is considered here to be an estimate of ρ. The t-test in (2.31) is distributed as a Student's t with $(n - 2)$ degrees of freedom. Thus, $H_0 : \rho = 0$ is rejected if (2.26) holds (or, equivalently, if (2.27) holds). Again if $H_0 : \rho - 0$ is rejected, it means that there is a statistically significant linear relationship between Y and X.

It is clear that if no linear relationship exists between Y and X, then $\beta_1 = 0$. Consequently, the statistical tests for $H_0 : \beta_1 = 0$ and $H_0 : \rho - 0$ should be identical. Although the statistics for testing these hypotheses given in (2.25) and (2.31) look different, it can be demonstrated that they are indeed algebraically equivalent.

2.7 CONFIDENCE INTERVALS

To construct confidence intervals for the regression parameters, we also need to assume that the ε's have a normal distribution, which will enable us to conclude that the sampling distributions of $\hat{\beta}_0$ and $\hat{\beta}_1$ are normal, as discussed in Section 2.6. Consequently, the $(1 - \alpha) \times 100\%$ confidence interval for β_0 is given by

$$\hat{\beta}_0 \pm t_{(n-2,\alpha/2)} \times s.e.(\hat{\beta}_0), \tag{2.32}$$

where $t_{(n-2,\alpha/2)}$ is the $(1 - \alpha/2)$ percentile of a t distribution with $(n - 2)$ degrees of freedom. Similarly, limits of the $(1 - \alpha) \times 100\%$ confidence interval for β_1 are given by

$$\hat{\beta}_1 \pm t_{(n-2,\alpha/2)} \times s.e.(\hat{\beta}_1). \tag{2.33}$$

The confidence interval in (2.33) has the usual interpretation, namely, if we were to take repeated samples of the same size at the same values of X and construct for example 95% confidence intervals for the slope parameter for each sample, then 95% of these intervals would be expected to contain the true value of the slope.

From Table 2.9 we see that a 95% confidence interval for β_1 is

$$15.509 \pm 2.18 \times 0.505 = (14.408, 16.610). \tag{2.34}$$

That is, the incremental time required for each broken unit is between 14 and 17 minutes. The calculation of confidence interval for β_0 in this example is left as an exercise for the reader.

Note that the confidence limits in (2.32) and (2.33) are constructed for each of the parameters β_0 and β_1, separately. This does not mean that a simultaneous (joint) confidence region for the two parameters is rectangular. Actually, the simultaneous confidence region is elliptical. This region is given for the general case of multiple regression in the Appendix to Chapter 3 in (A.13), of which the simultaneous confidence region for β_0 and β_1 is a special case.

2.8   PREDICTIONS

The fitted regression equation can be used for prediction. We distinguish between two types of predictions:

1. The prediction of the value of the response variable Y which corresponds to any chosen value, x_0, of the predictor variable, or

2. The estimation of the mean response μ_0, when $X = x_0$.

For the first case, the predicted value $\hat{y}_0$ is

$$\hat{y}_0 = \hat{\beta}_0 + \hat{\beta}_1 x_0. \tag{2.35}$$

The standard error of this prediction is

$$s.e.(\hat{y}_0) = \hat{\sigma}\sqrt{1 + \frac{1}{n} + \frac{(x_0 - \bar{x})^2}{\sum(x_i - \bar{x})^2}}. \tag{2.36}$$

Hence, the confidence limits for the predicted value with confidence coefficient $(1 - \alpha)$ are given by

$$\hat{y}_0 \pm t_{(n-2,\alpha/2)} \, s.e.(\hat{y}_0). \tag{2.37}$$

For the second case, the mean response μ_0 is estimated by

$$\hat{\mu}_0 = \hat{\beta}_0 + \hat{\beta}_1 x_0. \tag{2.38}$$

The standard error of this estimate is

$$s.e.(\hat{\mu}_0) = \hat{\sigma}\sqrt{\frac{1}{n} + \frac{(x_0 - \bar{x})^2}{\sum(x_i - \bar{x})^2}}, \tag{2.39}$$

from which it follows that the confidence limits for μ_0 with confidence coefficient $(1 - \alpha)$ are given by

$$\hat{\mu}_0 \pm t_{(n-2,\alpha/2)} \ s.e.(\hat{\mu}_0). \tag{2.40}$$

Note that the point estimate of μ_0 is identical to the predicted response $\hat{y}_0$. This can be seen by comparing (2.35) with (2.38). The standard error of $\hat{\mu}_0$ is, however, smaller than the standard error of $\hat{y}_0$ and can be seen by comparing (2.36) with (2.39). Intuitively, this makes sense. There is greater uncertainty (variability) in predicting one observation (the next observation) than in estimating the mean response when $X = x_0$. The averaging that is implied in the mean response reduces the variability and uncertainty associated with the estimate.

To distinguish between the limits in (2.37) and (2.40), the limits in (2.37) are sometimes referred to as the *prediction* or *forecast* limits, whereas the limits given in (2.40) are called the *confidence limits*.

Suppose that we wish to predict the length of a service call in which four components had to be repaired. If $\hat{y}_4$ denotes the predicted value, then from (2.35) we get

$$\hat{y}_4 = 4.162 + 15.509 \times 4 = 66.20,$$

with a standard error that is obtained from (2.36) as

$$s.e.(\hat{y}_4) = 5.392\sqrt{1 + \frac{1}{14} + \frac{(4-6)^2}{114}} = 5.67.$$

On the other hand, if the service department wishes to estimate the expected (mean) service time for a call that needed four components repaired, we would use (2.38) and (2.39), respectively. Denoting by μ_4, the expected service time for a call that needed four components to be repaired, we have:

$$\hat{\mu}_4 = 4.162 + 15.509 \times 4 = 66.20,$$

with a standard error

$$s.e.(\hat{\mu}_4) = 5.392\sqrt{\frac{1}{14} + \frac{(4-6)^2}{114}} = 1.76.$$

With these standard errors we can construct confidence intervals using (2.37) and (2.40), as appropriate.

As can be seen from (2.36), the standard error of prediction increases the farther the value of the predictor variable is from the center of the actual observations. Care should be taken when predicting the value of Minutes corresponding to a value for Units that does not lie close to the observed data. There are two dangers in such predictions. First, there is substantial uncertainty due to the large standard error. More important, the linear relationship that has been estimated may not hold outside the range of observations. Therefore, care should be taken in employing fitted regression lines for prediction far outside the range of observations. In our example we would not use the fitted equation to predict the service time for a service call which requires that 25 components be replaced or repaired. This value lies too far outside the existing range of observations.

2.9   MEASURING THE QUALITY OF FIT

After fitting a linear model relating Y to X, we are interested not only in knowing whether a linear relationship exits, but also in measuring the quality of the fit of the model to the data. The quality of the fit can be assessed by one of the following highly related (hence, somewhat redundant) ways:

1. When using the tests in (2.25) or (2.31), if H_0 is rejected, the magnitude of the values of the test (or the corresponding p-values) gives us information about the *strength* (not just the existence) of the linear relationship between Y and X. Basically, the larger the t (in absolute value) or the smaller the corresponding p-value, the stronger the linear relationship between Y and X. These tests are objective but they require all the assumptions stated earlier, specially the assumption of normality of the ε's.

2. The strength of the linear relationship between Y and X can also be assessed directly from the examination of the scatter plot of Y versus X together with the corresponding value of the correlation coefficient $Cor(Y, X)$ in (2.6). The closer the set of points to a straight line (the closer $Cor(Y, X)$ to 1 or -1), the stronger the linear relationship between Y and X. This approach is informal and subjective but it requires only the linearity assumption.

3. Examine the scatter plot of Y versus $\hat{Y}$. The closer the set of points to a straight line, the stronger the linear relationship between Y and X. One can measure the strength of the linear relationship in this graph by computing the correlation coefficient between Y and $\hat{Y}$, which is given by

$$Cor(Y, \hat{Y}) = \frac{\sum(y_i - \bar{y})(\hat{y}_i - \bar{\hat{y}})}{\sqrt{\sum(y_i - \bar{y})^2 \sum(\hat{y}_i - \bar{\hat{y}})^2}}, \qquad (2.41)$$

where $\bar{y}$ is the mean of the response variable Y and $\bar{\hat{y}}$ is the mean of the fitted values. In fact, the scatter plot of Y versus X and the scatter plot of Y versus $\hat{Y}$ are redundant because the patterns of points in the two graphs are identical. The two corresponding values of the correlation coefficient are related. In fact,

$$Cor(Y, \hat{Y}) = |Cor(Y, X)|. \qquad (2.42)$$

Note that $Cor(Y, \hat{Y})$ cannot be negative (why?), but $Cor(Y, X)$ can be positive or negative ($-1 \leq Cor(Y, X) \leq 1$). Therefore, in simple linear regression, the scatter plot of Y versus $\hat{Y}$ is redundant. However, in multiple regression, the scatter plot of Y versus $\hat{Y}$ is not redundant. The graph is very useful because, as we shall see in Chapter 3, it is used to assess the strength of the relationship between Y and the set of predictor variables $X_1, X_2, \ldots, X_p$.

4. Although scatter plots of Y versus $\hat{Y}$ and $Cor(Y, \hat{Y})$ are redundant in simple linear regression, they give us an indication of the quality of the fit in both simple and multiple regression. Furthermore, in both simple and multiple regressions, $Cor(Y, \hat{Y})$ is related to another useful measure of the quality of fit of the linear model to the observed data. This measure is developed as follows. After we compute the least squares estimates of the parameters of a linear model, let us compute the following quantities:

$$
\begin{aligned}
SST &= \sum (y_i - \bar{y})^2, \\
SSR &= \sum (\hat{y}_i - \bar{y})^2, \\
SSE &= \sum (y_i - \hat{y}_i)^2,
\end{aligned}
\tag{2.43}
$$

where SST stands for the total sum of squared deviations in Y from its mean $\bar{y}$, SSR denotes the sum of squares due to regression, and SSE represents the sum of squared residuals (errors). The quantities $(\hat{y}_i - \bar{y})$, $(\hat{y}_i - \bar{y})$, and $(y_i - \hat{y}_i)$ are depicted in Figure 2.7 for a typical point (x_i, y_i). The line $\hat{y}_i = \hat{\beta}_0 + \hat{\beta}_1 x_i$ is the fitted regression line based on all data points (not shown on the graph) and the horizontal line is drawn at $Y = \bar{y}$. Note that for every point (x_i, y_i), there are two points, $(x_i, \hat{y}_i)$, which lies on the fitted line, and $(x_i, \bar{y})$ which lies on the line $Y = \bar{y}$.

A fundamental equality, in both simple and multiple regressions, is given by

$$
SST = SSR + SSE.
\tag{2.44}
$$

This equation arises from the description of an observation as

$$
\begin{array}{ccccc}
y_i & = & \hat{y}_i & + & (y_i - \hat{y}_i) \\
\text{Observed} & = & \text{Fit} & + & \text{Deviation from fit.}
\end{array}
$$

Subtracting $\bar{y}$ from both sides, we obtain

$$
\begin{array}{ccccc}
y_i - \bar{y} & = & (\hat{y}_i - \bar{y}) & + & (y_i - \hat{y}_i) \\
\text{Deviation from mean} & = & \text{Deviation due to fit} & + & \text{Residual.}
\end{array}
$$

Accordingly, the total sum of squared deviations in Y can be decomposed into the sum of two quantities, the first, SSR, measures the quality of X

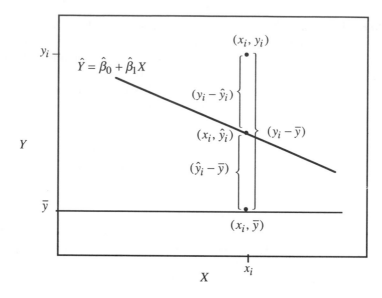

Fig. 2.7 A graphical illustration of various quantities computed after fitting a regression line to data.

as a predictor of Y, and the second, SSE, measures the error in this prediction. Therefore, the ratio $R^2 = SSR/SST$ can be interpreted as the proportion of the total variation in Y that is accounted for by the predictor variable X. Using (2.44), we can rewrite R^2 as

$$R^2 = \frac{SSR}{SST} = 1 - \frac{SSE}{SST}.$$ (2.45)

Additionally, it can be shown that

$$[Cor(Y, X)]^2 = [Cor(Y, \hat{Y})]^2 = R^2.$$ (2.46)

In simple linear regression, R^2 is equal to the square of the correlation coefficient between the response variable Y and the predictor X or to the square of the correlation coefficient between the response variable Y and the fitted values $\hat{Y}$. The definition given in (2.45) provides us with an alternative interpretation of the squared correlation coefficients. The *goodness-of-fit index*, R^2, may be interpreted as the proportion of the total variability in the response variable Y that is accounted for by the predictor variable X. Note that $0 \le R^2 \le 1$ because $SSE \le SST$. If R^2 is near 1, then X accounts for a large part of the variation in Y. For this reason, R^2 is known as the *coefficient of determination* because it gives us an idea of how the predictor variable X accounts for (determines) the response variable Y. The same interpretation of R^2 will carry over to the case of multiple regression.

Using the Computer Repair data, the fitted values, and the residuals in Table 2.7, the reader can verify that $Cor(Y, X) = Cor(Y, \hat{Y}) = 0.994$, from which it follows that $R^2 = (0.994)^2 = .987$. The same value of R^2 can be computed using (2.45). Verify that $SST = 27768.348$ and $SSE = 348.848$. So that

$$R^2 = 1 - \frac{SSE}{SST} = 1 - \frac{348.848}{27768.348} = 0.987.$$

The value $R^2 = 0.987$ indicates that nearly 99% of the total variability in the response variable (Minutes) is accounted for by the predictor variable (Units). The high value of R^2 indicates a strong linear relationship between servicing time and the number of units repaired during a service call.

We reemphasize that the regression assumptions should be checked before drawing statistical conclusions from the analysis (e.g., conducting tests of hypothesis or constructing confidence or prediction intervals) because the validity of these statistical procedures hinges on the validity of the assumptions. Chapter 4 presents a collection of graphical displays that can be used for checking the validity of the assumptions. We have used these graphs for the computer repair data and found no evidence that the underlying assumptions of regression analysis are not in order. In summary, the 14 data points in the Computer Repair data have given us an informative view of the repair time problem. Within the range of observed data, we are confident of the validity of our inferences and predictions.

2.10   REGRESSION LINE THROUGH THE ORIGIN

We have considered fitting the model

$$Y = \beta_0 + \beta_1 X + \varepsilon, \tag{2.47}$$

which is a regression line with an intercept. Sometimes, it may be necessary to fit the model

$$Y = \beta_1 X + \varepsilon, \tag{2.48}$$

a line passing through the origin. This model is also called the *no-intercept* model. The line may be forced to go through the origin because of subject matter or other external considerations. The choice between the models given in (2.47) and (2.48) has to be made with care. Rather than comparing R^2 in these cases, the goodness of fit should be judged by comparing the residual mean squares ($\hat{\sigma}^2$) produced by the two models. The R^2 value obtained from (2.47) is not strictly comparable with the R^2 value obtained from (2.48) because R^2 compares the deviations as measured from the sample mean in the first case, whereas in the second case the deviations are measured about zero. In fact, it is possible for R^2 to be negative in some cases from the fit of (2.48). The residual mean square, however, measures the closeness of the observed and predicted values for the two models.

2.11 TRIVIAL REGRESSION MODELS

In this section we give two examples of trivial regression models, that is, regression equations that have no regression coefficients. The first example arises when we wish to test for the mean μ of a single variable Y based on a random sample of n observations $y_1, y_2, \ldots, y_n$. Here we have $H_0 : \mu = 0$ against $H_1 : \mu \neq 0$. Assuming that Y is normally distributed with mean μ and variance σ^2, the well-known *one-sample t-test*

$$t = \frac{\bar{y} - 0}{s.e.(\bar{y})} = \frac{\bar{y}}{s_y/\sqrt{n}} , \tag{2.49}$$

can be used to test H_0, where s_y is sample standard deviation of Y. Alternatively, the above hypotheses can be formulated as

$$H_0(Model\ 1) : Y = \varepsilon \quad \text{against} \quad H_1(Model\ 2) : Y = \beta_0 + \varepsilon, \tag{2.50}$$

where $\beta_0 = \mu_0$. Thus, Model 1 indicates that $\mu = 0$ and Model 2 indicates that $\mu \neq 0$. The least squares estimate of β_0 in Model 2 is $\bar{y}$, the ith fitted value is $\hat{y}_i = \bar{y}$, and the ith residual is $e_i = y_i - \bar{y}$. It follows then that an estimate of σ^2 is

$$\hat{\sigma}^2 = \frac{SSE}{n-1} = \frac{\sum(y_i - \bar{y})^2}{n-1} = s_y^2, \tag{2.51}$$

which is the *sample variance* of Y. The standard error of $\hat{\beta}_0$ is then $\hat{\sigma}/\sqrt{n} = s_y/\sqrt{n}$, which is the familiar standard error of the sample mean $\bar{y}$. The t-test for testing Model 1 against Model 2 is

$$t_1 = \frac{\hat{\beta}_0 - 0}{s.e.(\hat{\beta}_0)} = \frac{\bar{y}}{s_y/\sqrt{n}} , \tag{2.52}$$

which is the same as the one-sample t-test in (2.49).

The second example occurs in connection with the *paired two-sample t-test*. For example, to test whether a given diet is effective in weight reduction, a random sample of n people is chosen and each person in the sample follows the diet for a specified period of time. Each person's weight is measured at the beginning of the diet and at the end of the period. Let Y_1 and Y_2 denote the weight at the beginning and at the end of diet period, respectively. Let $Y = Y_1 - Y_2$ be the difference between the two weights. Then Y is a random variable with mean μ and variance σ^2. Consequently, testing whether or not the diet is effective is the same as testing $H_0 : \mu = 0$ against $H_1 : \mu > 0$. With the definition of Y and assuming that Y is normally distributed, the well-known paired two-sample t-test is the same as the test in (2.49). This situation can be modeled as in (2.50) and the test in (2.52) can be used to test whether the diet is effective in weight reduction.

The above two examples show that the one-sample and the paired two-sample tests can be obtained as special cases using regression analysis.

2.12   BIBLIOGRAPHIC NOTES

The standard theory of regression analysis is developed in a number of good text books, some of which have been written to serve specific disciplines. Each provides a complete treatment of the standard results. The books by Snedecor and Cochran (1980), Fox (1984), and Kmenta (1986) develop the results using simple algebra and summation notation. The development in Searle (1971), Rao (1973), Seber (1977), Myers (1990), Sen and Srivastava (1990), Green (1993), Graybill and Iyer (1994), and Draper and Smith (1998) lean more heavily on matrix algebra.

EXERCISES

2.1   Using the data in Table 2.6:

(a)   Compute $Var(Y)$ and $Var(X)$.

(b)   Prove or verify that $\sum_{i=1}^{n}(y_i - \bar{y}) = 0$.

(c)   Prove or verify that any standardized variable has a mean of 0 and a standard deviation of 1.

(d)   Prove or verify that the three formulas for $Cor(Y, X)$ in (2.5), (2.6), and (2.7) are identical.

(e)   Prove or verify that the three formulas for $\hat{\beta}_1$ in (2.13) and (2.19) are identical.

2.2   Explain why you would or wouldn't agree with each of the following statements:

(a) $Cov(Y, X)$ and $Cor(Y, X)$ can take values between $-\infty$ and $+\infty$.

(b)   If $Cov(Y, X) = 0$ or $Cor(Y, X) = 0$, one can conclude that there is no relationship between Y and X.

(c)   The least squares line fitted to the points in the scatter plot of Y versus $\hat{Y}$ has a zero intercept and a unit slope.

2.3   Using the regression output in Table 2.9, test the following hypotheses using $\alpha = 0.1$:

(a) $H_0 : \beta_1 = 15$ versus $H_1 : \beta_1 \neq 15$

(b) $H_0 : \beta_1 = 15$ versus $H_1 : \beta_1 > 15$

(c) $H_0 : \beta_0 = 0$ versus $H_1 : \beta_0 \neq 0$

(d) $H_0 : \beta_0 = 5$ versus $H_1 : \beta_0 \neq 5$

2.4 Using the regression output in Table 2.9, construct the 99% confidence interval for β_0.

2.5 When fitting the simple linear regression model $Y = \beta_0 + \beta_1 X + \varepsilon$ to a set of data using the least squares method, each of the following statements can be proven to be true. Prove each statement mathematically or demonstrate its correctness numerically (using the data in Table 2.5):

(a) The sum of the ordinary least squares residuals is zero.

(b) The two tests in (2.25) and (2.31) are equivalent.

(c) The scatter plot of Y versus X and the scatter plot of Y versus $\hat{Y}$ have identical patterns.

(d) The correlation coefficient between Y and $\hat{Y}$ must be nonnegative.

2.6 Using the data in Table 2.5, and the fitted values and the residuals in Table 2.7, verify that:

(a) $Cor(Y, X) = Cor(Y, \hat{Y}) = 0.994$

(b) $SST = 27768.348$ (c) $SSE = 348.848$

2.7 Verify that the four data sets in Table 2.4 give identical results for the following quantities:.

(a) $\hat{\beta}_0$ and $\hat{\beta}_1$ (b) $Cor(Y, X)$

(c) R^2 (d) The t-test

2.8 When fitting a simple linear regression model $Y = \beta_0 + \beta_1 X + \varepsilon$ to a set of data using the least squares method, suppose that $H_0 : \beta_1 = 0$ was not rejected. This implies that the model can be written simply as: $Y = \beta_0 + \varepsilon$. The least squares estimate of β_0 is $\hat{\beta}_0 = \bar{y}$. (Can you prove that?)

(a) What are the ordinary least squares residuals in this case?

(b) Show that the ordinary least squares residuals sum up to zero.

2.9 Let Y and X denote the labor force participation rate of women in 1972 and 1968, respectively, in each of 19 cities in the United States. The regression output for this data set is shown in Table 2.10. It was also found that $SSR = 0.0358$ and $SSE = 0.0544$. Suppose that the model $Y = \beta_0 + \beta_1 X + \varepsilon$ satisfies the usual regression assumptions.

(a) Compute $Var(Y)$ and $Cor(Y, X)$.

(b) Suppose that the participation rate of women in 1968 in a given city is 45%. What is the estimated participation rate of women in 1972 for the same city?

(c) Suppose further that the mean and variance of the participation rate of women in 1968 are 0.5 and 0.005, respectively. Construct the 95% confidence interval for the estimate in Exercise 2.9b.

(d) Construct the 95% confidence interval for the slope of the true regression line, β_1.

(e) Test the hypothesis: $H_0 : \beta_1 = 1$ versus $H_1 : \beta_1 > 1$ at the 5% significance level.

Table 2.10 Regression Output When Y Is Regressed on X for the Labor Force Participation Rate of Women.

Variable	Coefficient	s.e.	t-test	p-value
Constant	0.203311	0.0976	2.08	0.0526
X	0.656040	0.1961	3.35	< 0.0038
$n = 19$	$R^2 = 0.397$	$R_a^2 = 0.362$	$\hat{\sigma} = 0.0566$	$d.f. = 17$

(f) If Y and X were reversed in the above regression, what would you expect R^2 to be?

2.10 One may wonder if people of similar heights tend to marry each other. For this purpose, a sample of newly married couples was selected. Let X be the height of the husband and Y be the height of the wife. The heights (in centimeters) of husbands and wives are found in Table 2.11. The data can also be found in the book's Web site.

(a) Compute the covariance between the heights of the husbands and wives.

(b) What would the covariance be if heights were measured in inches rather than in centimeters?

(c) Compute the correlation coefficient between the heights of the husband and wife.

(d) What would the correlation be if heights were measured in inches rather than in centimeters?

(e) What would the correlation be if every man married a woman exactly 5 centimeters shorter than him?

(f) We wish to fit a regression model relating the heights of husbands and wives. Which one of the two variables would you choose as the response variable? Justify your answer.

(g) Using your choice of the response variable in Exercise 2.10f, test the null hypothesis that the slope is zero.

(h) Using your choice of the response variable in Exercise 2.10f, test the null hypothesis that the intercept is zero.

(i) Using your choice of the response variable in Exercise 2.10f, test the null hypothesis that the both the intercept and the slope are zero.

(j) Which of the above hypotheses and tests would you choose to test whether people of similar heights tend to marry each other? What is your conclusion?

(k) If none of the above tests is appropriate for testing the hypothesis that people of similar heights tend to marry each other, which test would you use? What is your conclusion based on this test?

2.11 Consider fitting a simple linear regression model through the origin, $Y = \beta_1 X + \varepsilon$, to a set of data using the least squares method. The least squares estimate of β_1 in this case is $\hat{\beta}_1 = \sum(y_i x_i)/(\sum x_i^2)$. (Can you prove that?)

(a) Give an example of a situation where fitting the model (2.48) is justified by theoretical or other considerations.

(b) Show that the residuals $e_1, e_2, \ldots, e_n$ will not necessarily add up to zero.

(c) Give an example of a data set Y and X in which R^2 computed from fitting (2.48) to the data is negative.

(d) Which goodness of fit measures would you use to compare model (2.48) with model (2.47)?

2.12 In order to investigate the feasibility of starting a Sunday edition for a large metropolitan newspaper, information was obtained from a sample of 34 newspapers concerning their daily and Sunday circulations (in thousands) (*Source: Gale Directory of Publications*, 1994). The data are given in Table 2.12 and can be found in the book's Web site.

(a) Construct a scatter plot of Sunday circulation versus daily circulation. Does the plot suggest a linear relationship between Daily and Sunday circulation? Do you think this is a plausible relationship?

(b) Fit a regression line predicting Sunday circulation from Daily circulation.

(c) Obtain the 95% confidence intervals for β_0 and β_1.

(d) Is there a significant relationship between Sunday circulation and Daily circulation? Justify your answer by a statistical test. Indicate what hypothesis you are testing and your conclusion.

(e) What proportion of the variability in Sunday circulation is accounted for by Daily circulation?

(f) Provide an interval estimate (based on 95% level) for the true average Sunday circulation of newspapers with Daily circulation of 500,000.

(g) The particular newspaper that is considering a Sunday edition has a Daily circulation of 500,000. Provide an interval estimate (based on 95% level) for the predicted Sunday circulation of this paper. How does this interval differ from that given in Exercise 2.12f?

(h) Another newspaper being considered as a candidate for a Sunday edition has a Daily circulation of 2,000,000. Provide an interval estimate for the predicted Sunday circulation for this paper? How does this interval compare with the one given in Exercise 2.12g? Do you think it is likely to be accurate?

Table 2.11 Heights of Husband (H) and Wife (W) in (Centimeters).

Row	H	W	Row	H	W	Row	H	W
1	186	175	33	180	166	65	181	175
2	180	168	34	188	181	66	170	169
3	160	154	35	153	148	67	161	149
4	186	166	36	179	169	68	188	176
5	163	162	37	175	170	69	181	165
6	172	152	38	165	157	70	156	143
7	192	179	39	156	162	71	161	158
8	170	163	40	185	174	72	152	141
9	174	172	41	172	168	73	179	160
10	191	170	42	166	162	74	170	149
11	182	170	43	179	159	75	170	160
12	178	147	44	181	155	76	165	148
13	181	165	45	176	171	77	165	154
14	168	162	46	170	159	78	169	171
15	162	154	47	165	164	79	171	165
16	188	166	48	183	175	80	192	175
17	168	167	49	162	156	81	176	161
18	183	174	50	192	180	82	168	162
19	188	173	51	185	167	83	169	162
20	166	164	52	163	157	84	184	176
21	180	163	53	185	167	85	171	160
22	176	163	54	170	157	86	161	158
23	185	171	55	176	168	87	185	175
24	169	161	56	176	167	88	184	174
25	182	167	57	160	145	89	179	168
26	162	160	58	167	156	90	184	177
27	169	165	59	157	153	91	175	158
28	176	167	60	180	162	92	173	161
29	180	175	61	172	156	93	164	146
30	157	157	62	184	174	94	181	168
31	170	172	63	185	160	95	187	178
32	186	181	64	165	152	96	181	170

Table 2.12 Newspapers Data: Daily and Sunday Circulations (in Thousands).

Newspaper	Daily	Sunday
Baltimore Sun	391.952	488.506
Boston Globe	516.981	798.298
Boston Herald	355.628	235.084
Charlotte Observer	238.555	299.451
Chicago Sun Times	537.780	559.093
Chicago Tribune	733.775	1133.249
Cincinnati Enquirer	198.832	348.744
Denver Post	252.624	417.779
Des Moines Register	206.204	344.522
Hartford Courant	231.177	323.084
Houston Chronicle	449.755	620.752
Kansas City Star	288.571	423.305
Los Angeles Daily News	185.736	202.614
Los Angeles Times	1164.388	1531.527
Miami Herald	444.581	553.479
Minneapolis Star Tribune	412.871	685.975
New Orleans Times-Picayune	272.280	324.241
New York Daily News	781.796	983.240
New York Times	1209.225	1762.015
Newsday	825.512	960.308
Omaha World Herald	223.748	284.611
Orange County Register	354.843	407.760
Philadelphia Inquirer	515.523	982.663
Pittsburgh Press	220.465	557.000
Portland Oregonian	337.672	440.923
Providence Journal-Bulletin	197.120	268.060
Rochester Democrat & Chronicle	133.239	262.048
Rocky Mountain News	374.009	432.502
Sacramento Bee	273.844	338.355
San Francisco Chronicle	570.364	704.322
St. Louis Post-Dispatch	391.286	585.681
St. Paul Pioneer Press	201.860	267.781
Tampa Tribune	321.626	408.343
Washington Post	838.902	1165.567

3

Multiple Linear Regression

3.1 INTRODUCTION

In this chapter the general multiple linear regression model is presented. The presentation serves as a review of the standard results on regression analysis. The standard theoretical results are given without mathematical derivations, but illustrated by numerical examples. Readers interested in mathematical derivations are referred to the bibliographic notes at the end of Chapter 2, where a number of books that contain a formal development of multiple linear regression theory is given.

3.2 DESCRIPTION OF THE DATA AND MODEL

The data consist of n observations on a dependent or response variable Y and p predictor or explanatory variables, $X_1, X_2, \ldots, X_p$. The observations are usually represented as in Table 3.1. The relationship between Y and $X_1, X_2, \ldots, X_p$ is formulated as a linear model

$$Y = \beta_0 + \beta_1 X_1 + \beta_2 X_2 + \cdots + \beta_p X_p + \varepsilon, \tag{3.1}$$

where $\beta_0, \beta_1, \beta_2, \ldots, \beta_p$ are constants referred to as the model *partial* regression coefficients (or simply as the *regression coefficients*) and ε is a random disturbance or error. It is assumed that for any set of fixed values of $X_1, X_2, \ldots, X_p$ that fall within the range of the data, the linear equation (3.1) provides an acceptable approximation of the true relationship between Y and the X's (Y is approximately a linear function of the X's, and ε measures the discrepancy in that approximation). In particular, ε contains no systematic information for determining Y that is not already captured by the X's.

Table 3.1 Notation for the Data Used in Multiple Regression Analysis.

Observation	Response	Predictors			
Number	Y	X_1	X_2	$\ldots$	X_p
1	y_1	x_{11}	x_{12}	$\ldots$	x_{1p}
2	y_2	x_{21}	x_{22}	$\ldots$	x_{2p}
3	y_3	x_{31}	x_{32}	$\ldots$	x_{3p}
$\vdots$	$\vdots$	$\vdots$	$\vdots$	$\vdots$	$\vdots$
n	y_n	x_{n1}	x_{n2}	$\ldots$	x_{np}

According to (3.1), each observation in Table 3.1 can be written as

$$y_i = \beta_0 + \beta_1 x_{i1} + \ldots + \beta_p x_{ip} + \varepsilon_i, \quad i = 1, 2, \ldots, n, \qquad (3.2)$$

where y_i represents the ith value of the response variable Y, $x_{i1}, x_{i2}, \ldots, x_{ip}$ represent values of the predictor variables for the ith unit (the ith row in Table 3.1), and ε_i represents the error in the approximation of y_i.

Multiple linear regression is an extension (generalization) of simple linear regression. Thus, the results given here are essentially extensions of the results given in Chapter 2. One can similarly think of simple regression as a special case of multiple regression because all simple regression results can be obtained using the multiple regression results when the number of predictor variables $p = 1$. For example, when $p = 1$, (3.1) and (3.2) reduce to (2.9) and (2.10), respectively.

3.3 EXAMPLE: SUPERVISOR PERFORMANCE DATA

Throughout this chapter we use data from a study in *industrial psychology* (management) to illustrate some of the standard regression results. A recent survey of the clerical employees of a large financial organization included questions related to employee satisfaction with their supervisors. There was a question designed to measure the overall performance of a supervisor, as well as questions that were related to specific activities involving interaction between supervisor and employee. An exploratory study was undertaken to try to explain the relationship between specific supervisor characteristics and overall satisfaction with supervisors as perceived by the employees. Initially, six questionnaire items were chosen as possible explanatory variables. Table 3.2 gives the description of the variables in the study. As can be seen from the list, there are two broad types of variables included in the study. Variables

X_1, X_2, and X_5 relate to direct interpersonal relationships between employee and supervisor, whereas variables X_3 and X_4 are of a less personal nature and relate to the job as a whole. Variable X_6 is not a direct evaluation of the supervisor but serves more as a general measure of how the employee perceives his or her own progress in the company.

The data for the analysis were generated from the individual employee response to the items on the survey questionnaire. The response on any item ranged from 1 through 5, indicating very satisfactory to very unsatisfactory, respectively. A dichotomous index was created to each item by collapsing the response scale to two categories: {1,2}, to be interpreted as a favorable response, and {3,4,5}, representing an unfavorable response. The data were collected in 30 departments selected at random from the organization. Each department had approximately 35 employees and one supervisor. The data to be used in the analysis, given in Table 3.3, were obtained by aggregating responses for departments to get the proportion of favorable responses for each item for each department. The resulting data therefore consist of 30 observations on seven variables, one observation for each department. We refer to this data set as the *Supervisor Performance* data. The data set can also be found in the book's Web site.[1]

Table 3.2 Description of Variables in Supervisor Performance Data.

Variable	Description
Y	Overall rating of job being done by supervisor
X_1	Handles employee complaints
X_2	Does not allow special privileges
X_3	Opportunity to learn new things
X_4	Raises based on performance
X_5	Too critical of poor performance
X_6	Rate of advancing to better jobs

A linear model of the form

$$Y = \beta_0 + \beta_1 X_1 + \beta_2 X_2 + \cdots + \beta_6 X_6 + \varepsilon, \tag{3.3}$$

relating Y and the six explanatory variables, is assumed. Methods for the validation of this and other assumptions are presented in Chapter 4.

---

[1] http://www.ilr.cornell.edu/~hadi/RABE

Table 3.3 Supervisor Performance Data.

Row	Y	X_1	X_2	X_3	X_4	X_5	X_6
1	43	51	30	39	61	92	45
2	63	64	51	54	63	73	47
3	71	70	68	69	76	86	48
4	61	63	45	47	54	84	35
5	81	78	56	66	71	83	47
6	43	55	49	44	54	49	34
7	58	67	42	56	66	68	35
8	71	75	50	55	70	66	41
9	72	82	72	67	71	83	31
10	67	61	45	47	62	80	41
11	64	53	53	58	58	67	34
12	67	60	47	39	59	74	41
13	69	62	57	42	55	63	25
14	68	83	83	45	59	77	35
15	77	77	54	72	79	77	46
16	81	90	50	72	60	54	36
17	74	85	64	69	79	79	63
18	65	60	65	75	55	80	60
19	65	70	46	57	75	85	46
20	50	58	68	54	64	78	52
21	50	40	33	34	43	64	33
22	64	61	52	62	66	80	41
23	53	66	52	50	63	80	37
24	40	37	42	58	50	57	49
25	63	54	42	48	66	75	33
26	66	77	66	63	88	76	72
27	78	75	58	74	80	78	49
28	48	57	44	45	51	83	38
29	85	85	71	71	77	74	55
30	82	82	39	59	64	78	39

3.4 PARAMETER ESTIMATION

Based on the available data, we wish to estimate the parameters $\beta_0, \beta_1, \ldots, \beta_p$. As in the case of simple regression presented in Chapter 2, we use the least squares method, that is, we minimize the sum of squares of the errors. From (3.2), the errors can be written as

$$\varepsilon_i = y_i - \beta_0 - \beta_1 x_{i1} - \ldots - \beta_p x_{ip}, \quad i = 1, 2, \ldots, n. \tag{3.4}$$

The sum of squares of these errors is

$$S(\beta_0, \beta_1, \ldots, \beta_p) = \sum_{i=1}^{n} \varepsilon_i^2 = \sum_{i=1}^{n} (y_i - \beta_0 - \beta_1 x_{i1} - \ldots - \beta_p x_{ip})^2.$$

By a direct application of calculus, it can be shown that the least squares estimates $\hat{\beta}_0, \hat{\beta}_1, \ldots, \hat{\beta}_p$ which minimize $S(\beta_0, \beta_1, \ldots, \beta_p)$, are given by the solution of the following system of equations:

$$s_{11}\hat{\beta}_1 + s_{12}\hat{\beta}_2 + \cdots + s_{1p}\hat{\beta}_p = s_{y1}$$
$$s_{12}\hat{\beta}_1 + s_{22}\hat{\beta}_2 + \cdots + s_{2p}\hat{\beta}_p = s_{y2}$$
$$\vdots \tag{3.5}$$
$$s_{1p}\hat{\beta}_1 + s_{2p}\hat{\beta}_2 + \cdots + s_{pp}\hat{\beta}_p = s_{yp},$$

where

$$s_{ij} = \sum_{a=1}^{n} (x_{ai} - \bar{x}_i)(x_{aj} - \bar{x}_j), \quad i, j = 1, 2, \ldots, p;$$

$$s_{yj} = \sum_{a=1}^{n} (y_a - \bar{y})(x_{aj} - \bar{x}_j), \qquad j = 1, 2, \ldots, p;$$

$$\bar{x}_j = \frac{\sum_{a=1}^{n} x_{aj}}{n}, \quad \bar{y} = \frac{\sum_{a=1}^{n} y_a}{n},$$

and

$$\hat{\beta}_0 = \bar{y} - \hat{\beta}_1 \bar{x}_1 - \hat{\beta}_2 \bar{x}_2 - \cdots - \hat{\beta}_p \bar{x}_p.$$

The equations in the above system are called the *normal equations*. The estimate $\hat{\beta}_0$ is usually referred to as the *intercept* or *constant*, and $\hat{\beta}_j$ as the *estimate* of the (partial) regression coefficient of the predictor X_j.

We assume that the system of equations is solvable and has a unique solution. A closed-form formula for the solution is given in the Appendix of this chapter for readers who are familiar with matrix notation. We shall not say anything more about the actual process of solving the normal equations. We assume the availability of computer software that gives a numerically accurate solution.

Using the estimated regression coefficients $\hat{\beta}_0, \hat{\beta}_1, \ldots, \hat{\beta}_p$, we write the fitted least squares regression equation as

$$\hat{Y} = \hat{\beta}_0 + \hat{\beta}_1 X_1 + \ldots + \hat{\beta}_p X_p. \tag{3.6}$$

For each observation in our data we can compute

$$\hat{y}_i = \hat{\beta}_0 + \hat{\beta}_1 x_{i1} + \ldots + \hat{\beta}_p x_{ip}, \quad i = 1, 2, \ldots, n. \tag{3.7}$$

These are called the *fitted* values. The corresponding *ordinary* least squares residuals are given by

$$e_i = y_i - \hat{y}_i, \quad i = 1, 2, \ldots, n. \tag{3.8}$$

An unbiased estimate of σ^2 is given by

$$\hat{\sigma}^2 = \frac{SSE}{n - p - 1}, \tag{3.9}$$

where

$$SSE = \sum_{i=1}^{n} (y_i - \hat{y}_i)^2 = \sum_{i=1}^{n} e_i^2, \tag{3.10}$$

is the *sum of squared residuals*. The number $n - p - 1$ in the denominator of (3.9) is called the *degrees of freedom* $(d.f.)$. It is equal to the number of observations minus the number of estimated regression coefficients.

When certain assumptions hold, the least squares estimators have several desirable properties. Chapter 4 is devoted entirely to validation of the assumptions. We should note, however, that we have applied these validation procedures on the Supervisor Performance data that we use as illustrative numerical examples in this chapter and found no evidence for model misspecification. We will, therefore, continue with the presentation of multiple regression analysis in this chapter knowing that the required assumptions are valid for the Supervisor Performance data.

The properties of least squares estimators are presented in Section 3.6. Based on these properties, one can develop proper statistical inference procedures (e.g., confidence interval estimation, tests of hypothesis, and goodness-of-fit tests). These are presented in Sections 3.7 to 3.10.

3.5   INTERPRETATIONS OF REGRESSION COEFFICIENTS

The interpretation of the regression coefficients in a multiple regression equation is a source of common confusion. The simple regression equation represents a line, while the multiple regression equation represents a plane (in cases of two predictors) or a hyperplane (in cases of more than two predictors). In multiple regression, the coefficient β_0, called the *constant coefficient*, is the value of Y when $X_1 = X_2 = \ldots = X_p = 0$, as in simple regression. The regression coefficient $\beta_j, j = 1, 2, \ldots, p$, has several interpretations. It may be interpreted as the change in Y corresponding to a unit change in X_j when all other predictor variables are held constant. Magnitude of the change is not dependent on the values at which the other predictor variables are fixed. In practice, however, the predictor variables may be inherently related, and holding some of them constant while varying the others may not be possible.

The regression coefficient β_j is also called the *partial regression coefficient* because β_j represents the contribution of X_j to the response variable Y after it has been adjusted for the other predictor variables. What does "adjusted for" mean in multiple regression? Without loss of any generality, we address this question using the simplest multiple regression case where we have two predictor variables. When $p = 2$, the model is

$$Y = \beta_0 + \beta_1 X_1 + \beta_2 X_2 + \varepsilon. \tag{3.11}$$

We use the variables X_1 and X_2 from the Supervisor data to illustrate the concepts. A statistical package gives the estimated regression equation as

$$\hat{Y} = 15.3276 + 0.7803\ X_1 - 0.0502\ X_2. \tag{3.12}$$

The coefficient of X_1 suggests that each unit of X_1 adds 0.7803 to Y when the value of X_2 is held fixed. As we show below, this is also the effect of X_1 after adjusting for X_2. Similarly, the coefficient of X_2 suggests that each unit of X_2 subtracts about 0.0502 from Y when the value of X_1 is held fixed. This is also the effect of X_2 after adjusting for X_1.

This interpretation can be easily understood when we consider the fact that the multiple regression equation can be obtained from a series of simple regression equations. For example, the coefficient of X_2 in (3.12) can be obtained as follows:

1. Fit the simple regression model that relates Y to X_1. Let the residuals from this regression be denoted by $e_{Y \cdot X_1}$. This notation indicates that the variable that comes before the dot is treated as a response variable and the variable that comes after the dot is considered as a predictor. The fitted regression equation is:

$$\hat{Y} = 14.3763 + 0.754610\ X_1. \tag{3.13}$$

2. Fit the simple regression model that relates X_2 (considered temporarily here as a response variable) to X_1. Let the residuals from this regression be denoted by $e_{X_2 \cdot X_1}$. The fitted regression equation is:

$$\hat{X}_2 = 18.9654 + 0.513032\ X_1. \tag{3.14}$$

The residuals, $e_{Y \cdot X_1}$ and $e_{X_2 \cdot X_1}$ are given in Table 3.4.

Table 3.4 Partial Residuals.

Row	$e_{Y \cdot X_1}$	$e_{X_2 \cdot X_1}$	Row	$e_{Y \cdot X_1}$	$e_{X_2 \cdot X_1}$
1	−9.8614	−15.1300	16	−1.2912	−15.1383
2	0.3287	−0.7995	17	−4.5182	1.4269
3	3.8010	13.1224	18	5.3471	15.2527
4	−0.9167	−6.2864	19	−2.1990	−8.8776
5	7.7641	−2.9819	20	−8.1437	19.2787
6	−12.8799	1.8178	21	5.4393	−6.4867
7	−6.9352	−11.3385	22	3.5925	1.7397
8	0.0279	−7.4428	23	−11.1806	−0.8255
9	−4.2543	10.9660	24	−2.2969	4.0524
10	6.5925	−5.2604	25	7.8748	−4.6691
11	9.6294	6.8439	26	−6.4813	7.5311
12	7.3471	−2.7473	27	7.0279	0.5572
13	7.8379	6.2266	28	−9.3891	−4.2082
14	−9.0089	21.4529	29	6.4818	8.4269
15	4.5187	−4.4689	30	5.7457	−22.0340

3. Fit the simple regression model that relates the above two residuals. In this regression, the response variable is $e_{Y \cdot X_1}$ and the predictor variable is $e_{X_2 \cdot X_1}$. The fitted regression equation is:

$$\hat{e}_{Y \cdot X_1} = 0 - 0.0502\, e_{X_2 \cdot X_1}. \tag{3.15}$$

The interesting result here is that the coefficient of $e_{X_2 \cdot X_1}$ in this last regression is the same as the multiple regression coefficient of X_2 in (3.12). The two coefficients are equal to −0.0502. In fact, their standard errors are also the same. What's the intuition here? In the first step, we found the linear relationship between Y and X_1. The residual from this regression is Y after taking or partialling out the linear effects of X_1. In other words, the residual is that part of Y that is not linearly related to X_1. In the second step we do the same thing, replacing Y by X_2, so the residual is the part of X_2 that is not linearly related to X_1. In the third step we look for the linear relationship between the Y residual and the X_2 residual. The resultant regression coefficient represents the effect of X_2 on Y after taking out the effects of X_1 from both Y and X_2.

The regression coefficient β_j is the partial regression coefficient because it represents the contribution of X_j to the response variable Y after both variables have been linearly adjusted for the other predictor variables (see also Exercise 3.4).

Note that the estimated intercept in the regression equation in (3.15) is zero because the two sets of residuals have a mean of zero (they sum up to zero). The same procedures can be applied to obtain the multiple regression coefficient of X_1 in (3.12). Simply interchange X_2 by X_1 in the above three steps. This is left as an exercise for the reader.

From the above discussion we see that the simple and the multiple regression coefficients are not the same unless the predictor variables are uncorrelated. In nonexperimental, or observational data, the predictor variables are rarely uncorrelated. In an experimental setting, in contrast, the experimental design is often set up to produce uncorrelated explanatory variables because in an experiment the researcher sets the values of the predictor variables. So in samples derived from experiments it may be the case that the explanatory variables are uncorrelated and hence the simple and multiple regression coefficients in that sample would be the same.

3.6   PROPERTIES OF THE LEAST SQUARES ESTIMATORS

Under certain standard regression assumptions (to be stated in Chapter 4), the least squares estimators have the properties listed below. A reader familiar with matrix algebra will find concise statements of these properties employing matrix notation in the Appendix at the end of the chapter.

1. The estimator $\hat{\beta}_j, j = 0, 1, \ldots, p$, is an unbiased estimate of β_j and has a variance of $\sigma^2 c_{jj}$, where c_{jj} is the jth diagonal element of the inverse of a matrix known as the *corrected sums of squares and products* matrix. The covariance between $\hat{\beta}_i$ and $\hat{\beta}_j$ is $\sigma^2 c_{ij}$, where c_{ij} is the element in the ith row and jth column of the inverse of the corrected sums of squares and products matrix. For all unbiased estimates that are linear in the observations the least squares estimators have the smallest variance. Thus, the least squares estimators are said to be BLUE (*best linear unbiased estimators*).

2. The estimator $\hat{\beta}_j, j = 0, 1, \ldots, p$, is normally distributed with mean β_j and variance $\sigma^2 c_{jj}$.

3. $W = SSE/\sigma^2$ has a χ^2 distribution with $(n - p - 1)$ degrees of freedom, and $\hat{\beta}_j$'s and $\hat{\sigma}^2$ are distributed independently of each other.

4. The vector $\hat{\boldsymbol{\beta}} = (\hat{\beta}_0, \hat{\beta}_1, \ldots, \hat{\beta}_p)$ has a $(p+1)$-variate normal distribution with mean vector $\boldsymbol{\beta} = (\beta_0, \beta_1, \ldots, \beta_p)$ and variance-covariance matrix with elements $\sigma^2 c_{ij}$.

The results above enable us to test various hypotheses about individual regression parameters and to construct confidence intervals. These are discussed in Section 3.8.

3.7 MULTIPLE CORRELATION COEFFICIENT

After fitting the linear model to a given data set, an assessment is made of the adequacy of fit. The discussion given in Section 2.9 applies here. All the material extend naturally to multiple regression and will not be repeated here.

The strength of the linear relationship between Y and the set of predictors $X_1, X_2, \ldots, X_p$ can be assessed through the examination of the scatter plot of Y versus $\hat{Y}$ and the correlation coefficient between Y and $\hat{Y}$, which is given by

$$Cor(Y, \hat{Y}) = \frac{\sum (y_i - \bar{y})(\hat{y}_i - \bar{\hat{y}})}{\sqrt{\sum (y_i - \bar{y})^2 \sum (\hat{y}_i - \bar{\hat{y}})^2}}, \qquad (3.16)$$

where $\bar{y}$ is the mean of the response variable Y and $\bar{\hat{y}}$ is the mean of the fitted values. As in the simple regression case, the coefficient of determination $R^2 = [Cor(Y, \hat{Y})]^2$ is also given by

$$R^2 = \frac{SSR}{SST} = 1 - \frac{SSE}{SST} = 1 - \frac{\sum (y_i - \hat{y}_i)^2}{\sum (y_i - \bar{y})^2}, \qquad (3.17)$$

as in (2.45). Thus, R^2 may be interpreted as the proportion of the total variability in the response variable Y that can be accounted for by the set of predictor variables $X_1, X_2, \ldots, X_p$. In multiple regression, $R = \sqrt{R^2}$ is called the *multiple correlation coefficient* because it measures the relationship between one variable Y and a set of variables $X_1, X_2, \ldots, X_p$.

From Table 3.5, the value of R^2 for the Supervisor Performance data is 0.73, showing that about 73% of the total variation in the overall rating of the job being done by the supervisor can be accounted for by the six variables.

When the model fits the data well, it is clear that the value of R^2 is close to unity. With a good fit, the observed and predicted values will be close to each other, and $\sum (y_i - \hat{y}_i)^2$ will be small. Then R^2 will be near unity. On the other hand, if there is no linear relationship between Y and the predictor variables, $X_1, \ldots, X_p$, the linear model gives a poor fit, the best predicted value for an observation y_i would be $\bar{y}$; that is, in the absence of any relationship with the predictors, the best estimate of any value of Y is the sample mean, because the sample mean minimizes the sum of squared deviations. So in the absence of any linear relationship between Y and the X's, R^2 will be near zero. The value of R^2 is used as a summary measure to judge the fit of the linear model to a given body of data. As pointed out in Chapter 2, a large value of R^2 does not necessarily mean that the model fits the data well. As we outline in Section 3.9, a more detailed analysis is needed to ensure that the model adequately describes the data.

A quantity related to R^2, known as the *adjusted R-squared*, R_a^2, is also used for judging the goodness of fit. It is defined as

$$R_a^2 = 1 - \frac{SSE/(n - p - 1)}{SST/(n - 1)}, \qquad (3.18)$$

which is obtained from R^2 in (3.17) after dividing SSE and SST by their respective degrees of freedom. From (3.18) and (3.17) it follows that

$$R_a^2 = 1 - \frac{n-1}{n-p-1}(1 - R^2). \tag{3.19}$$

R_a^2 is sometimes used to compare models having different numbers of predictor variables. (This is described in Chapter 11.) In comparing the goodness of fit of models with different numbers of explanatory variables, R_a^2 tries to "adjust" for the unequal number of variables in the different models. Unlike R^2, R_a^2 cannot be interpreted as the proportion of total variation in Y accounted for by the predictors. Many regression packages provide values for both R^2 and R_a^2.

3.8 INFERENCE FOR INDIVIDUAL REGRESSION COEFFICIENTS

Using the properties of the least squares estimators discussed in Section 3.6, one can make statistical inference regarding the regression coefficients. The statistic for testing $H_0 : \beta_j = \beta_j^0$ versus $H_1 : \beta_j \neq \beta_j^0$, where β_j^0 is a constant chosen by the investigator, is

$$t_j = \frac{\hat{\beta}_j - \beta_j^0}{s.e.(\hat{\beta}_j)} , \tag{3.20}$$

which has a Student's t-distribution with $(n - p - 1)$ degrees of freedom. The test is carried out by comparing the observed value with the appropriate critical value $t_{(n-p-1,\alpha/2)}$, which is obtained from the t-table given in the Appendix to this book (see Table A.2), where α is the significance level. Note that we divide the significance level α by 2 because we have a two-sided alternative hypothesis. Accordingly, H_0 is to be rejected at the significance level α if

$$|t_j| \geq t_{(n-p-1,\alpha/2)}, \tag{3.21}$$

where $|t_j|$ denotes the absolute value of t_j. A criterion equivalent to that in (3.21) is to compare the p-value of the test with α and reject H_0 if

$$p(|t_j|) \leq \alpha, \tag{3.22}$$

where $p(|t_j|)$, is the p-value of the test, which is the probability that a random variable having a Student t-distribution, with $(n - p - 1)$, is greater than $|t_j|$ (the absolute value of the observed value of the t-test); see Figure 2.6. The p-value is usually computed and supplied as part of the regression output by many statistical packages.

The usual test is for $H_0 : \beta_j^0 = 0$, in which case the t-test reduces to

$$t_j = \frac{\hat{\beta}_j}{s.e.(\hat{\beta}_j)} , \tag{3.23}$$

which is the ratio of $\hat{\beta}_j$ to its standard error, $s.e.(\hat{\beta}_j)$ given in the Appendix to this chapter, in (A.8). The standard errors of the coefficients are computed by the statistical packages as part of their standard regression output.

Note that the rejection of H_0: $\beta_j = 0$ would mean that β_j is likely to be different from 0, and hence the predictor variable X_j is a statistically significant predictor of the response variable Y after adjusting for the other predictor variables.

As another example of statistical inference, the confidence limits for β_j with confidence coefficient α are given by

$$\hat{\beta}_j \pm \hat{\sigma}\sqrt{c_{jj}}\, t_{(n-p-1,\alpha/2)}, \tag{3.24}$$

where $t_{(n-p-1,\alpha)}$ is the $(1-\alpha)$ percentile point of the t-distribution with $(n-p-1)$ degrees of freedom. The confidence interval in (3.24) is for the individual coefficient β_j. A joint confidence region of all regression coefficients is given in the Appendix to this chapter in (A.13).

Note that when $p = 1$ (simple regression), the t-test in (3.23) and the criteria in (3.21) and (3.22) reduce to the t-test in (2.25) and the criteria in (2.26) and (2.27), respectively, illustrating the fact that simple regression results can be obtained from the multiple regression results by setting $p = 1$.

Many other statistical inference situations arise in practice in connection with multiple regression. These will be considered in the following sections.

Example: Supervisor Performance Data (Cont.)

Let us now illustrate the above t-tests using the Supervisor Performance data described earlier in this chapter. The results of fitting a linear model relating Y and the six explanatory variables are given in Table 3.5. The fitted regression equation is

$$\hat{Y} = 10.787 + 0.613X_1 - 0.073X_2 + 0.320X_3 + 0.081X_4 + 0.038X_5 - 0.217X_6. \tag{3.25}$$

The t-values in Table 3.5 test the null hypothesis $H_0 : \beta_j = 0, j = 0, 1, \ldots, p$, against an alternative $H_1 : \beta_j \neq 0$. From Table 3.5 it is seen that only the regression coefficient of X_1 is significantly different from zero and X_3 has a regression coefficient that approach being significantly different from zero. The other variables have insignificant t-tests. The construction of confidence intervals for the individual parameters is left as an exercise for the reader.

It should be noted here that the constant in the above model is statistically not significant (t-value of 0.93 and p-value of 0.3616). In any regression model, unless there is strong theoretical reason, a constant should always be included even if the term is statistically not significant. The constant represents the base or background level of the response variable. Insignificant predictors should not be in general retained but a constant should be retained.

Table 3.5   Regression Output for the Supervisor Performance Data.

Variable	Coefficient	*s.e.*	*t*-test	*p*-value
Constant	10.787	11.5890	0.93	0.3616
X_1	0.613	0.1610	3.81	0.0009
X_2	−0.073	0.1357	−0.54	0.5956
X_3	0.320	0.1685	1.90	0.0699
X_4	0.081	0.2215	0.37	0.7155
X_5	0.038	0.1470	0.26	0.7963
X_6	−0.217	0.1782	−1.22	0.2356
$n = 30$	$R^2 = 0.73$	$R_a^2 = 0.66$	$\hat{\sigma} = 7.068$	$d.f. = 23$

3.9   TESTS OF HYPOTHESES IN A LINEAR MODEL

In addition to looking at hypotheses about individual β's, several different hypotheses are considered in connection with the analysis of linear models. The most commonly investigated hypotheses are:

1. All the regression coefficients associated with the predictor variables are zero.

2. Some of the regression coefficients are zero.

3. Some of the regression coefficients are equal to each other.

4. The regression parameters satisfy certain specified constraints.

The different hypotheses about the regression coefficients can all be tested in the same way by a unified approach. Rather than describing the individual tests, we first describe the general unified approach, then illustrate specific tests using the Supervisor Performance data.

The model given in (3.1) will be referred to as the *full model* (FM). The null hypothesis to be tested specifies values for some of the regression coefficients. When these values are substituted in the full model, the resulting model is called the *reduced model* (RM). The number of *distinct* parameters to be estimated in the reduced model is smaller than the number of parameters to be estimated in the full model. Accordingly, we wish to test

H_0 : Reduced model is adequate   against H_1 : Full model is adequate.

Note that the reduced model is *nested.* A set of models are said to be nested if they can be obtained from a larger model as special cases. The test for these nested hypotheses involves a comparison of the goodness of fit that is obtained when using the full model, to the goodness of fit that results using the reduced model specified by the null hypothesis. If the reduced model gives as good a fit as the full model, the null hypothesis, which defines the reduced model (by specifying some values of β_j), is not rejected. This procedure is described formally as follows.

Let $\hat{y}_i$ and $\hat{y}_i^*$ be the values predicted for y_i by the full model and the reduced model, respectively. The lack of fit in the data associated with the full model is the sum of the squared residuals obtained when fitting the full model to the data. We denote this by SSE(FM), the sum of squares due to error associated with the full model,

$$SSE(FM) = \sum (y_i - \hat{y}_i)^2. \tag{3.26}$$

Similarly, the lack of fit in the data associated with the reduced model is the sum of the squared residuals obtained when fitting the reduced model to the data. This quantity is denoted by SSE(RM), the sum of squares due to error associated with the reduced model,

$$SSE(RM) = \sum (y_i - \hat{y}_i^*)^2. \tag{3.27}$$

In the full model there are $p + 1$ regression parameters $(\beta_0, \beta_1, \beta_2, \ldots, \beta_p)$ to be estimated. Let us suppose that for the reduced model there are k distinct parameters. Note that $SSE(RM) \geq SSE(FM)$ because the additional parameters (variables) in the full model cannot increase the residual sum of squares. Note also that the difference $SSE(RM) - SSE(FM)$ represents the increase in the residual sum of squares due to fitting the reduced model. If this difference is large, the reduced model is inadequate. To see whether the reduced model is adequate, we use the ratio

$$F = \frac{[SSE(RM) - SSE(FM)]/(p + 1 - k)}{SSE(FM)/(n - p - 1)}. \tag{3.28}$$

This ratio is referred to as the F-test. Note that we divide $SSE(RM) - SSE(FM)$ and $SSE(FM)$ in the above ratio by their respective degrees of freedom to compensate for the different number of parameters involved in the two models as well as to ensure that the resulting test statistic has a standard statistical distribution. The full model has $p + 1$ parameters, hence $SSE(FM)$ has $n - p - 1$ degrees of freedom. Similarly, the reduced model has k parameters and $SSE(RM)$ has $n - k$ degrees of freedom. Consequently, the difference $SSE(RM) - SSE(FM)$ has $(n - k) - (n - p - 1) = p + 1 - k$ degrees of freedom. Therefore, the observed F ratio in (3.28) has F distribution with $(p + 1 - k)$ and $(n - p - 1)$ degrees of freedom.

If the observed F value is large in comparison to the tabulated value of F with $(p + 1 - k)$ and $(n - p - 1)$ d.f., the result is significant at level α; that is, the reduced model is unsatisfactory and the null hypothesis, with its suggested values of β's in the full model is rejected. The reader interested in proofs of the statements above is referred to Rao (1973), Searle (1971), Seber (1977), or Graybill (1976).

Accordingly, H_0 is rejected if

$$F \geq F_{(p+1-k, n-p-1; \alpha)}, \tag{3.29}$$

or, equivalently, if

$$p(F) \leq \alpha, \tag{3.30}$$

where F is the observed value of the F-test in (3.28), $F_{(p+1-k, n-p-1; \alpha)}$ is the appropriate critical value obtained from the F table given in the Appendix to this book (see Tables A.4 and A.5), α is the significance level, and $p(F)$ is the *p-value* for the F-test, which is the probability that a random variable having an F distribution, with $(p+1-k)$ and $(n-p-1)$ degrees of freedom, is greater than the observed F-test in (3.28). The p-value is usually computed and supplied as part of the regression output by many statistical packages.

In the rest of this section, we give several special cases of the general F-test in (3.28) with illustrative numerical examples using the Supervisor Performance data.

3.9.1 Testing All Regression Coefficients Equal to Zero

An important special case of the F-test in (3.28) is obtained when we test the hypothesis that all predictor variables under consideration have no explanatory power and that all their regression coefficients are zero. In this case, the reduced and full models become:

$$\text{RM}: \quad H_0 : Y = \beta_0 \qquad\qquad\qquad + \varepsilon, \tag{3.31}$$
$$\text{FM}: \quad H_1 : Y = \beta_0 + \beta_1 X_1 + \cdots + \beta_p X_p + \varepsilon. \tag{3.32}$$

The residual sum of squares from the full model is $SSE(FM) = SSE$. Because the least squares estimate of β_0 in the reduced model is $\bar{y}$, the residual sum of squares from the reduced model is $SSE(RM) = \sum(y_i - \bar{y})^2 = SST$. The reduced model has one regression parameter and the full model has $p+1$ regression parameters. Therefore, the F-test in (3.28) reduces to

$$
\begin{aligned}
F &= \frac{[SSE(RM) - SSE(FM)]/(p+1-k)}{SSE(FM)/(n-p-1)} \\
&= \frac{[SST - SSE]/p}{SSE/(n-p-1)},
\end{aligned} \tag{3.33}
$$

Because $SST = SSR + SSE$, we can replace $SST - SSE$ in the above formula by SSR and obtain

$$F = \frac{SSR/p}{SSE/(n-p-1)} = \frac{MSR}{MSE}, \tag{3.34}$$

where MSR is the *mean square due to regression* and MSE is the *mean square due to error*. The F-test in (3.34) can be used for testing the hypothesis that

Table 3.6   The Analysis of Variance (ANOVA) Table in Multiple Regression.

Source	Sum of Squares	d.f.	Mean Square	F-test
Regression	SSR	p	$MSR = \frac{SSR}{p}$	$F = \frac{MSR}{MSE}$
Residuals	SSE	$n - p - 1$	$MSE = \frac{SSE}{n-p-1}$	

the regression coefficients of all predictor variables (excluding the constant) are zero.

The F-test in (3.34) can also be expressed directly in terms of the sample multiple correlation coefficient. The null hypothesis which tests whether all the population regression coefficients are zero is equivalent to the hypothesis that states that the population multiple correlation coefficient is zero. Let R_p denote the sample multiple correlation coefficient, which is obtained from fitting a model to n observations in which there are p predictor variables (i.e., we estimate p regression coefficients and one intercept). The appropriate F for testing $H_0 : \beta_1 = \beta_2 = \cdots = \beta_p = 0$ in terms of R_p is

$$F = \frac{R_p^2/p}{(1 - R_p^2)/(n - p - 1)} \,, \tag{3.35}$$

with p and $n - p - 1$ degrees of freedom.

The values involved in the above F-test are customarily computed and compactly displayed in a table called the *analysis of variance* (ANOVA) table. The ANOVA table is given in Table 3.6. The first column indicates that there are two sources of variability in the response variable Y. The total variability in Y, $SST = \sum (y_i - \bar{y})^2$, can be decomposed into two sources: the *explained* variability, $SSR = \sum (\hat{y}_i - \bar{y})^2$, which is the variability in Y that can be accounted for by the predictor variables, and the *unexplained* variability, $SSE = \sum (y_i - \hat{y}_i)^2$. This is the same decomposition $SST = SSR + SSE$. This decomposition is given under the column heading Sum of Squares. The third column gives the degrees of freedom $(d.f.)$ associated with the sum of squares in the second column. The fourth column is the Mean Square (MS), which is obtained by dividing each sum of squares by its respective degrees of freedom. Finally, the F-test in (3.34) is reported in the last column of the table. Some statistical packages also give an additional column containing the corresponding p-value, $p(F)$.

Returning now to the Supervisor Performance data, although the t-tests for the regression coefficients have already indicated that some of the regression coefficients (β_1 and β_3) are significantly different from zero, we will, for illustrative purposes, test the hypothesis that all six predictor variables have no explanatory power, that is, $\beta_1 = \beta_2 = \ldots = \beta_6 = 0$. In this case, the

Table 3.7 Supervisor Performance Data: The Analysis of Variance (ANOVA) Table.

Source	Sum of Squares	d.f.	Mean Square	F-test
Regression	3147.97	6	524.661	10.5
Residuals	1149.00	23	49.9565	

reduced and full models in (3.31) and (3.32) become:

$$\text{RM}: \quad H_0: Y = \beta_0 \qquad\qquad\qquad + \varepsilon, \qquad (3.36)$$
$$\text{FM}: \quad H_1: Y = \beta_0 + \beta_1 X_1 + \cdots + \beta_6 X_6 + \varepsilon. \qquad (3.37)$$

For the full model we have to estimate seven parameters, six regression coefficients, and an intercept term β_0. The ANOVA table is given in Table 3.7. The sum of squares due to error in the full model is $SSE(FM) = SSE = 1149$. Under the null hypothesis, where all the β's are zero, the number of parameters estimated for the reduced model is therefore 1 (β_0). Consequently, the sum of squares of the residuals in the reduced model is

$$SSE(RM) = SST = SSR + SSE = 3147.97 + 1149 = 4296.97.$$

Note that this is the same quantity obtained by $\sum(y_i - \bar{y})^2$. The observed F ratio is 10.5. In our present example the numerical equivalence of (3.34) and (3.35) is easily seen for

$$F = \frac{R_p^2/p}{(1 - R_p^2)/(n - p - 1)} = \frac{0.7326/6}{(1 - 0.7326)/23} = 10.50.$$

This F-value has an F distribution with 6 and 23 degrees of freedom. The 1% F value with 6 and 23 degrees of freedom is found in Table A.5 to be 3.71. (Note that the value of 3.71 is obtained in this case by interpolation.) Since the observed F value is larger than this value, the null hypothesis is rejected; not all the β's can be taken as zero. This, of course, comes as no surprise, because of the large values of some of the t-tests.

If any of the t-tests for the individual regression coefficients prove significant, the F for testing all the regression coefficients zero will usually be significant. A more puzzling case can, however, arise when none of the t-values for testing the regression coefficients are significant, but the F-test given in (3.35) is significant. This implies that although none of the variables individually have significant explanatory power, the entire set of variables taken collectively explain a significant part of the variation in the dependent variable. This situation, when it occurs, should be looked at very carefully, for it may indicate a problem with the data analyzed, namely, that some of the explanatory variables may be highly correlated, a situation commonly called *multicollinearity*. We discuss this problem in Chapter 9.

3.9.2 Testing a Subset of Regression Coefficients Equal to Zero

We have so far attempted to explain Y in the Supervisor Performance data, in terms of six variables, $X_1, X_2, \ldots X_6$. The F-test in (3.34) indicates that all the regression coefficients cannot be taken as zero, hence one or more of the predictor variables is related to Y. The question of interest now is: Can Y be explained adequately by fewer variables? An important goal in regression analysis is to arrive at adequate descriptions of observed phenomenon in terms of as few meaningful variables as possible. This economy in description has two advantages. First, it enables us to isolate the most important variables, and second, it provides us with a simpler description of the process studied, thereby making it easier to understand the process. *Simplicity of description* or the *principle of parsimony*, as it is sometimes called, is one of the important guiding principles in regression analysis.

To examine whether the variable Y can be explained in terms of fewer variables, we look at a hypothesis that specifies that some of the regression coefficients are zero. If there are no overriding theoretical considerations as to which variables are to be included in the equation, preliminary t-tests, like those given in Table 3.5, are used to suggest the variables. In our current example, suppose it was desired to explain the overall rating of the job being done by the supervisor by means of two variables, one taken from the group of personal employee-interaction variables X_1, X_2, X_5, and another taken from the group of variables X_3, X_4, X_6, which are of a less personal nature. From this point of view X_1 and X_3 suggest themselves because they have significant t-tests. Suppose then that we wish to determine whether Y can be explained by X_1 and X_3 as adequately as the full set of six variables. The reduced model in this case is

$$\text{RM} : Y = \beta_0 + \beta_1 X_1 + \beta_3 X_3 + \varepsilon. \tag{3.38}$$

This model corresponds to hypothesis

$$H_0 : \beta_2 = \beta_4 = \beta_5 = \beta_6 = 0. \tag{3.39}$$

The regression output from fitting this model is given in Table 3.8, which includes both the ANOVA and the coefficients tables.

The residual sum of squares in this output is the residual sum of squares for the reduced model, which is $SSE(RM) = 1254.65$. From Table 3.7, the residual sum of squares from the full model is $SSE(FM) = 1149.00$. Hence the F-test in (3.28) is

$$F = \frac{[1254.65 - 1149]/4}{1149/23} = 0.528, \tag{3.40}$$

with 4 and 23 degrees of freedom.

The corresponding tabulated value for this test is $F_{(4,23,0.05)} = 2.8$. The value of F is not significant and the null hypothesis is not rejected. The

Table 3.8 Regression Output From the Regression of Y on X_1 and X_3.

ANOVA Table				
Source	Sum of Squares	d.f.	Mean Square	F-test
Regression	3042.32	2	1521.1600	32.7
Residuals	1254.65	27	46.4685	

Coefficients Table				
Variable	Coefficient	s.e.	t-test	p-value
Constant	9.8709	7.0610	1.40	0.1735
X_1	0.6435	0.1185	5.43	< 0.0001
X_3	0.2112	0.1344	1.57	0.1278
$n = 30$	$R^2 = 0.708$	$R_a^2 = 0.686$	$\hat{\sigma} = 6.817$	$d.f. = 27$

variables X_1 and X_3 together explain the variation in Y as adequately as the full set of six variables. At this stage residual plots similar to those described in Chapter 4 are examined to see if the deletion of variables X_2, X_4, X_5, X_6 has caused any violations of the model assumptions. In our present example, the residual plots appear satisfactory, and we conclude that deletion of X_2, X_4, X_5, X_6 does not adversely affect the explanatory power of the model.

We conclude this section with a few remarks:

1. The F-test in this case can also be expressed in terms of the sample multiple correlation coefficients. Let R_p denote the sample multiple correlation coefficient that is obtained when the full model with all the p variables in it is fitted to the data. Let R_q denote the sample multiple correlation coefficient when the model is fitted with q specific variables: that is, the null hypothesis states that $(p - q)$ specified variables have zero regression coefficients. The F-test for testing the above hypothesis is

$$F = \frac{(R_p^2 - R_q^2)/(p - q)}{(1 - R_p^2)/(n - p - 1)}, \quad d.f. = p - q, n - p - 1. \tag{3.41}$$

In our present example, from Tables 3.7 and 3.8, we have $n = 30, p = 6, q = 2, R_6^2 = 0.7326$, and $R_2^2 = 0.7080$. Substituting these in (3.41) we get an F value of 0.528, as before.

2. When the reduced model has only one coefficient (predictor variable) less than the full model, say β_j, then the F-test in (3.28) has 1 and $n - p - 1$ degrees of freedom. In this case, it can be shown that the F-test in (3.28) is equivalent to the t-test in (3.21). More precisely, we have

$$F = t_j^2, \tag{3.42}$$

Table 3.9   The Analysis of Variance (ANOVA) Table in Simple Regression.

Source	Sum of Squares	d.f.	Mean Square	F-test
Regression	SSR	1	$MSR = SSR$	$F = \frac{MSR}{MSE}$
Residuals	SSE	$n - 2$	$MSE = \frac{SSE}{n-2}$	

which indicates that an F-value with 1 and $n - p - 1$ degrees of freedom is equal to the square of a t-value with $n - p - 1$ degrees of freedom, a result which is well-known in statistical theory. (Check the t- and F-Tables A.2, A.4, and A.5 in the Appendix to this book to see that $F(1, v) = t^2(v)$.)

3. In simple regression the number of predictors is $p = 1$. Replacing p by one in the multiple regression ANOVA table (Table 3.6) we obtain the simple regression ANOVA table (Table 3.9). The F-test in Table 3.9 tests the null hypothesis that the predictor variable X_1 has no explanatory power, that is, its regression coefficient is zero. But this is the same hypothesis tested by the t_1 test introduced in Section 2.6 and defined in (2.25) as

$$t_1 = \frac{\hat{\beta}_1}{s.e.(\hat{\beta}_1)}.$$
(3.43)

Therefore in simple regression, the F and t_1 tests are equivalent, they are related by

$$F = t_1^2.$$
(3.44)

3.9.3   Testing the Equality of Regression Coefficients

By the general method outlined in Section 3.9, it is possible to test the equality of two regression coefficients in the same model. In the present example we test whether the regression coefficient of the variables X_1 and X_3 can be treated as equal. The test is performed assuming that it has already been established that the regression coefficients for $X_2, X_4, X_5,$ and X_6 are zero. The null hypothesis to be tested is

$$H_0 : \beta_1 = \beta_3 \mid (\beta_2 = \beta_4 = \beta_5 = \beta_6 = 0).$$
(3.45)

The full model assuming that $\beta_2 = \beta_4 = \beta_5 = \beta_6 = 0$ is

$$Y = \beta_0 + \beta_1 X_1 + \beta_3 X_3 + \varepsilon.$$
(3.46)

Under the null hypothesis, where $\beta_1 = \beta_3 = \beta_1'$, say, the reduced model is

$$Y = \beta_0' + \beta_1'(X_1 + X_3) + \varepsilon.$$
(3.47)

A simple way to carry out the test is to fit the model given by (3.46) to the data. The resulting regression output has been given in Table 3.8. We next fit the reduced model given in (3.47). This can be done quite simply by generating a new variable $W = X_1 + X_3$ and fitting the model

$$Y = \beta_0' + \beta_1'\, W + \varepsilon. \tag{3.48}$$

The least squares estimates of β_0', β_1' and the sample multiple correlation coefficient (in this case it is the simple correlation coefficient between Y and W since we have only one variable) are obtained. The fitted equation is

$$\hat{Y} = 9.988 + 0.444W$$

with $R_1^2 = 0.6685$. The appropriate F for testing the null hypothesis, defined in (3.41), becomes

$$F = \frac{(R_p^2 - R_q^2)/(p-q)}{(1 - R_p^2)/(n-p-1)} = \frac{(0.7080 - 0.6685)/(2-1)}{(1 - 0.7080)/(30-2-1)} = 3.65,$$

with 1 and 27 degrees of freedom. The corresponding tabulated value is $F_{(1,27,0.05)} = 4.21$. The resulting F is not significant; the null hypothesis is not rejected. The distribution of the residuals for this equation (not given here) was found satisfactory.

The equation

$$\hat{Y} = 9.988 + 0.444\,(X_1 + X_3)$$

is not inconsistent with the given data. We conclude then that X_1 and X_3 have the same incremental effect in determining employee satisfaction with a supervisor. This test could also be performed by using a t-test, given by

$$t = \frac{\hat{\beta}_1 - \hat{\beta}_3}{s.e.(\hat{\beta}_1 - \hat{\beta}_3)}$$

with 27 degrees of freedom.[2] The conclusions are identical and follow from the fact that F with 1 and p degrees of freedom is equal to the square of t with p degrees of freedom.

In this example we have discussed a sequential or step-by-step approach to model building. We have discussed the equality of β_1 and β_3 under the assumption that the other regression coefficients are equal to zero. We can, however, test a more complex null hypothesis which states that β_1 and β_3 are equal and $\beta_2, \beta_4, \beta_5, \beta_6$ are all equal to zero. This null hypothesis H_0' is formally stated as

$$H_0' : \beta_1 = \beta_3, \beta_2 = \beta_4 = \beta_5 = \beta_6 = 0. \tag{3.49}$$

---

[2] The $s.e.(\hat{\beta}_i - \hat{\beta}_j) = \sqrt{Var(\hat{\beta}_i) + Var(\hat{\beta}_j) - 2Cov(\hat{\beta}_i, \hat{\beta}_j)}$. These quantities are defined in the Appendix to this chapter.

The difference between (3.45) and (3.49) is that in (3.45), $\beta_2, \beta_4, \beta_5$ and β_6 are assumed to be zero, whereas in (3.49) this is under test. The null hypothesis (3.49) can be tested quite easily. The reduced model under H_0' is (3.47), but this model is not compared to the model of equation (3.46), as in the case of H_0, but with the full model with all six variables in the equation. The F-test for testing H_0' is, therefore,

$$F = \frac{(0.7326 - 0.6685)/5}{0.2674/23} = 1.10, \quad d.f. = 5, 23.$$

The result is insignificant as before. The first test is more sensitive for detecting departures from equality of the regression coefficients than the second test. (Why?)

3.9.4   Estimating and Testing of Regression Parameters Under Constraints

Sometimes in fitting regression equations to a given body of data it is desired to impose some constraints on the values of the parameters. A common constraint is that the regression coefficients sum to a specified value, usually unity. The constraints often arise because of some theoretical or physical relationships that may connect the variables. Although no such relationships are obvious in our present example, we consider $\beta_1 + \beta_3 = 1$ for the purpose of demonstration. Assuming that the model in (3.46) has already been accepted, we may further argue that if each of X_1 and X_3 is increased by a fixed amount, Y should increase by that same amount. Formally, we are led to the null hypothesis H_0 which states that

$$H_0 : \beta_1 + \beta_3 = 1 \mid (\beta_2 = \beta_4 = \beta_5 = \beta_6 = 0). \tag{3.50}$$

Since $\beta_1 + \beta_3 = 1$, or equivalently, $\beta_3 = 1 - \beta_1$, then under H_0 the reduced model is

$$H_0 : Y = \beta_0 + \beta_1 X_1 + (1 - \beta_1)X_3 + \varepsilon.$$

Rearranging terms we obtain

$$H_0 : Y - X_3 = \beta_0 + \beta_1(X_1 - X_3) + \varepsilon,$$

which can be written as

$$H_0 : Y' = \beta_0 + \beta_1 V + \varepsilon,$$

where $Y' = Y - X_3$ and $V = X_1 - X_3$. The least squares estimates of the parameters, β_1 and β_3 under the constraint are obtained by fitting a regression equation with Y' as response variable and V as the predictor variable. The fitted equation is

$$\hat{Y}' = 1.166 + 0.694 \, V,$$

from which it follows that the fitted equation for the reduced model is

$$\hat{Y} = 1.166 + 0.694\,X_1 + 0.306\,X_3$$

with $R^2 = 0.6905$.

The test for H_0 is given by

$$F = \frac{(0.7080 - 0.6905)/1}{0.2920/27} = 1.62, \quad d.f. = 1, 27,$$

which is not significant. The data support the proposition that the sum of the partial regression coefficients of X_1 and X_3 equal unity.

Recall that we have now tested two separate hypotheses about β_1 and β_3, one which states that they are equal and the other that they sum to unity. Since both hypotheses hold, it is implied that both coefficients can be taken to be 0.5. A test of this null hypothesis, $\beta_1 = \beta_3 = 0.5$, may be performed directly by applying the methods we have outlined.

The previous example, in which the equality of β_1 and β_3 was investigated, can be considered as a special case of constrained problem in which the constraint is $\beta_1 - \beta_3 = 0$. The tests for the full set or subsets of regression coefficients being zero can also be thought of as examples of testing regression coefficients under constraints.

From the above discussion it is clear that several models may describe a given body of data adequately. Where several descriptions of the data are available, it is important that they all be considered. Some descriptions may be more meaningful than others (meaningful being judged in the context of the application and considerations of subject matter), and one of them may be finally adopted. Looking at alternative descriptions of the data provides insight that might be overlooked in focusing on a single description.

The question of which variables to include in a regression equation is very complex and is taken up in detail in Chapter 11. We make two remarks here that will be elaborated on in later chapters.

1. The estimates of regression coefficients that do not significantly differ from zero are most commonly replaced by zero in the equation. The replacement has two advantages: a simpler model and a smaller prediction variance.

2. A variable or a set of variables may sometimes be retained in an equation because of their theoretical importance in a given problem, even though the sample regression coefficients are statistically insignificant. That is, sample coefficients which are not significantly different from zero are not replaced by zero. The variables so retained should give a meaningful process description, and the coefficients help to assess the contributions of the X's to the value of the dependent variable Y.

3.10 PREDICTIONS

The fitted multiple regression equation can be used to predict the value of the response variable using a set of specific values of the predictor variables, $\mathbf{x}_0 = (x_{01}, x_{02}, \ldots, x_{0p})$. The predicted value, $\hat{y}_0$, corresponding to $\mathbf{x}_0$ is given by

$$\hat{y}_0 = \hat{\beta}_0 + \hat{\beta}_1 x_{01} + \hat{\beta}_2 x_{02} + \cdots + \hat{\beta}_p x_{0p}, \tag{3.51}$$

and its standard error, $s.e.(\hat{y}_0)$, is given, in the Appendix to this chapter, in (A.10) for readers who are familiar with matrix notation. The standard error is usually computed by many statistical packages. Confidence limits for $\hat{y}_0$ with confidence coefficient α are

$$\hat{y}_0 \pm t_{(n-p-1, \alpha/2)} \ s.e.(\hat{y}_0).$$

As already mentioned in connection with simple regression, instead of predicting the response Y corresponding to an observation $\mathbf{x}_0$ we may want to estimate the mean response corresponding to that observation. Let us denote the mean response at $\mathbf{x}_0$ by μ_0 and its estimate by $\hat{\mu}_0$. Then

$$\hat{\mu}_0 = \hat{\beta}_0 + \hat{\beta}_1 x_{01} + \hat{\beta}_2 x_{02} + \cdots + \hat{\beta}_p x_{0p},$$

as in (3.51), but its standard error, $s.e.(\hat{\mu}_0)$, is given, in the Appendix to this chapter, in (A.12) for readers who are familiar with matrix notation. Confidence limits for $\hat{\mu}_0$ with confidence coefficient α are

$$\hat{\mu}_0 \pm t_{(n-p-1, \alpha/2)} s.e.(\hat{\mu}_0).$$

3.11 SUMMARY

We have illustrated the testing of various hypotheses in connection with the linear model. Rather than describing individual tests we have outlined a general procedure by which they can be performed. It has been shown that the various tests can also be described in terms of the appropriate sample multiple correlation coefficients. It is to be emphasized here, that before starting on any testing procedure, the adequacy of the model assumptions should always be examined. As we shall see in Chapter 4, residual plots provide a very convenient graphical way of accomplishing this task. The test procedures are not valid if the assumptions on which the tests are based do not hold. If a new model is chosen on the basis of a statistical test, residuals from the new model should be examined before terminating the analysis. It is only by careful attention to detail that a satisfactory analysis of data can be carried out.

EXERCISES

3.1 Using the Supervisor data, verify that the coefficient of X_1 in the fitted equation $\hat{Y} = 15.3276 + 0.7803\, X_1 - 0.0502\, X_2$ in (3.12) can be obtained from a series of simple regression equations, as outlined in Section 3.5 for the coefficient of X_2.

3.2 Construct a small data set consisting of one response and two predictor variables so that the regression coefficient of X_1 in the following two fitted equations are equal: $\hat{Y} = \hat{\beta}_0 + \hat{\beta}_1 X_1$ and $\hat{Y} = \hat{\alpha}_0 + \hat{\alpha}_1 X_1 + \hat{\alpha}_2 X_2$. Hint: The two predictor variables should be uncorrelated.

3.3 Table 3.10 shows the scores in the final examination F and the scores in two preliminary examinations P_1 and P_2 for 22 students in a statistics course. The data can be found in the book's Web site.

(a) Fit each of the following models to the data:

$$
\begin{aligned}
Model\ 1: \quad & F = \beta_0 + \beta_1 P_1 && + \varepsilon \\
Model\ 2: \quad & F = \beta_0 && + \beta_2 P_2 + \varepsilon \\
Model\ 3: \quad & F = \beta_0 + \beta_1 P_1 + \beta_2 P_2 + \varepsilon
\end{aligned}
$$

(b) Test whether $\beta_0 = 0$ in each of the three models.

(c) Which variable individually, P_1 or P_2, is a better predictor of F?

(d) Which of the three models would you use to predict the final examination scores for a student who scored 78 and 85 on the first and second preliminary examinations, respectively? What is your prediction in this case?

3.4 The relationship between the simple and the multiple regression coefficients can be seen when we compare the following regression equations:

$$
\begin{aligned}
\hat{Y} &= \hat{\beta}_0 + \hat{\beta}_1 X_1 + \hat{\beta}_2 X_2, && (3.52) \\
\hat{Y} &= \hat{\beta}_0' + \hat{\beta}_1' X_1, && (3.53) \\
\hat{Y} &= \hat{\beta}_0'' \qquad + \hat{\beta}_2' X_2, && (3.54) \\
\hat{X}_1 &= \hat{\alpha}_0 \qquad + \hat{\alpha}_2 X_2, && (3.55) \\
\hat{X}_2 &= \hat{\alpha}_0' + \hat{\alpha}_1 X_1. && (3.56)
\end{aligned}
$$

Using the Examination Data in Table 3.10 with $Y = F$, $X_1 = P_1$ and $X_2 = P_2$, verify that:

(a) $\hat{\beta}_1' = \hat{\beta}_1 + \hat{\beta}_2 \hat{\alpha}_1$, that is, the simple regression coefficient of Y on X_1 is the multiple regression coefficient of X_1 plus the multiple regression coefficient of X_2 times the coefficient from the regression of X_2 on X_1.

(b) $\hat{\beta}_2' = \hat{\beta}_2 + \hat{\beta}_1 \hat{\alpha}_2$, that is, the simple regression coefficient of Y on X_2 is the multiple regression coefficient of X_2 plus the multiple

Table 3.10 Examination Data: Scores in the Final (F), First Preliminary (P_1), and Second Preliminary (P_2) Examinations.

Row	F	P_1	P_2	Row	F	P_1	P_2
1	68	78	73	12	75	79	75
2	75	74	76	13	81	89	84
3	85	82	79	14	91	93	97
4	94	90	96	15	80	87	77
5	86	87	90	16	94	91	96
6	90	90	92	17	94	86	94
7	86	83	95	18	97	91	92
8	68	72	69	19	79	81	82
9	55	68	67	20	84	80	83
10	69	69	70	21	65	70	66
11	91	91	89	22	83	79	81

Table 3.11 Regression Output When Y Is Regressed on X_1 for 20 Observations.

ANOVA Table				
Source	Sum of Squares	d.f.	Mean Square	F-test
Regression	1848.76	–	–	–
Residuals	–	–	–	

Coefficients Table				
Variable	Coefficient	s.e.	t-test	p-value
Constant	-23.4325	12.74	–	0.0824
X_1	–	0.1528	8.32	< 0.0001
$n = -$	$R^2 = -$	$R_a^2 = -$	$\hat{\sigma} = -$	d.f. $= -$

regression coefficient of X_1 times the coefficient from the regression of X_1 on X_2.

3.5 Table 3.11 shows the regression output, with some numbers erased, when a simple regression model relating a response variable Y to a predictor variable X_1 is fitted based on twenty observations. Complete the 13 missing numbers, then compute $Var(Y)$ and $Var(X_1)$.

3.6 Table 3.12 shows the regression output, with some numbers erased, when a simple regression model relating a response variable Y to a predictor variable X_1 is fitted based on eighteen observations. Complete the 13 missing numbers, then compute $Var(Y)$ and $Var(X_1)$.

3.7 Construct the 95% confidence intervals for the individual parameters β_1 and β_2 using the regression output in Table 3.5.

Table 3.12   Regression Output When Y Is Regressed on X_1 for 18 Observations.

ANOVA Table				
Source	Sum of Squares	d.f.	Mean Square	F-test
Regression	–	–	–	–
Residuals	–	–	–	

Coefficients Table				
Variable	Coefficient	s.e.	t-test	p-value
Constant	3.43179	–	0.265	0.7941
X_1	–	0.1421	–	< 0.0001
$n = -$	$R^2 = 0.716$	$R_a^2 = -$	$\hat{\sigma} = 7.342$	$d.f. = -$

3.8 Explain why the test for testing the hypothesis H_0 in (3.45) is more sensitive for detecting departures from equality of the regression coefficients than the test for testing the hypothesis H_0' in (3.49).

3.9 Using the Supervisor Performance data, test the hypothesis $H_0 : \beta_1 = \beta_3 = 0.5$ in each of the following models:
(a) $Y = \beta_0 + \beta_1 X_1 + \beta_3 X_3 + \varepsilon$.
(b) $Y = \beta_0 + \beta_1 X_1 + \beta_2 X_2 + +\beta_3 X_3 + \varepsilon$.

3.10 One may wonder if people of similar heights tend to marry each other. For this purpose, a sample of newly married couples was selected. Let X be the height of the husband and Y be the height of the wife. The heights (in centimeters) of husbands and wives are found in Table 2.11. The data can also be found in the book's Web site.
(a) Using your choice of the response variable in Exercise 2.10f, test the null hypothesis that both the intercept and the slope are zero.

3.11 To decide whether a company is discriminating against women, the following data were collected from the company's records: Salary is the annual salary in thousands of dollars, Qualification is an index of employee qualification, and Sex (1, if the employee is a man, and 0, if the employee is a woman). Two linear models were fit to the data and the regression outputs are shown in Table 3.13. Suppose that the usual regression assumptions hold.
(a) Are men paid more than equally qualified women?
(b) Are men less qualified than equally paid women?
(c) Do you detect any inconsistency in the above results? Explain.
(d) Which model would you advocate if you were the defense lawyer? Explain.

Table 3.13 Regression Outputs for the Salary Discriminating Data.

Model 1: Dependent variable is: Salary				
Variable	Coefficient	s.e.	t-test	p-value
Constant	20009.5	0.8244	24271	< 0.0001
Qualification	0.935253	0.0500	18.7	< 0.0001
Sex	0.224337	0.4681	0.479	0.6329

Model 2: Dependent variable is: Qualification				
Variable	Coefficient	s.e.	t-test	p-value
Constant	−16744.4	896.4	−18.7	< 0.0001
Sex	0.850979	0.4349	1.96	0.0532
Salary	0.836991	0.0448	18.7	< 0.0001

Table 3.14 Regression Output When Salary Is Related to Four Predictor Variables.

ANOVA Table				
Source	Sum of Squares	d.f.	Mean Square	F-test
Regression	23665352	4	5916338	22.98
Residuals	22657938	88	257477	

Coefficients Table				
Variable	Coefficient	s.e.	t-test	p-value
Constant	3526.4	327.7	10.76	0.000
Sex	722.5	117.8	6.13	0.000
Education	90.02	24.69	3.65	0.000
Experience	1.2690	0.5877	2.16	0.034
Months	23.406	5.201	4.50	0.000
$n = 93$	$R^2 = 0.515$	$R_a^2 = 0.489$	$\hat{\sigma} = 507.4$	$d.f. = 88$

3.12 Table 3.14 shows the regression output of a multiple regression model relating the beginning salaries in dollars of employees in a given company to the following predictor variables:

Sex   An indicator variable (1 = man and 0 = woman)
Education   Years of schooling at the time of hire
Experience   Number of months of previous work experience
Months   Number of months with the company.

In (a)–(b) below, specify the null and alternative hypotheses, the test used, and your conclusion using a 5% level of significance.

Table 3.15   ANOVA Table When the Beginning Salary Is Regressed on Education.

ANOVA Table				
Source	Sum of Squares	d.f.	Mean Square	F-test
Regression	7862535	1	7862535	18.60
Residuals	38460756	91	422646	

(a) Conduct the F-test for the overall fit of the regression.

(b) Is there a *positive* linear relationship between Salary and Experience, after accounting for the effect of the variables Sex, Education, and Months?

(c) What salary would you forecast for a man with 12 years of education, 10 months of experience, and 15 months with the company?

(d) What salary would you forecast, on average, for men with 12 years of education, 10 months of experience, and 15 months with the company?

(e) What salary would you forecast, on average, for women with 12 years of education, 10 months of experience, and 15 months with the company?

3.13 Consider the regression model that generated the output in Table 3.14 to be a full model. Now consider the reduced model in which Salary is regressed on only Education. The ANOVA table obtained when fitting this model is shown in Table 3.15. Conduct a single test to compare the full and reduced models. What conclusion can be drawn from the result of the test? (Use $\alpha = 0.05$.)

3.14 Cigarette Consumption Data: A national insurance organization wanted to study the consumption pattern of cigarettes in all 50 states and the District of Columbia. The variables chosen for the study are given in Table 3.16. The data from 1970 are given in Table 3.17. The states are given in alphabetical order. The data can be found in the book's Web site.

In (a)–(b) below, specify the null and alternative hypotheses, the test used, and your conclusion using a 5% level of significance.

(a) Test the hypothesis that the variable Female is not needed in the regression equation relating Sales to the six predictor variables.

(b) Test the hypothesis that the variables Female and HS are not needed in the above regression equation.

(c) Compute the 95% confidence interval for the true regression coefficient of the variable Income.

Table 3.16 Variables in the Cigarette Consumption Data in Table 3.17.

Variable	Definition
Age	Median age of a person living in a state
HS	Percentage of people over 25 years of age in a state who had completed high school
Income	Per capita personal income for a state (income in dollars)
Black	Percentage of blacks living in a state
Female	Percentage of females living in a state
Price	Weighted average price (in cents) of a pack of cigarettes in a state
Sales	Number of packs of cigarettes sold in a state on a per capita basis

(d) What percentage of the variation in Sales can be accounted for when Income is removed from the above regression equation? Explain.

(e) What percentage of the variation in Sales can be accounted for by the three variables: Price, Age, and Income? Explain.

(f) What percentage of the variation in Sales that can be accounted for by the variable Income, when Sales is regressed on only Income? Explain.

3.15 Consider the two models:

$$\text{RM}: \quad H_0 : Y = \qquad\qquad\qquad \varepsilon,$$
$$\text{FM}: \quad H_1 : Y = \beta_0 + \beta_1 X_1 + \cdots + \beta_p X_p + \varepsilon.$$

(a) Develop an F-test for testing the above hypotheses.

(b) Let $p = 1$ (simple regression) and construct a data set Y and X_1 such that H_0 is not rejected at the 5% significance level.

(c) What does the null hypothesis indicate in this case?

(d) Compute the appropriate value of R^2 that relates the above two models.

Appendix

We present the standard results of multiple regression analysis in matrix notation. Let us define the following matrices:

$$
\mathbf{Y} = \begin{bmatrix} y_1 \\ y_2 \\ \vdots \\ y_n \end{bmatrix}, \quad
\mathbf{X} = \begin{bmatrix} x_{10} & x_{11} & \cdots & x_{1p} \\ x_{20} & x_{21} & \cdots & x_{2p} \\ \vdots & \vdots & & \vdots \\ x_{n0} & x_{n1} & \cdots & x_{np} \end{bmatrix}, \quad
\beta = \begin{bmatrix} \beta_0 \\ \beta_1 \\ \vdots \\ \beta_p \end{bmatrix}, \quad
\varepsilon = \begin{bmatrix} \varepsilon_1 \\ \varepsilon_2 \\ \vdots \\ \varepsilon_n \end{bmatrix}.
$$

Table 3.17   Cigarette Consumption Data (1970).

State	Age	HS	Income	Black	Female	Price	Sales
AL	27.0	41.3	2948.0	26.2	51.7	42.7	89.8
AK	22.9	66.7	4644.0	3.0	45.7	41.8	121.3
AZ	26.3	58.1	3665.0	3.0	50.8	38.5	115.2
AR	29.1	39.9	2878.0	18.3	51.5	38.8	100.3
CA	28.1	62.6	4493.0	7.0	50.8	39.7	123.0
CO	26.2	63.9	3855.0	3.0	50.7	31.1	124.8
CT	29.1	56.0	4917.0	6.0	51.5	45.5	120.0
DE	26.8	54.6	4524.0	14.3	51.3	41.3	155.0
DC	28.4	55.2	5079.0	71.1	53.5	32.6	200.4
FL	32.3	52.6	3738.0	15.3	51.8	43.8	123.6
GA	25.9	40.6	3354.0	25.9	51.4	35.8	109.9
HI	25.0	61.9	4623.0	1.0	48.0	36.7	82.1
ID	26.4	59.5	3290.0	0.3	50.1	33.6	102.4
IL	28.6	52.6	4507.0	12.8	51.5	41.4	124.8
IN	27.2	52.9	3772.0	6.9	51.3	32.2	134.6
IA	28.8	59.0	3751.0	1.2	51.4	38.5	108.5
KS	28.7	59.9	3853.0	4.8	51.0	38.9	114.0
KY	27.5	38.5	3112.0	7.2	50.9	30.1	155.8
LA	24.8	42.2	3090.0	29.8	51.4	39.3	115.9
ME	28.	54.7	3302.0	0.3	51.3	38.8	128.5
MD	27.1	52.3	4309.0	17.8	51.1	34.2	123.5
MA	29.0	58.5	4340.0	3.1	52.2	41.0	124.3
MI	26.3	52.8	4180.0	11.2	51.0	39.2	128.6
MN	26.8	57.6	3859.0	0.9	51.0	40.1	104.3
MS	25.1	41.0	2626.0	36.8	51.6	37.5	93.4
MO	29.4	48.8	3781.0	10.3	51.8	36.8	121.3
MT	27.1	59.2	3500.0	0.3	50.0	34.7	111.2
NB	28.6	59.3	3789.0	2.7	51.2	34.7	108.1
NV	27.8	65.2	4563.0	5.7	49.3	44.0	189.5
NH	28.0	57.6	3737.0	0.3	51.1	34.1	265.7
NJ	30.1	52.5	4701.0	10.8	51.6	41.7	120.7
NM	23.9	55.2	3077.0	1.9	50.7	41.7	90.0
NY	30.3	52.7	4712.0	11.9	52.2	41.7	119.0
NC	26.5	38.5	3252.0	22.2	51.0	29.4	172.4
ND	26.4	50.3	3086.0	0.4	49.5	38.9	93.8
OH	27.7	53.2	4020.0	9.1	51.5	38.1	121.6
OK	29.4	51.6	3387.0	6.7	51.3	39.8	108.4
OR	29.0	60.0	3719.0	1.3	51.0	29.0	157.0
PA	30.7	50.2	3971.0	8.0	52.0	44.7	107.3
RI	29.2	46.4	3959.0	2.7	50.9	40.2	123.9
SC	24.8	37.8	2990.0	30.5	50.9	34.3	103.6
SD	27.4	53.3	3123.0	0.3	50.3	38.5	92.7
TN	28.1	41.8	3119.0	15.8	51.6	41.6	99.8
TX	26.4	47.4	3606.0	12.5	51.0	42.0	106.4
UT	23.1	67.3	3227.0	0.6	50.6	36.6	65.5
VT	26.8	57.1	3468.0	0.2	51.1	39.5	122.6
VA	26.8	47.8	3712.0	18.5	50.6	30.2	124.3
WA	27.5	63.5	4053.0	2.1	50.3	40.3	96.7
WV	30.0	41.6	3061.0	3.9	51.6	41.6	114.5
WI	27.2	54.5	3812.0	2.9	50.9	40.2	106.4
WY	27.2	62.9	3815.0	0.8	50.0	34.4	132.2

The linear model in (3.1) can be expressed in terms of the above matrices as

$$\mathbf{Y} = \mathbf{X}\boldsymbol{\beta} + \boldsymbol{\varepsilon}, \tag{A.1}$$

where $x_{i0} = 1$ for all i. The assumptions made about ε for least squares estimation are

$$E(\boldsymbol{\varepsilon}) = \mathbf{0}, \text{ and } Var(\boldsymbol{\varepsilon}) = E(\boldsymbol{\varepsilon}\boldsymbol{\varepsilon}^T) = \sigma^2 \mathbf{I}_n,$$

where $E(\boldsymbol{\varepsilon})$ is the expected value (mean) of $\boldsymbol{\varepsilon}$, $\mathbf{I}_n$ is the identity matrix of order n, and $\boldsymbol{\varepsilon}^T$ is the transpose of $\boldsymbol{\varepsilon}$. Accordingly, ε_i's are independent and have zero mean and constant variance. This implies that

$$E(\mathbf{Y}) = \mathbf{X}\boldsymbol{\beta}.$$

The least squares estimator $\hat{\boldsymbol{\beta}}$ of $\boldsymbol{\beta}$ is obtained by minimizing the sum of squared deviations of the observations from their expected values. Hence the least squares estimators are obtained by minimizing $S(\boldsymbol{\beta})$, where

$$S(\boldsymbol{\beta}) = \boldsymbol{\varepsilon}^T \boldsymbol{\varepsilon} = (\mathbf{Y} - \mathbf{X}\boldsymbol{\beta})^T (\mathbf{Y} - \mathbf{X}\boldsymbol{\beta}).$$

Minimization of $S(\boldsymbol{\beta})$ leads to the system of equations

$$(\mathbf{X}^T\mathbf{X})\hat{\boldsymbol{\beta}} = \mathbf{X}^T\mathbf{Y}.$$

This is the system of normal equations in (3.5). Assuming that $(\mathbf{X}^T\mathbf{X})$ has an inverse, the least squares estimates $\hat{\boldsymbol{\beta}}$ can be written explicitly as

$$\hat{\boldsymbol{\beta}} = (\mathbf{X}^T\mathbf{X})^{-1}\mathbf{X}^T\mathbf{Y},$$

from which it can be seen that $\hat{\boldsymbol{\beta}}$ is a linear function of $\mathbf{Y}$. The vector of fitted values $\hat{\mathbf{Y}}$ corresponding to the observed $\mathbf{Y}$ is

$$\hat{\mathbf{Y}} = \mathbf{X}\hat{\boldsymbol{\beta}} = \mathbf{P}\mathbf{Y}, \tag{A.2}$$

where

$$\mathbf{P} = \mathbf{X}(\mathbf{X}^T\mathbf{X})^{-1}\mathbf{X}^T, \tag{A.3}$$

is known as the *hat* or *projection* matrix. The vector of residuals is given by

$$\mathbf{e} = \mathbf{Y} - \hat{\mathbf{Y}} = \mathbf{Y} - \mathbf{P}\mathbf{Y} = (\mathbf{I}_n - \mathbf{P})\mathbf{Y}. \tag{A.4}$$

The properties of the least squares estimators are:

1. $\hat{\boldsymbol{\beta}}$ is an unbiased estimator of $\boldsymbol{\beta}$ (that is, $E(\hat{\boldsymbol{\beta}}) = \boldsymbol{\beta}$) with variance-covariance matrix $Var(\hat{\boldsymbol{\beta}})$, which is

$$Var(\hat{\boldsymbol{\beta}}) = E(\hat{\boldsymbol{\beta}} - \boldsymbol{\beta})(\hat{\boldsymbol{\beta}} - \boldsymbol{\beta})^T = \sigma^2(\mathbf{X}^T\mathbf{X})^{-1} = \sigma^2\mathbf{C},$$

where

$$C = (X^T X)^{-1}. \tag{A.5}$$

Of all unbiased estimators of β that are linear in the observations, the least squares estimator has minimum variance. For this reason, $\hat{\beta}$ is said to be the *best linear unbiased estimator* (BLUE) of β.

2. The residual sum of squares can be expressed as

$$e^T e = Y^T(I_n - P)^T(I_n - P)Y = Y^T(I_n - P)Y. \tag{A.6}$$

The last equality follows because $(I_n - P)$ is a symmetric idempotent matrix.

3. An unbiased estimator of σ^2 is

$$\hat{\sigma}^2 = \frac{e^T e}{n - p - 1} = \frac{Y^T(I_n - P)Y}{n - p - 1}. \tag{A.7}$$

With the added assumption that the ε_i's are normally distributed we have the following additional results:

4. The vector $\hat{\beta}$ has a $(p+1)$-variate normal distribution with mean vector β and variance-covariance matrix $\sigma^2 C$. The marginal distribution of $\hat{\beta}_j$ is normal with mean β_j and variance $\sigma^2 c_{jj}$, where c_{jj} is the jth diagonal element of C in (A.5). Accordingly, the standard error of $\hat{\beta}_j$ is

$$s.e.(\hat{\beta}_j) = \hat{\sigma}\sqrt{c_{jj}}, \tag{A.8}$$

and the covariance of $\hat{\beta}_i$ and $\hat{\beta}_j$ is $Cov(\hat{\beta}_i, \hat{\beta}_j) = \sigma^2 c_{ij}$.

5. The quantity $W = e^T e/\sigma^2$ has an χ^2 distribution with $(n-p-1)$ degrees of freedom.

6. $\hat{\beta}$ and $\hat{\sigma}^2$ are distributed independently of one another.

7. The vector of fitted values $\hat{Y}$ has a singular n-variate normal distribution with mean $E(\hat{Y}) = X\beta$ and variance-covariance matrix $Var(\hat{Y}) = \sigma^2 P$.

8. The residual vector e has a singular n-variate normal distribution with mean $E(e) = 0$ and variance-covariance matrix $Var(e) = \sigma^2(I_n - P)$.

9. The predicted value $\hat{y}_0$ corresponding to an observation vector $x_0 = (x_{00}, x_{01}, x_{02}, \ldots, x_{0p})^T$, with $x_{00} = 1$ is

$$\hat{y}_0 = x_0^T \hat{\beta} \tag{A.9}$$

and its standard error is

$$s.e.(\hat{y}_0) = \hat{\sigma}\sqrt{1 + x_0^T(X^T X)^{-1}x_0}. \tag{A.10}$$

The mean response μ_0^T corresponding to $\mathbf{x}_0^T$ is

$$\hat{\mu}_0 = \mathbf{x}_0^T \hat{\boldsymbol{\beta}} \tag{A.11}$$

with a standard error

$$s.e.(\hat{\mu}_0) = \hat{\sigma} \sqrt{\mathbf{x}_0^T (\mathbf{X}^T \mathbf{X})^{-1} \mathbf{x}_0}. \tag{A.12}$$

10. The $100(1 - \alpha)\%$ joint confidence region for the regression parameters $\boldsymbol{\beta}$ is given by

$$\left\{ \boldsymbol{\beta} : \frac{(\boldsymbol{\beta} - \hat{\boldsymbol{\beta}})^T (\mathbf{X}^T \mathbf{X})(\boldsymbol{\beta} - \hat{\boldsymbol{\beta}})}{\hat{\sigma}^2 (p + 1)} \leq F_{(p+1, n-p-1, \alpha)} \right\}, \tag{A.13}$$

which is an ellipsoid centered at $\hat{\boldsymbol{\beta}}$.

4

$$\mathcal{4}$$

Regression Diagnostics: Detection of Model Violations

4.1 INTRODUCTION

We have stated the basic results that are used for making inferences about simple and multiple linear regression models in Chapters 2 and 3. The results are based on summary statistics that are computed from the data. In fitting a model to a given body of data, we would like to ensure that the fit is not overly determined by one or few observations. The distribution theory, confidence intervals, and tests of hypotheses outlined in Chapters 2 and 3 are valid and have meaning only if the standard regression assumptions are satisfied. These assumptions are stated in this chapter (Section 4.2). When these assumptions are violated the standard results quoted previously do not hold and an application of them may lead to serious error. We reemphasize that the prime focus of this book is on the detection and correction of violations of the basic linear model assumptions as a means of achieving a thorough and informative analysis of the data. This chapter presents methods for checking these assumptions. We will rely mainly on graphical methods as opposed to applying rigid numerical rules to check for model violations.

4.2 THE STANDARD REGRESSION ASSUMPTIONS

In the previous two chapters we have given the least squares estimates of the regression parameters and stated their properties. The properties of least squares estimators and the statistical analysis presented in Chapters 2 and 3 are based on the following assumptions:

1. **Assumptions about the form of the model**: The model that relates the response Y to the predictors $X_1, X_2, \ldots, X_p$ is assumed to be linear in the regression parameters $\beta_0, \beta_1, \ldots, \beta_p$, namely,

$$Y = \beta_0 + \beta_1 X_1 + \ldots + \beta_p X_p + \varepsilon, \tag{4.1}$$

which implies that the ith observation can be written as

$$y_i = \beta_0 + \beta_1 x_{i1} + \ldots + \beta_p x_{ip} + \varepsilon_i, \; i = 1, 2, \ldots, n. \tag{4.2}$$

We refer to this as the *linearity* assumption. Checking the linearity assumption in simple regression is easy because the validity of this assumption can be determined by examining the scatter plot of Y versus X. A linear scatter plot ensures linearity. Checking the linearity in multiple regression is more difficult due to the high dimensionality of the data. Some graphs that can be used for checking the linearity assumption in multiple regression are given later in this chapter. When the linearity assumption does not hold, transformation of the data can sometimes lead to linearity. Data transformation is discussed in Chapter 6.

2. **Assumptions about the errors**: The errors $\varepsilon_1, \varepsilon_2, \ldots, \varepsilon_n$ in (4.2) are assumed to be *independently and identically distributed* (iid) normal random variables each with mean zero and a common variance σ^2. Note that this implies four assumptions:

   - The error $\varepsilon_i, i = 1, 2, \ldots, n$, has a normal distribution. We refer to this as the *normality assumption*. The normality assumption is not as easily validated especially when the values of the predictor variables are not replicated. The validity of the normality assumption can be assessed by examination of appropriate graphs of the residuals, as we describe later in this chapter.

   - The errors $\varepsilon_1, \varepsilon_2, \ldots, \varepsilon_n$ have mean zero.

   - The errors $\varepsilon_1, \varepsilon_2, \ldots, \varepsilon_n$ have the same (but unknown) variance σ^2. This is the *constant variance assumption*. It is also known by other names such as the *homogeneity* or the *homoscedasticity* assumption. When this assumption does not hold, the problem is called the *heterogeneity* or the *heteroscedasticity* problem. This problem is considered in Chapter 7.

   - The errors $\varepsilon_1, \varepsilon_2, \ldots, \varepsilon_n$ are independent of each other (their pairwise covariances are zero). We refer to this as the *independent-errors assumption*. When this assumption does not hold, we have the *autocorrelation* problem. This problem is considered in Chapter 8.

3. **Assumptions about the predictors**: There are three assumptions concerning the predictor variables:

- The predictor variables $X_1, X_2, \ldots, X_p$ are nonrandom, that is, the values $x_{1j}, x_{2j}, \ldots, x_{nj}; j = 1, 2, \ldots, p$, are assumed fixed or selected in advance. This assumption is satisfied only when the experimenter can set the values of the predictor variables at predetermined levels. It is clear that under nonexperimental or observational situations this assumption will not be satisfied. The theoretical results that are presented in Chapters 2 and 3 will continue to hold, but their interpretation has to be modified. When the predictors are random variables, all inferences are conditional, conditioned on the observed data. It should be noted that this conditional aspect of the inference is consistent with the approach to data analysis presented in this book. Our main objective is to extract the maximum amount of information from the available data.

- The values $x_{1j}, x_{2j}, \ldots, x_{nj}; j = 1, 2, \ldots, p$, are measured without error. This assumption is hardly ever satisfied. The errors in measurement will affect the residual variance, the multiple correlation coefficient, and the individual estimates of the regression coefficients. The exact magnitude of the effects will depend on several factors, the most important of which are the standard deviation of the errors of measurement and the correlation structure among the errors. The effect of the measurement errors will be to increase the residual variance and reduce the magnitude of the observed multiple correlation coefficient. The effects of measurement errors on individual regression coefficients are more difficult to assess. The estimate of the regression coefficient for a variable is affected not only by its own measurement errors, but also by the measurement errors of other variables included in the equation.

  Correction for measurement errors on the estimated regression coefficients, even in the simplest case where all the measurement errors are uncorrelated, requires a knowledge of the ratio between the variances of the measurement errors for the variables and the variance of the random error. Since these quantities are seldom, if ever, known (particularly in the social sciences, where this problem is most acute), we can never hope to remove completely the effect of measurement errors from the estimated regression coefficients. If the measurement errors are not large compared to the random errors, the effect of measurement errors is slight. In interpreting the coefficients in such an analysis, this point should be remembered. Although there is some problem in the estimation of the regression coefficients when the variables are in error, the regression equation may still be used for prediction. However, the presence of errors in the predictors decreases the accuracy of predictions. For a more extensive discussion of this problem, the reader is referred to Fuller (1987), Chatterjee and Hadi (1988), and Chi-Lu and Van Ness (1999).

- The predictor variables $X_1, X_2, \ldots, X_p$ are assumed to be linearly independent of each other. This assumption is needed to guarantee the uniqueness of the least squares solution (the solution of the normal equations in (3.5)). If this assumption is violated, the problem is referred to as the *collinearity* problem This problem is considered in Chapters 9 and 10.

The first two of the above assumptions about the predictors cannot be validated, so they do not play a major role in the analysis. However, they do influence interpretation of the regression results.

4. **Assumptions about the observations**: All observations are equally reliable and have approximately equal role in determining the regression results and in influencing conclusions.

A feature of the method of least squares is that small or minor violations of the underlying assumptions do not invalidate the inferences or conclusions drawn from the analysis in a major way. Gross violations of the model assumptions can, however, seriously distort conclusions. Consequently, it is important to investigate the structure of the residuals and the data pattern through graphs.

4.3 VARIOUS TYPES OF RESIDUALS

A simple and effective method for detecting model deficiencies in regression analysis is the examination of residual plots. Residual plots will point to serious violations in one or more of the standard assumptions when they exist. Of more importance, the analysis of residuals may lead to suggestions of structure or point to information in the data that might be missed or overlooked if the analysis is based only on summary statistics. These suggestions or cues can lead to a better understanding and possibly a better model of the process under study. A careful graphical analysis of residuals may often prove to be the most important part of the regression analysis.

As we have seen in Chapters 2 and 3, when fitting the linear model in (4.1) to a set of data by least squares, we obtain the fitted values,

$$\hat{y}_i = \hat{\beta}_0 + \hat{\beta}_1 x_{i1} + \ldots + \hat{\beta}_p x_{ip}, \ i = 1, 2, \ldots, n, \tag{4.3}$$

and the corresponding *ordinary* least squares residuals,

$$e_i = y_i - \hat{y}_i, \ i = 1, 2, \ldots, n. \tag{4.4}$$

The fitted values in (4.3) can also be written in an alternative form as

$$\hat{y}_i = p_{i1}y_1 + p_{i2}y_2 + \ldots + p_{in}y_n, \ i = 1, 2, \ldots, n, \tag{4.5}$$

where the p_{ij}'s are quantities that depend only on the values of the predictor variables (they do not involve the response variable). Equation (4.5) shows directly the relationship between the observed and predicted values. In simple regression, p_{ij} is given by

$$p_{ij} = \frac{1}{n} + \frac{(x_i - \bar{x})(x_j - \bar{x})}{\sum(x_i - \bar{x})^2} . \qquad (4.6)$$

In multiple regression the p_{ij}'s are elements of a matrix known as the *hat* or *projection* matrix, which is defined in (A.3) in the Appendix to Chapter 3.

When $i = j$, p_{ii} is the ith diagonal element of the projection matrix **P**. In simple regression,

$$p_{ii} = \frac{1}{n} + \frac{(x_i - \bar{x})^2}{\sum(x_i - \bar{x})^2} . \qquad (4.7)$$

The value p_{ii} is called the *leverage* value for the ith observation because, as can be seen from (4.5), $\hat{y}_i$ is a weighted sum of all observations in Y and p_{ii} is the weight (leverage) given to y_i in determining the ith fitted value $\hat{y}_i$ (Hoaglin and Welsch, 1978). Thus, we have n leverage values and they are denoted by

$$p_{11}, p_{22}, \ldots, p_{nn}. \qquad (4.8)$$

The leverage values play an important role in regression analysis and we shall often encounter them.

When the assumptions stated in Section 4.2 hold, the ordinary residuals, $e_1, e_2, \ldots, e_n$, defined in (4.4), will sum to zero, but they will not have the same variance because

$$Var(e_i) = \sigma^2(1 - p_{ii}), \qquad (4.9)$$

where p_{ii} is the ith leverage value in (4.8), which depends on $x_{i1}, x_{i2}, \ldots, x_{ip}$. To overcome the problem of unequal variances, we standardize the ith residual e_i by dividing it by its standard deviation and obtain

$$z_i = \frac{e_i}{\sigma\sqrt{1 - p_{ii}}} . \qquad (4.10)$$

This is called the ith *standardized residual* because it has mean zero and standard deviation 1. The standardized residuals depend on σ, the unknown standard deviation of ε. An unbiased estimate of σ^2 is given by

$$\hat{\sigma}^2 = \frac{\sum e_i^2}{n-p-1} = \frac{\sum(y_i - \hat{y}_i)^2}{n-p-1} = \frac{SSE}{n-p-1} , \qquad (4.11)$$

where SSE is the sum of squares of the residuals. The number $n - p - 1$ in the denominator of (4.11) is called the *degrees of freedom* (*d.f.*). It is equal to the number of observations, n, minus the number of estimated regression coefficients, $p + 1$.

An alternative unbiased estimate of σ^2 is given by

$$\hat{\sigma}_{(i)}^2 = \frac{SSE_{(i)}}{(n-1)-p-1} = \frac{SSE_{(i)}}{n-p-2} , \qquad (4.12)$$

where $SSE_{(i)}$ is the sum of squared residuals when we fit the model to the $(n-1)$ observations obtained by omitting the ith observation. Both $\hat{\sigma}^2$ and $\hat{\sigma}_{(i)}^2$ are unbiased estimates of σ^2.

Using $\hat{\sigma}$ as an estimate of σ in (4.10), we obtain

$$r_i = \frac{e_i}{\hat{\sigma}\sqrt{1-p_{ii}}} , \qquad (4.13)$$

whereas using $\hat{\sigma}_{(i)}$ as an estimate of σ, we obtain

$$r_i^* = \frac{e_i}{\hat{\sigma}_{(i)}\sqrt{1-p_{ii}}} . \qquad (4.14)$$

The form of residual in (4.13) is called the *internally studentized residual,* and the residual in (4.14) is called the *externally studentized residual,* because e_i is not involved in (external to) $\hat{\sigma}_{(i)}$. For simplicity of terminology and presentation, however, we shall refer to the studentized residuals as the standardized residuals.

The standardized residuals do not sum to zero, but they all have the same variance. The externally standardized residuals follow a t-distribution with $n - p - 2$ degrees of freedom, but the internally standardized residuals do not. However, with a moderately large sample, these residuals should approximately have a standard normal distribution. The residuals are not strictly independently distributed, but with a large number of observations, the lack of independence may be ignored.

The two forms of residuals are related by

$$r_i^* = r_i\sqrt{\frac{n-p-2}{n-p-1-r_i^2}} , \qquad (4.15)$$

hence one is a monotone transformation of the other. Therefore, for the purpose of residuals plots, it makes little difference as to which of the two forms of the standardized residuals is used. From here on, we shall use the internally standardized residuals in the graphs. We need not make any distinction between the internally and externally standardized residuals in our residual plots. Several graphs of the residuals are used for checking the regression assumptions.

4.4 GRAPHICAL METHODS

Graphical methods play an important role in data analysis. It is of particular importance in fitting linear models to data. As Chambers et al. (1983, page 1) put it, "There is no single statistical tool that is as powerful as a well-chosen graph." Graphical methods can be regarded as exploratory tools. They are also an integral part of confirmatory analysis or statistical inference. Huber (1991) says, "Eye-balling can give diagnostic insights no formal diagnostics will ever provide." One of the best examples that illustrates this is the Anscombe's quartet, the four data sets given in Chapter 2 (Table 2.4). The four data sets are constructed by Anscombe (1973) in such a way that all pairs (Y, X) have identical values of descriptive statistics (same correlation coefficients, same regression lines, same standard errors, etc.), yet their pairwise scatter plots (reproduced in Figure 4.1 for convenience) give completely different scatters.

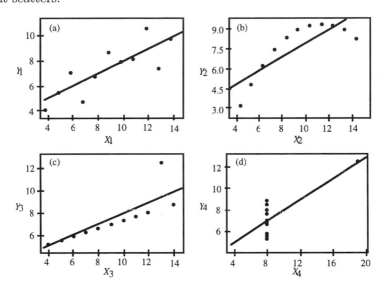

Fig. 4.1   Plot of the data (X, Y) with the least squares fitted line for the Anscombe's quartet.

The scatter plot in Figure 4.1(a) indicates that a linear model may be reasonable, whereas the one in Figure 4.1(b) suggests a (possibly linearizable) nonlinear model. Figure 4.1(c) shows that the data follow a linear model closely except for one point which is clearly off the line. This point may be an outlier, hence it should be examined before conclusions can be drawn from the data. Figure 4.1(d) indicates either a deficient experimental design or a bad sample. For the point at $X = 19$, the reader can verify that (a) the residual at this point is always zero (with a variance of zero) no matter how large or small its corresponding value of Y and (b) if the point is removed, the least squares estimates based on the remaining points are no longer unique (except the vertical line, any line that passes through the average of the remaining points is a least squares line!). Observations which unduly influence regression results are called *influential observations*. The point at $X = 19$ is therefore extremely influential because it alone determines both the intercept and the slope of the fitted line.

We have used the scatter plot here as an exploratory tool, but one can also use graphical methods to complement numerical methods in a confirmatory analysis. Suppose we wish to test whether there is a positive correlation between Y and X or, equivalently, if Y and X can be fitted by a positively sloped regression line. The reader can verify that the correlation coefficients are the same in all four data sets $(Cor(Y, X) = 0.80)$ and all four data sets also have the same regression line $(Y = 3 + 0.5\ X)$ with the same standard errors of the coefficients. Thus, based on these numerical summaries, one would reach the erroneous conclusion that all four data sets can be described by the same model. The underlying assumption here is that the relationship between Y and X is linear and this assumption does not hold here, for example, for the data set in Figure 4.1(b). Hence the test is invalid. The test for linear relationship, like other statistical methods, is based on certain underlying assumptions. Thus conclusions based on these methods are valid only when the underlying assumptions hold. It is clear from the above example that if analyses were solely based on numerical results, wrong conclusions will be reached.

Graphical methods can be useful in many ways. They can be used to:

1. Detect errors in the data (e.g., an outlying point may be a result of a typographical error),

2. Recognize patterns in the data (e.g., clusters, outliers, gaps, etc.),

3. Explore relationships among variables,

4. Discover new phenomena,

5. Confirm or negate assumptions,

6. Assess the adequacy of a fitted model,

7. Suggest remedial actions (e.g., transform the data, redesign the experiment, collect more data, etc.), and

8. Enhance numerical analyses in general.

This chapter presents some graphical displays useful in regression analysis. The graphical displays we discuss here can be classified into two (not mutually exclusive) classes:

- Graphs before fitting a model. These are useful, for example, in correcting errors in data and in selecting a model.

- Graphs after fitting a model. These are particularly useful for checking the assumptions and for assessing the goodness of the fit.

Our presentation draws heavily from Hadi (1993) and Hadi and Son (1997). Before examining a specific graph, consider what the graph should look like when the assumptions hold. Then examine the graph to see whether it is consistent with expectations. This will then confirm or disprove the assumption.

4.5 GRAPHS BEFORE FITTING A MODEL

The form of a model that represents the relationship between the response and predictor variables should be based on the theoretical background or the hypothesis to be tested. But if no prior information about the form of the model is available, the data may be used to suggest the model. The data should be examined thoroughly before a model is fitted. The graphs that one examines before fitting a model to the data serve as exploratory tools. Four possible groups of graphs are:

1. One-dimensional graphs,

2. Two-dimensional graphs,

3. Rotating plots, and

4. Dynamic graphs.

4.5.1 One-Dimensional Graphs

Data analysis usually begins with the examination of each variable in the study. The purpose is to have a general idea about the distribution of each individual variable. One of the following graphs may be used for examining a variable:

- Histogram

- Stem-and-leaf display

- Dot Plot

- Box Plot

The one-dimensional graphs serve two major functions. They indicate the distribution of a particular variable, whether the variable is symmetric or skewed. When a variable is very skewed it should be transformed. For a highly skewed variable a logarithmic transformation is recommended. Univariate graphs provide guidance on the question as to whether one should work with the original or with the transformed variables.

Univariate graphs also point out the presence of outliers in the variables. Outliers should be checked to see if they are due to transcription errors. No observation should be deleted at this stage. They should be noted as they may show up as troublesome points later.

4.5.2 Two-Dimensional Graphs

Ideally, when we have multidimensional data, we should examine a graph of the same dimension as that of the data. Obviously, this is feasible only when the number of variables is small. However, we can take the variables in pairs and look at the scatter plots of each variable versus each other variable in the data set. The purposes of these pairwise scatter plots are to explore the relationships between each pair of variables and to identify general patterns.

When the number of variables is small, it may be possible to arrange these pairwise scatter plots in a matrix format, sometimes referred to as the *draftsman's* plot or the *plot* matrix. Figure 4.2 is an example of a plot matrix for one response and two predictor variables. The pairwise scatter plots are given in the upper triangular part of the plot matrix. We can also arrange the corresponding correlation coefficients in a matrix. The corresponding correlation coefficients are given in the lower triangular part of the plot matrix. These arrangements facilitate the examination of the plots. The pairwise correlation coefficients should always be interpreted in conjunction with the corresponding scatter plots. The reason for this is two-fold: (a) the correlation coefficient measures only linear relationships, and (b) the correlation coefficient is non-robust, that is, its value can be substantially influenced by one or two observations in the data.

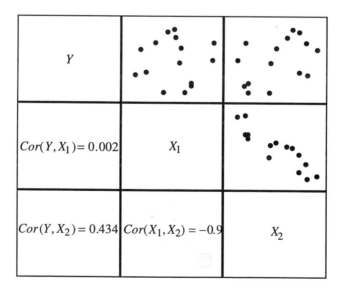

Fig. 4.2 The plot matrix for Hamilton's data.

What do we expect each of the graphs in the plot matrix to look like? In simple regression, the plot of Y versus X is expected to show a linear pattern. In multiple regression, however, the scatter plots of Y versus each predictor variable may or may not show linear patterns. Where the presence of a linear

Table 4.1 Hamilton's (1987) Data.

Y	X_1	X_2	Y	X_1	X_2
12.37	2.23	9.66	12.86	3.04	7.71
12.66	2.57	8.94	10.84	3.26	5.11
12.00	3.87	4.40	11.20	3.39	5.05
11.93	3.10	6.64	11.56	2.35	8.51
11.06	3.39	4.91	10.83	2.76	6.59
13.03	2.83	8.52	12.63	3.90	4.90
13.13	3.02	8.04	12.46	3.16	6.96
11.44	2.14	9.05			

pattern is reassuring, the absence of such a pattern does not imply that our linear model is incorrect. An example is given below.

Example: Hamilton's Data

Hamilton (1987) generates sets of data in such a way that Y depends on the predictor variables collectively but not individually. One such data set is given in Table 4.1. It can be seen from the plot matrix of this data (Figure 4.2) that no linear relationships exist in the plot of Y versus X_1 ($R^2 = 0$) and Y versus X_2 ($R^2 = 0.19$). Yet, when Y is regressed on X_1 and X_2 simultaneously, we obtain an almost perfect fit. The reader can verify that the following fitted equations are obtained:

$$\hat{Y} = 11.989 + 0.004\ X_1; \qquad t\text{-test} = 0.009; \qquad R^2 = 0.0,$$
$$\hat{Y} = 10.632 + 0.195\ X_2; \qquad t\text{-test} = 1.74; \qquad R^2 = 0.188,$$
$$\hat{Y} = -4.515 + 3.097\ X_1 + 1.032\ X_2; \quad F\text{-test} = 39222; \quad R^2 = 1.0.$$

The first two equations indicate that Y is related to neither X_1 nor X_2 individually, yet X_1 and X_2 predict Y almost perfectly. Incidentally, the first equation produces a negative value for the adjusted R^2, $R_a^2 = -0.08$.

The scatter plots that should look linear in the plot matrix are the plots of Y versus each predictor variable after adjusting for all other predictor variables (that is, taking the linear effects of all other predictor variables out). Two types of these graphs known as the *added-variable plot* and the *residual plus component plot*, are presented in Section 4.12.1.

The pairwise scatter plot of the predictors should show no linear pattern (ideally, we should see no discernible pattern, linear or otherwise) because the predictors are assumed to be linearly independent. In Hamilton's data, this assumption does not hold because there is a clear linear pattern in the scatter plot of X_1 versus X_2 (Figure 4.2). We should caution here that the absence of linear relationships in these scatter plots does not imply that the entire set of predictors are linearly independent. The linear relationship may involve more than two predictor variables. Pairwise scatter plots will fail to detect such a multivariate relationship. This multicollinearity problem will be dealt with in Chapters 9 and 10.

4.5.3   Rotating Plots

Recent advances in computer hardware and software have made it possible to plot data of three or more dimensions. The simplest of these plots is the three-dimensional rotating plot. The rotating plot is a scatter plot of three variables in which the points can be rotated in various directions so that the three-dimensional structure becomes apparent. Describing rotating plots in words does not do them justice. The real power of rotation can be felt only when one watches a rotating plot in motion on a computer screen. The motion can be stopped when one sees an interesting view of the data. For example, in the Hamilton's data we have seen that X_1 and X_2 predict Y almost perfectly. This finding is confirmed in the rotating plot of Y against X_1 and X_2. When this plot is rotated, the points fall on an almost perfect plane. The plot is rotated until an interesting direction is found. Figure 4.3 shows one such direction, where the plane is viewed from an angle that makes the scatter of points seem to fall on a straight line.

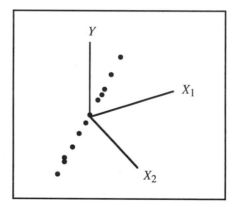

Fig. 4.3   Rotating plot for Hamilton's data.

4.5.4   Dynamic Graphs

Dynamic graphics are an extraordinarily useful tool for exploring the structure and relationships in multivariate data. In a dynamic graphics environment the data analyst can go beyond just looking at a static graph. The graphs can be manipulated and the changes can be seen instantaneously on the computer screen. For example, one can make two or more three-dimensional rotating plots then use dynamic graphical techniques to explore the structure and relationships in more than three dimensions. Articles and books have been written about the subject, and many statistical software programs include dynamic graphical tools (e.g., rotating, brushing, linking, etc.). We refer the interested reader to Becker, Cleveland, and Wilks (1987), and Velleman (1999).

4.6   GRAPHS AFTER FITTING A MODEL

The graphs presented in the previous section are useful in data checking and the model formulation steps. The graphs after fitting a model to the data help in checking the assumptions and in assessing the adequacy of the fit of a given model. These graphs can be grouped into the following classes:

1. Graphs for checking the linearity and normality assumptions,

2. Graphs for the detection of outliers and influential observations, and

3. Diagnostic plots for the effect of variables.

4.7   CHECKING LINEARITY AND NORMALITY ASSUMPTIONS

When the number of variables is small, the assumption of linearity can be checked by interactively and dynamically manipulating the plots discussed in the previous section. The task of checking the linearity assumption becomes difficult when the number of variables is large. However, one can check the linearity and normality assumptions by examining the residuals after fitting a given model to the data.

The following plots of the standardized residuals can be used to check the linearity and normality assumptions:

1. *Normal probability plot of the standardized residuals:* This is a plot of the ordered standardized residuals versus the so-called *normal scores*. The normal scores are what we would expect to obtain if we take a sample of size n from a standard normal distribution. If the residuals are normally distributed, the ordered residuals should be approximately the same as the ordered normal scores. Under normality assumption, this plot should resemble a (nearly) straight line with an intercept of zero and a slope of one (these are the mean and the standard deviation of the standardized residuals, respectively).

2. *Scatter plots of the standardized residual against each of the predictor variables:* Under the standard assumptions, the standardized residuals are uncorrelated with each of the predictor variables. If the assumptions hold, this plot should be a random scatter of points. Any discernible pattern in this plot may indicate violation of some assumptions. If the linearity assumption does not hold, one may observe a plot like the one

given in Figure 4.4(a). In this case a transformation of the Y and/or the particular predictor variable may be necessary to achieve linearity. A plot that looks like Figure 4.4(b), may indicate heterogeneity of variance. In this case a transformation of the data that stabilizes the variance may be needed. Several types of transformations for the corrections of some model deficiencies are described in Chapter 6.

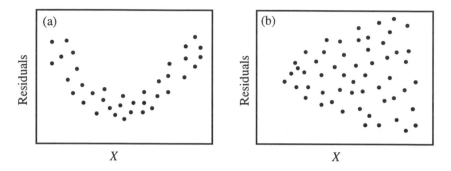

Fig. 4.4 Two scatter plots of residuals versus X illustrating violations of model assumptions: (a) a pattern indicating nonlinearity; and (b) a pattern indicating heterogeneity.

3. *Scatter plot of the standardized residual versus the fitted values:* Under the standard assumptions, the standardized residuals are also uncorrelated with the fitted values; therefore, this plot should also be a random scatter of points. In simple regression, the plots of standardized residuals against X and against the fitted values are identical.

4. *Index plot of the standardized residuals:* In this diagnostic plot we display the standardized residuals versus the observation number. If the order in which the observations were taken is immaterial, this plot is not needed. However, if the order is important (e.g., when the observations are taken over time or there is a spatial ordering), a plot of the residuals in serial order may be used to check the assumption of independence of the errors. Under the assumption of independent errors, the points should be scattered randomly within a horizontal band around zero.

4.8 LEVERAGE, INFLUENCE, AND OUTLIERS

In fitting a model to a given body of data, we would like to ensure that the fit is not overly determined by one or few observations. Recall, for example, that in the Anscombe quartet, the straight line for the data set in Figure 4.1(d) is determined entirely by one point. If the extreme point were to be removed, a very different line would result. When we have several variables, it is not

Table 4.2 New York Rivers Data: The t-tests for the Individual Coefficients.

Test	Observations Deleted		
	None	Neversink	Hackensack
t_0	1.40	1.21	2.08
t_1	0.39	0.92	0.25
t_2	-0.93	-0.74	-1.45
t_3	-0.21	-3.15	4.08
t_4	1.86	4.45	0.66

possible to detect such a situation graphically. We would, however, like to know the existence of such points. It should be pointed out that looking at residuals in this case would be of no help, because the residual for this point is zero! The point is therefore not an outlier because it does not have a large residual, but it is a very influential point.

A point is an *influential* point if its deletion, singly or in combination with others (two or three), causes substantial changes in the fitted model (estimated coefficients, fitted values, t-tests, etc.). Deletion of any point will in general cause changes in the fit. We are interested in detecting those points whose deletion cause large changes (i.e., they exercise undue influence). This point is illustrated by an example.

Example: New York Rivers Data

Consider the New York Rivers data described in Section 1.3.5 and given in Table 1.9. Let us fit a linear model relating the mean nitrogen concentration, Y, and the four predictor variables representing land use:

$$Y = \beta_0 + \beta_1 X_1 + \beta_2 X_2 + \beta_3 X_3 + \beta_4 X_4 + \varepsilon. \qquad (4.16)$$

Table 4.2 shows the regression coefficients and the t-tests for testing the significance of the coefficients for three subsets of the data. The second column in Table 4.2 gives the regression results based on all 20 observations (rivers). The third column gives the results after deleting the Neversink river (number 4). The fourth column gives the results after deleting the Hackensack river (number 5).

Note the striking difference among the regression outputs of three data sets that differ from each other by only one observation! Observe, for example, the values of the t-test for β_3. Based on all data, the test is insignificant, based on the data without the Neversink river, it is significantly negative, and based on the data without the Hackensack river, it is significantly positive. Only one observation can lead to substantially different results and conclusions! The Neversink and Hackensack rivers are called influential observations because they influence the regression results substantially more than other observations in the data. Examining the raw data in Table 1.9, one can easily identify the Hackensack river because it has an unusually large value for X_3 (percentage of residential land) relative to the other values for X_3. The reason for this large value is that the Hackensack river is the only urban river in the data due to its geographic proximity to New York City with its high population density. The other rivers are in rural areas. Although the Neversink river is influential (as can be seen from Table 4.2), it is not obvious from the raw data that it is different from the other rivers in the data.

It is therefore important to identify influential observations if they exist in data. We describe methods for the detection of influential observations. Influential observations are usually outliers in either the response variable Y or the predictor variable (the X-space).

4.8.1   Outliers in the Response Variable

Observations with large standardized residuals are outliers in the response variable because they lie far from the fitted equation in the Y-direction. Since the standardized residuals are approximately normally distributed with mean zero and a standard deviation 1, points with standardized residuals larger than 2 or 3 standard deviation away from the mean (zero) are called *outliers*. Outliers may indicate a model failure for those points. They can be identified using formal testing procedures (see, e.g., Hawkins (1980), Barnett and Lewis (1994), Hadi and Simonoff (1993), and Hadi and Velleman (1997) and the references therein) or through appropriately chosen graphs of the residuals, the approach we adopt here. The pattern of the points is more important than the numerical values of the residuals. Graphs of residuals will often expose gross model violations when they are present. Studying residual plots is one of the main tools in our analysis.

4.8.2   Outliers in the Predictors

Outliers can also occur in the predictor variables (the X-space). They can also affect the regression results. The leverage values p_{ii}, described earlier, can be used to measure outlyingness in the X-space. This can be seen from an examination of the formula for p_{ii} in the simple regression case given in (4.7), which shows that the farther a point is from $\bar{x}$, the larger the corresponding value of p_{ii}. This is also true in multiple regression. Therefore, p_{ii} can be used as a measure of outlyingness in the X-space because observations with large values of p_{ii} are outliers in the X-space (i.e., compared to other points in the space of the predictors). Observations that are outliers in the X-space (e.g., the point with the largest value of X_4 in Figure 4.1(d)) are known as *high leverage* points to distinguish them from observations that are outliers in the response variable (those with large standardized residuals).

The leverage values possess several interesting properties (see Dodge and Hadi (1999) and Chatterjee and Hadi (1988), Chapter 2, for a comprehensive discussion). For example, they lie between zero and 1 and their average value is $(p+1)/n$. Points with p_{ii} greater than $2(p+1)/n$ (twice the average value) are generally regarded as points with high leverage (Hoaglin and Welsch, 1978).

In any analysis, points with high leverage should be flagged and then examined to see if they are also influential. A plot of the leverage values (e.g., index plot, dot plot, or a box plot) will reveal points with high leverage if they exist.

4.8.3 Masking and Swamping Problems

The standardized residuals provide valuable information for validating linearity and normality assumptions and for the identification of outliers. However, analyses that are based on residuals alone may fail to detect outliers and influential observations for the following reasons:

1. *The presence of high leverage points*: The ordinary residuals, e_i, and leverage values, p_{ii} are related by

$$p_{ii} + \frac{e_i^2}{SSE} \le 1, \tag{4.17}$$

where SSE is the residual sum of squares. This inequality indicates that high leverage points (points with large values of p_{ii}) tend to have small residuals. For example, the point at $X = 19$ in Figure 4.1(d) is extremely influential even though its residual is identically zero. Therefore, in addition to an examination of the standardized residuals for outliers, an examination of the leverage values is also recommended for the identification of troublesome points.

2. *The masking and swamping problems*: Masking occurs when the data contain outliers but we fail to detect them. This can happen because some of the outliers may be hidden by other outliers in the data. Swamping occurs when we wrongly declare some of the non-outlying points as outliers. This can occur because outliers tend to pull the regression equation toward them, hence make other points lie far from the fitted equation. Thus, masking is a false negative decision whereas swamping is a false positive. An example of a data set in which masking and swamping problems are present is given below. Methods which are less susceptible to the masking and swamping problems than the standardized residuals and leverage values are given in Hadi and Simonoff (1993) and the references therein.

For the above reasons, additional measures of the influence of observations are needed. Before presenting these methods, we illustrate the above concepts using a real-life example.

Example: New York Rivers Data

Consider the New York rivers data, but now for illustrative purpose, let us consider fitting the simple regression model

$$Y = \beta_0 + \beta_4 X_4 + \varepsilon, \tag{4.18}$$

relating the mean nitrogen concentration, Y, to the percentage of land area in either industrial or commercial use, X_4. The scatter plot of Y versus X_4 together with the corresponding least squares fitted line are given in Figure 4.5. The corresponding standardized residuals, r_i, and the leverage values, p_{ii}, are given in Table 4.3 and their respective index plots are shown in Figure 4.6. In the index plot of the standardized residuals all the residuals are small indicating that there are no outliers in the data. This is a wrong conclusion because there are two clear outliers in the data as can be seen in the scatter plot in Figure 4.5. Thus masking has occurred! Because of the relationship between leverage and residual in (4.17), the Hackensack river with its large value of $p_{ii} = 0.67$, has a small residual. While a small value of the residual is desirable, the reason for the small value of the residual here is not due to a good fit; it is due to the fact that observation 5 is a high-leverage point and, in collaboration with observation 4, they pull the regression line toward them.

A commonly used cutoff value for p_{ii} is $2(p + 1)/n = 0.2$ (Hoaglin and Welsch, 1978). Accordingly, two points (Hackensack, $p_{ii} = 0.67$, and Neversink, $p_{ii} = 0.25$) that we have seen previously stand out in the scatter plot of points in Figure 4.5, are flagged as high leverage points as can be seen in the index plot of p_{ii} in Figure 4.6(b), where the two points are far from the other points. This example shows clearly that looking solely at residual plots is inadequate.

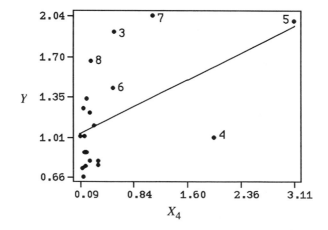

Fig. 4.5 New York Rivers Data: Scatter plot of Y versus X_4.

Table 4.3 New York Rivers Data: The Standardized Residuals, r_i, and the Leverage Values, p_{ii}, From Fitting Model 4.18.

Row	r_i	p_{ii}	Row	r_i	p_{ii}
1	0.03	0.05	11	0.75	0.06
2	−0.05	0.07	12	−0.81	0.06
3	1.95	0.05	13	−0.83	0.06
4	−1.85	0.25	14	−0.83	0.05
5	0.16	0.67	15	−0.94	0.05
6	0.67	0.05	16	−0.48	0.06
7	1.92	0.08	17	−0.72	0.06
8	1.57	0.06	18	−0.50	0.06
9	−0.10	0.06	19	−1.03	0.06
10	0.38	0.06	20	0.57	0.06

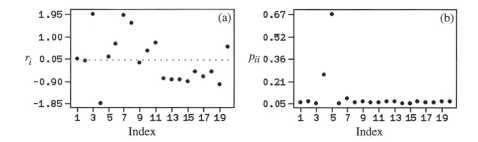

Fig. 4.6 New York Rivers Data: Index plots of (a) the standardized residuals, r_i, and (b) the leverage values p_{ii}.

4.9 MEASURES OF INFLUENCE

The influence of an observation is measured by the effects it produces on the fit when it is deleted in the fitting process. This deletion is almost always done one point at a time. Let $\hat{\beta}_{0(i)}$, $\hat{\beta}_{1(i)}$, ..., $\hat{\beta}_{p(i)}$ denote the regression coefficients obtained when the ith observation is deleted ($i = 1, 2, \ldots, n$). Similarly, let $\hat{y}_{1(i)}, \hat{y}_{2(i)}, \ldots, \hat{y}_{n(i)}$, and $\hat{\sigma}^2_{(i)}$ be the predicted values and residual mean square when we drop the ith observation. Note that

$$\hat{y}_{m(i)} = \hat{\beta}_{0(i)} + \hat{\beta}_{1(i)}x_{m1} + \ldots + \hat{\beta}_{p(i)}x_{mp} \qquad (4.19)$$

is the fitted value for observation m when the fitted equation is obtained with the ith observation deleted. Influence measures look at differences produced in quantities such as $(\hat{\beta}_j - \hat{\beta}_{j(i)})$ or $(\hat{y}_j - \hat{y}_{j(i)})$. There are numerous measures of influence in the literature, and the reader is referred to one of the books for details: Belsley, Kuh, and Welsch (1980), Cook and Weisberg (1982), Atkinson (1985), and Chatterjee and Hadi (1988). Here we give three of these measures.

4.9.1   Cook's Distance

An influence measure proposed by Cook (1977) is widely used. *Cook's distance* measures the difference between the regression coefficients obtained from the full data and the regression coefficients obtained by deleting the ith observation, or equivalently, the difference between the fitted values obtained from the full data and the fitted values obtained by deleting the ith observation. Accordingly, Cook's distance measures the influence of the ith observation by

$$C_i = \frac{\sum_{j=1}^{n}(\hat{y}_j - \hat{y}_{j(i)})^2}{\hat{\sigma}^2(p+1)}, \quad i = 1, 2, \ldots, n. \tag{4.20}$$

It can be shown that C_i can be expressed as

$$C_i = \left(\frac{r_i^2}{p+1}\right)\left(\frac{p_{ii}}{1-p_{ii}}\right), \quad i = 1, 2, \ldots, n. \tag{4.21}$$

Thus, Cook's distance is a multiplicative function of two basic quantities. The first is the square of the standardized residual, r_i, defined in (4.13) and the second is the so-called *potential* function $p_{ii}/(1 - p_{ii})$, where p_{ii} is the leverage of the ith observation introduced previously. If a point is influential, its deletion causes large changes and the value of C_i will be large. Therefore, a large value of C_i indicates that the point is influential. It has been suggested that points with C_i values greater than the 50% point of the F distribution with $p+1$ and $(n-p-1)$ degrees of freedom be classified as influential points. A practical operational rule is to classify points with C_i values greater than 1 as being influential. Rather than using a rigid cutoff rule, we suggest that all C_i values be examined graphically. A dot plot or an index plot of C_i is a useful graphical device. When the C_i values are all about the same, no action need be taken. On the other hand, if there are data points with C_i values that stand out from the rest, these points should be flagged and examined. The model may then be refitted without the offending points to see the effect of these points.

4.9.2   Welsch and Kuh Measure

A measure similar to Cook's distance has been proposed by Welsch and Kuh (1977) and named DFITS. It is defined as

$$DFITS_i = \frac{\hat{y}_i - \hat{y}_{i(i)}}{\hat{\sigma}_{(i)}\sqrt{p_{ii}}}, \; i = 1, 2, \dots, n. \tag{4.22}$$

Thus, $DFITS_i$ is the scaled difference between the ith fitted value obtained from the full data and the ith fitted value obtained by deleting the ith observation. The difference is scaled by $\hat{\sigma}_{(i)}\sqrt{p_{ii}}$. It can be shown that $DFITS_i$ can be written as

$$DFITS_i = r_i^* \sqrt{\frac{p_{ii}}{1 - p_{ii}}}, \; i = 1, 2, \dots, n, \tag{4.23}$$

where r_i^* is the standardized residual defined in (4.14). $DFITS_i$ corresponds to $\sqrt{C_i}$ when the normalization is done by using $\hat{\sigma}_{(i)}$ instead of $\hat{\sigma}$. Points with $|DFITS_i|$ larger than $2\sqrt{(p+1)/(n-p-1)}$ are usually classified as influential points. Again, instead of having a strict cutoff value, we use the measure to sort out points of abnormally high influence relative to other points on a graph such as the index plot, the dot plot, or the box plot. There is not much to choose between C_i and $DFITS_i$ – both give similar answers because they are functions of the residual and leverage values. Most computer software will give one or both of the measures, and it is sufficient to look at only one of them.

4.9.3   Hadi's Influence Measure

Hadi (1992) proposed a measure of the influence of the ith observation based on the fact that influential observations are outliers in either the response variable or in the predictors, or both. Accordingly, the influence of the ith observation can be measured by

$$H_i = \frac{p_{ii}}{1 - p_{ii}} + \frac{p+1}{1 - p_{ii}} \frac{d_i^2}{1 - d_i^2}, \; i = 1, 2, \dots, n, \tag{4.24}$$

where $d_i = e_i/\sqrt{SSE}$ is the so-called *normalized residual*. The first term on the right-hand-side of (4.24) is the potential function which measures outlyingness in the X-space. The second term is a function of the residual, which measures outlyingness in the response variable. It can be seen that observations will have large values of H_i if they are outliers in the response and/or the predictor variables, that is, if they have large values of r_i, p_{ii}, or both. The measure H_i does not focus on a specific regression result, but it can be thought of as an overall general measure of influence which depicts observations that are influential on at least one regression result.

Table 4.4 New York Rivers Data. Influence Measures From Fitting Model 4.18: Cook's Distance, C_i, Welsch and Kuh Measure, $DFITS_i$, and Hadi's Influence Measure H_i.

Row	C_i	$DFITS_i$	H_i	Row	C_i	$DFITS_i$	H_i
1	0.00	0.01	0.06	11	0.02	0.19	0.13
2	0.00	−0.01	0.07	12	0.02	−0.21	0.14
3	0.10	0.49	0.58	13	0.02	−0.22	0.15
4	0.56	−1.14	0.77	14	0.02	−0.19	0.13
5	0.02	0.22	2.04	15	0.02	−0.22	0.16
6	0.01	0.15	0.10	16	0.01	−0.12	0.09
7	0.17	0.63	0.60	17	0.02	−0.18	0.12
8	0.07	0.40	0.37	18	0.01	−0.12	0.09
9	0.00	−0.02	0.07	19	0.04	−0.27	0.19
10	0.00	0.09	0.08	20	0.01	0.15	0.11

Note that C_i and $DFITS_i$ are multiplicative functions of the residuals and leverage values, whereas H_i is an additive function. The influence measure H_i can best be examined graphically in the same way as Cook's distance and Welsch and Kuh Measure.

Example: New York Rivers Data

Consider again fitting the simple regression model in (4.18), which relates the mean nitrogen concentration, Y, to the percentage of land area in commercial/industrial use, X_4. The scatter plot of Y versus X_4 and the corresponding least-squares regression are given in Figure 4.5. Observations 4 (the Neversink river) and 5 (the Hackensack river) are located far from the bulk of other data points in Figure 4.5. Also observations 7, 3, 8, and 6 are somewhat sparse in the upper-left region of the graph. The three influence measures discussed above which result from fitting model (4.18) are shown in Table 4.4, and the corresponding index plots are shown in Figure 4.7. No value of C_i exceeds it cutoff value of 1. However, the index plot of C_i in Figure 4.7(a) shows clearly that observation number 4 (Neversink) should be flagged as an influential observation. This observation also exceeds its $DFITS_i$ cutoff value of $2\sqrt{(p+1)/(n-p-1)} = 2/3$. As can be seen from Figure 4.7, observation number 5 (Hackensack) was not flagged by C_i or by $DFITS_i$. This is due to the small value of the residual because of its high leverage and to the multiplicative nature of the measure. The index plot of H_i in Figure 4.7(c) indicates that observation number 5 (Hackensack) is the most influential one, followed by observation number 4 (Neversink), which is consistent with the scatter plot in Figure 4.5.

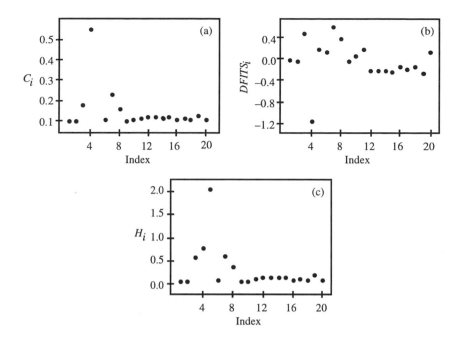

Fig. 4.7 New York Rivers data: Index plots of influence measures (a) Cook's distance, C_i, (b) Welsch and Kuh measure, $DFITS_i$, and (c) Hadi's influence measure H_i.

4.10   THE POTENTIAL-RESIDUAL PLOT

The formula for H_i in (4.24) suggests a simple graph to aid in classifying unusual observations as high-leverage points, outliers, or a combination of both. The graph is called the *potential-residual* (P-R) plot (Hadi, 1992) because it is the scatter plot of

Potential Function   Residual Function

$$\frac{p_{ii}}{1 - p_{ii}} \qquad \text{versus} \qquad \frac{p+1}{1 - p_{ii}} \frac{d_i^2}{1 - d_i^2} .$$

The P-R plot is related to the L-R (*leverage-residual*) plot suggested by Gray (1986) and McCulloch and Meeter (1983). The L-R plot is a scatter plot of p_{ii} versus d_i^2. For a comparison between the two plots, see Hadi (1992).

As an illustrative example, the P-R plot obtained from fitting model (4.18) is shown in Figure 4.8. Observation 5, which is a high-leverage point, is located by itself in the upper left corner of the plot. Four outlying observations (3, 7, 4, and 8) are located in the lower right area of the graph.

It is clear now that some individual data points may be flagged as outliers, leverage points, or influential points. The main usefulness of the leverage and

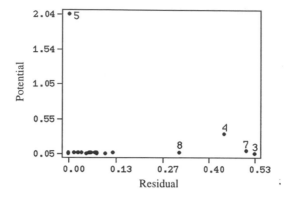

Fig. 4.8 New York Rivers data: Potential-Residual plot.

influence measures is that they give the analyst a complete picture of the role played by different points in the entire fitting process. Any point falling in one of these categories should be carefully examined for accuracy (gross error, transcription error), relevancy (whether it belongs to the data set), and special significance (abnormal condition, unique situation). Outliers should always be scrutinized carefully. Points with high leverage that are not influential do not cause problems. High leverage points that are influential should be investigated because these points are outlying as far as the predictor variables are concerned and also influence the fit. To get an idea of the sensitivity of the analysis to these points, the model should be fitted without the offending points and the resulting coefficients examined.

4.11 WHAT TO DO WITH THE OUTLIERS?

Outliers and influential observations should not routinely be deleted or automatically down-weighted because they are not necessarily bad observations. On the contrary, if they are correct, they may be the most informative points in the data. For example, they may indicate that the data did not come from a normal population or that the model is not linear. To illustrate that outliers and influential observations can be the most informative points in the data, we use the exponential growth data described in the following example.

Example: Exponential Growth Data

Figure 4.9 is the scatter plot of two variables, the size of a certain population, Y, and time, X. As can be seen from the scatter of points, the majority of the points resemble a linear relationship between population size and time as indicated by the straight line in Figure 4.9. According to this model the two points 22 and 23 in the upper right corner are outliers. If these points, however, are correct, they are the only observations in the data set that indicate that the data follow a nonlinear (e.g., exponential) model, such as the one shown in the graph. Think of this as a population of bacteria which increases very slowly over a period of time. After a critical point in time, however, the population explodes.

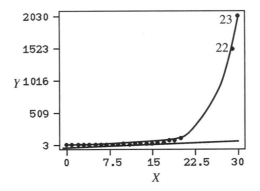

Fig. 4.9 A scatter plot of population size, Y, versus time, X. The curve is obtained by fitting an exponential function to the full data. The straight line is the least squares line when observations 22 and 23 are are deleted.

What to do with outliers and influential observations once they are identified? Because outliers and influential observations can be the most informative observations in the data set, they should not be automatically discarded without justification. Instead, they should be examined to determine why they are outlying or influential. Based on this examination, appropriate corrective actions can then be taken. These corrective actions include: correction of error in the data, deletion or down-weighing outliers, transforming the data, considering a different model, and redesigning the experiment or the sample survey, collecting more data.

4.12   ROLE OF VARIABLES IN A REGRESSION EQUATION

As we have indicated, successive variables are introduced sequentially into a regression equation. A question that arises frequently in practice is: Given a regression model which currently contains p predictor variables, what are the effects of deleting (or adding) one of the variables from (or to) the model? Frequently, the answer is to compute the t-test for each variable in the model. If the t-test is large in absolute value, the variable is retained, otherwise the variable is omitted. This is valid only if the underlying assumptions hold. Therefore, the t-test should be interpreted in conjunction with appropriate graphs of the data. Two plots have been proposed that give this information visually and are often very illuminating. They can be used to complement the t-test in deciding whether one should retain or remove a variable in a regression equation. The first graph is called the *added-variable plot* and the second is the *residual plus component plot*.

4.12.1 Added-Variable Plot

The added-variable plot, introduced by Mosteller and Tukey (1977), enables us graphically to see the magnitude of the regression coefficient of the new variable that is being considered for inclusion. The slope of the least squares line representing the points in the plot is equal to the estimated regression coefficient of the new variable. The plot also shows data points which play key roles in determining this magnitude. We can construct an added-variable plot for each predictor variable X_j. The added-variable plot for X_j is essentially a graph of two different sets of residuals. The first is the residuals when Y is regressed on all predictor variables except X_j. We call this set the Y-residuals. The second set of residuals are obtained when we regress X_j (treated temporarily as a response variable) on all other predictor variables. We refer to this set as the X_j-residuals. Thus, the added-variable plot for X_j is simply a scatter plot of the

Y-residuals versus X_j-residuals.

Therefore, if we have p predictor variables available, we can construct p added-variable plots, one for each predictor.

Note that the Y-residuals in the added-variable plot for X_j represent the part of Y not explained by all predictors other than X_j. Similarly, the X_j-residuals represent the part of X_j that is not explained by the other predictor variables. If a least squares regression line were fitted to the points in the added-variable plot for X_j, the slope of this line is equal to $\hat{\beta}_j$, the estimated regression coefficient of X_j when Y is regressed on all the predictor variables including X_j. This is an illuminating but equivalent interpretation of the partial regression coefficient as we have seen in Section 3.5.

The slope of the points in the plot gives the magnitude of the regression coefficient of the variable if it were brought into the equation. Thus, the stronger the linear relationship in the added-variable plot is, the more important the additional contribution of X_j to the regression equation already containing the other predictors. If the scatter of the points shows no marked slope, the variable is unlikely to be useful in the model. The scatter of the points will also indicate visually which of the data points are most influential in determining this slope and its corresponding t-test. The added-variable plot is also known as the *partial regression plot.*. We remark in passing that it is not actually necessary to carry out this fitting. These residuals can be obtained very simply from computations done in fitting Y on the full set of predictors. For a detailed discussion, see Velleman and Welsch (1981) and Chatterjee and Hadi (1988).

4.12.2 Residual Plus Component Plot

The residual plus component plot, introduced by Ezekiel (1924), is one of the earliest graphical procedures in regression analysis. It was revived by Larsen and McCleary (1972), who called it a *partial residual plot*. We are calling it a residual plus component plot, after Wood (1973), because this name is more self-explanatory.

The residual plus component plot for X_j is a scatter plot of

$$(e + \hat{\beta}_j X_j) \text{ versus } X_j,$$

where e is the ordinary least squares residuals when Y is regressed on all predictor variables and $\hat{\beta}_j$ is the coefficient of X_j in this regression. Note that $\hat{\beta}_j X_j$ is the contribution (component) of the jth predictor to the fitted values. Like in the added-variable plot, the slope of the points in this plot is $\hat{\beta}_j$, the regression coefficient of X_j. Besides indicating the slope graphically, this plot indicates whether any nonlinearity is present in the relationship between Y and X_j. The plot can therefore suggest possible transformations for linearizing the data. The indication of nonlinearity is, however, not present in the added-variable plot because the horizontal scale in the plot is not the variable itself. Both plots are useful, but the residual plus component plot is more sensitive than the added-variable plot in detecting nonlinearities in the variable being considered for introduction in the model. The added-variable plot is, however, easier to interpret and points out the influential observations.

Example: The Scottish Hills Races Data

The Scottish Hills Races data consist of a response variable (record times, in seconds) and two explanatory variables (the distance in miles, and the climb in feet) for 35 races in Scotland in 1984. The data set is given in Table 4.5. Since this data set is three-dimensional, let us first examine a three-dimensional rotating plot of the data as an exploratory tool. An interesting direction in this rotating plot is shown in Figure 4.10. Four observations are marked in this plot. Clearly, observations 7 and 18 are outliers, they lie far away (in the direction of Time) from the plane suggested by the majority of other points. Observation 7 lies far away in the direction of Climb. Observations 33 and 31 are also outliers in the graph but to a lesser extent. While observations 11 and 31 are near the plane suggested by the majority of other points, they are located far from the rest of the points on the plane. (Observation 11 is far mainly in the direction of Distance and observation 31 is in the direction of Climb.) The rotating plot clearly shows that the data contain unusual points (outliers, high leverage points, and/or influential observations).

The fitted equation is

$$Time = -539.483 + 373.073 \ Distance + 0.662888 \ Climb. \qquad (4.25)$$

Table 4.5   Scottish Hills Races Data.

Row	Race	Time	Distance	Climb
1	Greenmantle New Year Dash	965	2.5	650
2	Carnethy	2901	6	2500
3	Craig Dunain	2019	6	900
4	Ben Rha	2736	7.5	800
5	Ben Lomond	3736	8	3070
6	Goatfell	4393	8	2866
7	Bens of Jura	12277	16	7500
8	Cairnpapple	2182	6	800
9	Scolty	1785	5	800
10	Traprain Law	2385	6	650
11	Lairig Ghru	11560	28	2100
12	Dollar	2583	5	2000
13	Lomonds of Fife	3900	9.5	2200
14	Cairn Table	2648	6	500
15	Eildon Two	1616	4.5	1500
16	Cairngorm	4335	10	3000
17	Seven Hills of Edinburgh	5905	14	2200
18	Knock Hill	4719	3	350
19	Black Hill	1045	4.5	1000
20	Creag Beag	1954	5.5	600
21	Kildoon	957	3	300
22	Meall Ant-Suiche	1674	3.5	1500
23	Half Ben Nevis	2859	6	2200
24	Cow Hill	1076	2	900
25	North Berwick Law	1121	3	600
26	Creag Dubh	1573	4	2000
27	Burnswark	2066	6	800
28	Largo	1714	5	950
29	Criffel	3030	6.5	1750
30	Achmony	1257	5	500
31	Ben Nevis	5135	10	4400
32	Knockfarrel	1943	6	600
33	Two Breweries Fell	10215	18	5200
34	Cockleroi	1686	4.5	850
35	Moffat Chase	9590	20	5000

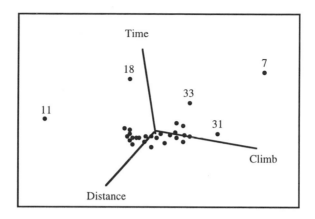

Fig. 4.10   Rotating plot for the Scottish Hills Races data.

We wish to address the question: Does each of the predictor variables contribute significantly when the other variable is included in the model? The t-test for the two predictors are 10.3 and 5.39, respectively, indicating very high significance. This implies that the answer to the above question is in the affirmative for both variables. The validity of this conclusion can be enhanced by examining the corresponding added-variable and residual plus component plots. These are given in Figures 4.11 and 4.12, respectively. For example, in the added-variable plot for Distance in Figure 4.11(a), the quantities plotted on the ordinate axis are the residuals obtained from the regression of Time on Climb (the other predictor variable), and the quantities plotted on the abscissa are the residuals obtained from the regression of Distance on Climb. Similarly for the added-variable plot for Climb, the quantities plotted are the residuals obtained from the regression of Time on Distance and the residuals obtained from the regression of Climb on Distance.

It can be seen that there is strong linear trend in all four graphs supporting the conclusions reached by the above t-tests. The graphs, however, indicate the presence of some points that may influence our results and conclusions. Races 7, 11, and 18 clearly stand out. These points are marked on the graphs by their numbers. Races 31 and 33 are also suspects but to a lesser extent. An examination of the P-R plot obtained from the above fitted equation (Figure 4.13) classifies Race 11 as a high leverage point, Race 18 as an outlier, and Race 7 as a combination of both. These points should be scrutinized carefully before continuing with further analysis.

4.13   EFFECTS OF AN ADDITIONAL PREDICTOR

We discuss in general terms the effect of introducing a new variable in a regression equation. Two questions should be addressed: (a) Is the regression

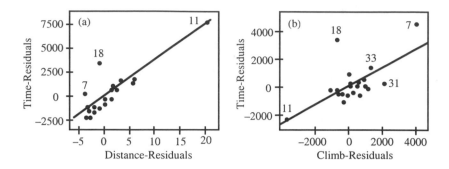

Fig. 4.11 The Scottish Hills Races Data: Added-variable plots for (a) Distance and (b) Climb.

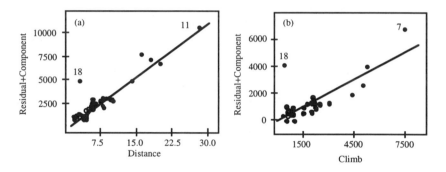

Fig. 4.12 The Scottish Hills Races Data: Residual plus component plots for (a) Distance and (b) Climb.

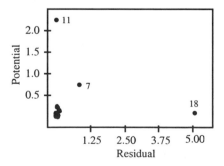

Fig. 4.13 The Scottish Hills Races Data: Potential-Residual plot.

coefficient of the new variable significant? and (b) Does the introduction of the new variable substantially change the regression coefficients of the variables already in the regression equation? When a new variable is introduced in a regression equation, four possibilities result, depending on the answer to each of the above questions:

- **Case A:** The new variable has an insignificant regression coefficient and the remaining regression coefficients do not change substantially from their previous values. Under these conditions the new variable should not be included in the regression equation, unless some other external conditions (e.g., theory or subject matter considerations) dictate its inclusion.

- **Case B:** The new variable has a significant regression coefficient, and the regression coefficients for the previously introduced variables are changed in a substantial way. In this case the new variable should be retained, but an examination of collinearity[1] should be carried out. If there is no evidence of collinearity, the variable should be included in the equation and other additional variables should be examined for possible inclusion. On the other hand, if the variables show collinearity, corrective actions, as outlined in Chapter 10 should be taken.

- **Case C:** The new variable has a significant regression coefficient, and the coefficients of the previously introduced variables do not change in any substantial way. This is the ideal situation and arises when the new variable is uncorrelated with the previously introduced variables. Under these conditions the new variable should be retained in the equation.

- **Case D:** The new variable has an insignificant regression coefficient, but the regression coefficients of the previously introduced variables are substantially changed as a result of the introduction of the new variable. This is a clear evidence of collinearity, and corrective actions have to be taken before the question of the inclusion or exclusion of the new variable in the regression equation can be resolved.

It is apparent from this discussion that the effect a variable has on the regression equation determines its suitability for being included in the fitted equation. The results presented in this chapter influence the formulation of different strategies devised for variable selection. Variable selection procedures are presented in Chapter 11.

---

[1]Collinearity occurs when the predictor variables are highly correlated. This problem is discussed in Chapters 9 and 10.

4.14   ROBUST REGRESSION

Another approach (not discussed here), useful for the identification of outliers and influential observations, is *robust regression*; a method of fitting that gives less weight to points of high leverage. There is a vast amount of literature on robust regression. The interested reader is referred, for example, to the books by Huber (1981), Hampel et al. (1986), Rousseeuw and Leroy (1987), Staudte and Sheather (1990), Birkes and Dodge (1993). We must also mention the papers by Krasker and Welsch (1982), Coakley and Hettmansperger (1993), Chatterjee and Mächler (1997), and Billor, Chatterjee, and Hadi (1999), which incorporate ideas of bounding influence and leverage in fitting.

EXERCISES

4.1   Check to see whether or not the standard regression assumptions are valid for each of the following data sets:

(a)   The Milk Production data described in Section 1.3.1.

(b)   The Right-To-Work Laws data described in Section 1.3.2 and given in Table 1.3.

(c)   The Egyptian Skulls data described in Section 1.3.3.

(d)   The Domestic Immigration data described in Section 1.3.4.

(e)   The New York Rivers data described in Section 1.3.5 and given in Table 1.9.

4.2   Find a data set where regression analysis can be used to answer a question of interest. Then:

(a)   Check to see whether or not the usual multiple regression assumptions are valid.

(b)   Analyze the data using the regression methods presented thus far, and answer the question of interest.

4.3   Consider the computer repair problem discussed in Section 2.3. In a second sampling period, 10 more observations on the variables Minutes and Units were obtained. Since all observations were collected by the same method from a fixed environment, all 24 observations were pooled to form one data set. The data appear in Table 4.6.

(a)   Fit a linear regression model relating Minutes to Units.

(b)   Check each of the standard regression assumptions and indicate which assumption(s) seems to be violated.

4.4   In an attempt to find unusual points in a regression data set, a data analyst examines the P-R plot (shown in Figure 4.14). Classify each of the unusual points in the above graph according to type.

4.5   Name one or more graphs that can be used to validate each of the following assumptions. For each graph, sketch an example where the

Table 4.6 Expanded Computer Repair Times Data: Length of Service Calls (Minutes) and Number of Units Repaired (Units).

Row	Units	Minutes	Row	Units	Minutes
1	1	23	13	10	154
2	2	29	14	10	166
3	3	49	15	11	162
4	4	64	16	11	174
5	4	74	17	12	180
6	5	87	18	12	176
7	6	96	19	14	179
8	6	97	20	16	193
9	7	109	21	17	193
10	8	119	22	18	195
11	9	149	23	18	198
12	9	145	24	20	205

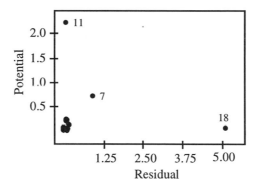

Fig. 4.14 P-R plot used in Exercise 4.4.

corresponding assumption is valid and an example where the assumption is clearly invalid.

(a) There is a linear relationship between the response and predictor variables.

(b) The observations are independent of each other.

(c) The error terms have constant variance.

(d) The error terms are uncorrelated.

(e) The error terms are normally distributed.

(f) The observations are equally influential on least squares results.

4.6   The following graphs are used to verify some of the assumptions of the ordinary least squares regression of Y on $X_1, X_2, \ldots, X_p$:

1. The scatter plot of Y versus each predictor X_j.
2. The scatter plot matrix of the variables $X_1, X_2, \ldots, X_p$.
3. The normal probability plot of the internally standardized residuals.
4. The residuals versus fitted values.
5. The potential-residual plot.
6. Index plot of Cook's distance.
7. Index plot of Hadi's influence measure.

For each of these graphs:

(a) What assumption can be verified by the graph?

(b) Draw an example of the graph where the assumption does not seem to be violated.

(c) Draw an example of the graph which indicates the violation of the assumption.

4.7   Consider again the Cigarette Consumption data described in Exercise 3.14 and given in Table 3.17.

(a) What would you expect the relationship between Sales and each of the other explanatory variables to be (i.e., positive, negative)? Explain.

(b) Compute the pairwise correlation coefficients matrix and construct the corresponding scatter plot matrix.

(c) Are there any disagreements between the pairwise correlation coefficients and the corresponding scatter plot matrix?

(d) Is there any difference between your expectations in part (a) and what you see in the pairwise correlation coefficients matrix and the corresponding scatter plot matrix?

(e) Regress Sales on the six predictor variables. Is there any difference between your expectations in part (a) and what you see in the regression coefficients of the predictor variables? Explain inconsistencies if any.

    (f) How would you explain the difference in the regression coefficients and the pairwise correlation coefficients between Sales and each of the six predictor variables?

    (g) Is there anything wrong with the tests you made and the conclusions you reached in Exercise 3.14?

4.8   Consider again the Salary data described in Exercise 3.12 and given in Table 3.14.

    (a) Draw the P-R plot. Identify all unusual observations (by number) and classify each as outlier, high leverage point, and/or influential observation.

    (b) Is there anything wrong with the tests you made and the conclusions you reached in Exercise 3.12?

4.9   Either prove each of the following statements mathematically or demonstrate its correctness numerically using the Cigarette Consumption data described in Exercise 3.14 and given in Table 3.17:

    (a) The sum of the ordinary least squares residuals is zero.

    (b) The relationship between $\hat{\sigma}^2$ and $\hat{\sigma}^2_{(i)}$ is

$$\hat{\sigma}^2_{(i)} = \hat{\sigma}^2 \left[\frac{n - p - 1 - r_i^2}{n - p - 2} \right]. \tag{4.26}$$

4.10 Identify unusual observations for the data set in Table 4.7

Table 4.7 Data for Exercise 4.10.

Row	Y	X	Row	Y	X
1	8.11	0	7	9.60	19
2	11.00	5	8	10.30	20
3	8.20	15	9	11.30	21
4	8.30	16	10	11.40	22
5	9.40	17	11	12.20	23
6	9.30	18	12	12.90	24

4.11 Consider the Scottish Hills Races data in Table 4.5. Choose an observation index i (e.g., $i = 33$, which corresponds to the outlying observation number 33) and create an indicator (dummy) variable U_i, where all the values of U_i are zero except for its ith value which is one. Now consider comparing the following models:

$$H_0 : Time \;=\; \beta_0 + \beta_1 \, Distance + \beta_2 \, Climb + \varepsilon, \tag{4.27}$$
$$H_1 : Time \;=\; \beta_0 + \beta_1 \, Distance + \beta_2 \, Climb + \beta_3 U_i + \varepsilon. \tag{4.28}$$

Let r_i^* be the ith externally standardized residual obtained from fitting model (4.27). Show (or verify using an example) that

(a) The t-test for testing $\beta_3 = 0$ in Model (4.28) is the same as the ith externally standardized residual obtained from Model (4.27), that is, $t_3 = r_i^*$.

(b) The F-test for testing Model (4.27) versus (4.28) reduces to the square of the ith externally standardized residual, that is, $F = r_i^{*2}$.

(c) Fit Model (4.27) to the Scottish Hills Races data without the ith observation.

(d) Show that the estimates of β_0, β_1, and β_2 in Model (4.28) are the same as those obtained in 4.11c. Hence adding an indicator variable for the ith observation is equivalent to deleting the corresponding observation!

4.12 Consider the data set in Table 4.8, which consist of a response variable Y and six predictor variables. The data can be obtained from the book's Web site. Consider first fitting a linear model relating Y to all six X-variables.

(a) What least squares assumptions (if any) seem to be violated?

(b) Compute r_i, C_i, $DFITS_i$, and H_i.

(c) Construct the index plots of r_i, C_i, $DFITS_i$, and H_i as well as the Potential-Residual plot.

(d) Identify all unusual observations in the data and classify each according to type (i.e., outliers, leverage points, etc.).

4.13 Consider again the data set in Table 4.8. Suppose now that we fit a linear model relating Y to the first three X-variables. Justify your answer to each of the following questions with the appropriate added-variable plot:

(a) Should we add X_4 to the above model? If yes, keep X_4 in the model.

(b) Should we add X_5 to the above model? If yes, keep X_5 in the model.

(c) Should we add X_6 to the above model?

(d) Which model(s) would you recommend as the best possible description of Y? Use the above results and/or perform additional analysis if needed.

4.14 Consider fitting the model $Y = \beta_0 + \beta_1 X_1 + \beta_2 X_2 + \beta_3 X_3 + \varepsilon$, to the data set in Table 4.8. Now let u be the residuals obtained from regressing Y on X_1 and X_2, and v be the residuals obtained from regressing X_3 on X_1. Show (or verify using the data set in Table 4.8 as an example) that:

(a) $\hat{\beta}_3 = \sum_{i=1}^{n} u_i v_i / \sum_{i=1}^{n} v_i^2$

(b) The standard error of $\hat{\beta}_3$ is $\hat{\sigma} / \sqrt{\sum_{i=1}^{n} v_i^2}$.

Table 4.8 Data for Exercises 4.12–4.14.

Row	Y	X_1	X_2	X_3	X_4	X_5	X_6
1	443	49	79	76	8	15	205
2	290	27	70	31	6	6	129
3	676	115	92	130	0	9	339
4	536	92	62	92	5	8	247
5	481	67	42	94	16	3	202
6	296	31	54	34	14	11	119
7	453	105	60	47	5	10	212
8	617	114	85	84	17	20	285
9	514	98	72	71	12	−1	242
10	400	15	59	99	15	11	174
11	473	62	62	81	9	1	207
12	157	25	11	7	9	9	45
13	440	45	65	84	19	13	195
14	480	92	75	63	9	20	232
15	316	27	26	82	4	17	134
16	530	111	52	93	11	13	256
17	610	78	102	84	5	7	266
18	617	106	87	82	18	7	276
19	600	97	98	71	12	8	266
20	480	67	65	62	13	12	196
21	279	38	26	44	10	8	110
22	446	56	32	99	16	8	188
23	450	54	100	50	11	15	205
24	335	53	55	60	8	0	170
25	459	61	53	79	6	5	193
26	630	60	108	104	17	8	273
27	483	83	78	71	11	8	233
28	617	74	125	66	16	4	265
29	605	89	121	71	8	8	283
30	388	64	30	81	10	10	176
31	351	34	44	65	7	9	143
32	366	71	34	56	8	9	162
33	493	88	30	87	13	0	207
34	648	112	105	123	5	12	340
35	449	57	69	72	5	4	200
36	340	61	35	55	13	0	152
37	292	29	45	47	13	13	123
38	688	82	105	81	20	9	268
39	408	80	55	61	11	1	197
40	461	82	88	54	14	7	225

Source: Chatterjee and Hadi (1988)

5

Qualitative Variables as Predictors

5.1 INTRODUCTION

Qualitative or categorical variables can be very useful as predictor variables in regression analysis. Qualitative variables such as sex, marital status, or political affiliation can be represented by *indicator* or *dummy* variables. These variables take on only two values, usually 0 and 1. The two values signify that the observation belongs to one of two possible categories. The numerical values of indicator variables are not intended to reflect a quantitative ordering of the categories, but only serve to identify category or class membership. For example, an analysis of salaries earned by computer programmers may include variables such as education, years of experience, and sex as predictor variables. The sex variable could be quantified, say, as 1 for female and 0 for male. Indicator variables can also be used in a regression equation to distinguish among three or more groups as well as among classifications across various types of groups. For example, the regression described above may also include an indicator variable to distinguish whether the observation was for a systems or applications programmer. The four conditions determined by sex and type of programming can be represented by combining the two variables, as we shall see in this chapter.

Indicator variables can be used in a variety of ways and may be considered whenever there are qualitative variables affecting a relationship. We shall illustrate some of the applications with examples and suggest some additional applications. It is hoped that the reader will recognize the general applicability of the technique from the examples. In the first example, we look at data on a

Table 5.1   Salary Survey Data.

Row	S	X	E	M	Row	S	X	E	M
1	13876	1	1	1	24	22884	6	2	1
2	11608	1	3	0	25	16978	7	1	1
3	18701	1	3	1	26	14803	8	2	0
4	11283	1	2	0	27	17404	8	1	1
5	11767	1	3	0	28	22184	8	3	1
6	20872	2	2	1	29	13548	8	1	0
7	11772	2	2	0	30	14467	10	1	0
8	10535	2	1	0	31	15942	10	2	0
9	12195	2	3	0	32	23174	10	3	1
10	12313	3	2	0	33	23780	10	2	1
11	14975	3	1	1	34	25410	11	2	1
12	21371	3	2	1	35	14861	11	1	0
13	19800	3	3	1	36	16882	12	2	0
14	11417	4	1	0	37	24170	12	3	1
15	20263	4	3	1	38	15990	13	1	0
16	13231	4	3	0	39	26330	13	2	1
17	12884	4	2	0	40	17949	14	2	0
18	13245	5	2	0	41	25685	15	3	1
19	13677	5	3	0	42	27837	16	2	1
20	15965	5	1	1	43	18838	16	2	0
21	12336	6	1	0	44	17483	16	1	0
22	21352	6	3	1	45	19207	17	2	0
23	13839	6	2	0	46	19346	20	1	0

salary survey, such as the one mentioned above, and use indicator variables to adjust for various categorical variables that affect the regression relationship. The second example uses indicator variables for analyzing and testing for equality of regression relationships in various subsets of a population.

We continue to assume that the response variable is a quantitative continuous variable, but the predictor variables can be quantitative and/or categorical. The case where the response variable is an indicator variable is dealt with in Chapter 12.

5.2   SALARY SURVEY DATA

The Salary Survey data set was developed from a salary survey of computer professionals in a large corporation. The objective of the survey was to identify and quantify those variables that determine salary differentials. In addition, the data could be used to determine if the corporation's salary administration guidelines were being followed. The data appear in Table 5.1 and can be obtained from the book's Web site.[1] The response variable is salary (S) and the predictors are: (1) experience (X), measured in years; (2) education (E), coded as 1 for completion of a high school (H.S.) diploma, 2 for completion

---

[1] http://www.ilr.cornell.edu/~hadi/RABE

of a bachelor degree (B.S.), and 3 for the completion of an advanced degree; and (3) management (M), which is coded as 1 for a person with management responsibility and 0 otherwise. We shall try to measure the effects of these three variables on salary using regression analysis.

A linear relationship will be used for salary and experience. We shall assume that each additional year of experience is worth a fixed salary increment. Education may also be treated in a linear fashion. If the education variable is used in the regression equation in raw form, we would be assuming that each step up in education is worth a fixed increment to salary. That is, with all other variables held constant, the relationship between salary and education is linear. That interpretation is possible but may be too restrictive. Instead, we shall view education as a categorical variable and define two indicator variables to represent the three categories. These two variables allow us to pick up the effect of education on salary whether or not it is linear. The management variable is also an indicator variable designating the two categories, 1 for management positions and 0 for regular staff positions.

Note that when using indicator variables to represent a set of categories, the number of these variables required is one less than the number of categories. For example, in the case of the education categories above, we create two indicator variables E_1 and E_2, where

$$E_{i1} = \begin{cases} 1, & \text{if the } i\text{th person is in the H.S. category,} \\ 0, & \text{otherwise,} \end{cases}$$

and

$$E_{i2} = \begin{cases} 1, & \text{if the } i\text{th person is in the B.S. category,} \\ 0, & \text{otherwise.} \end{cases}$$

As stated above, these two variables taken together uniquely represent the three groups. For H.S., $E_1 = 1, E_2 = 0$; for B.S., $E_1 = 0, E_2 = 1$; and for advanced degree, $E_1 = 0, E_2 = 0$. Furthermore, if there were a third variable, E_{i3}, defined to be 1 or 0 depending on whether or not the ith person is in the advanced degree category, then for each person we have $E_1 + E_2 + E_3 = 1$. Then $E_3 = 1 - E_1 - E_2$, showing clearly that one of the variables is superfluous.[2] Similarly, there is only one indicator variable required to distinguish the two management categories. The category that is not represented by an indicator variable is referred to as the *base category* or the *control group* because the regression coefficients of the indicator variables are interpreted relative to the control group.

In terms of the indicator variables described above, the regression model is

$$S = \beta_0 + \beta_1 X + \gamma_1 E_1 + \gamma_2 E_2 + \delta_1 M + \varepsilon. \tag{5.1}$$

---

[2]Had E_1, E_2, and E_3 been used, there would have been a perfect linear relationship among the predictors, which is an extreme case of multicollinearity, described in Chapter 9.

Table 5.2   Regression Equations for the Six Categories of Education and Management.

Category	E	M	Regression Equation
1	1	0	$S = (\beta_0 + \gamma_1) \qquad + \beta_1 X + \varepsilon$
2	1	1	$S = (\beta_0 + \gamma_1 + \delta_1) + \beta_1 X + \varepsilon$
3	2	0	$S = (\beta_0 + \gamma_2) \qquad + \beta_1 X + \varepsilon$
4	2	1	$S = (\beta_0 + \gamma_2 + \delta_1) + \beta_1 X + \varepsilon$
5	3	0	$S = \beta_0 \qquad\qquad\; + \beta_1 X + \varepsilon$
6	3	1	$S = (\beta_0 + \delta_1) \qquad + \beta_1 X + \varepsilon$

Table 5.3   Regression Analysis of Salary Survey Data.

Variable	Coefficient	s.e.	t-test	p-value
Constant	11031.800	383.2	28.80	< 0.0001
X	546.184	30.5	17.90	< 0.0001
E_1	−2996.210	411.8	−7.28	< 0.0001
E_2	147.825	387.7	0.38	0.7049
M	6883.530	313.9	21.90	< 0.0001
$n = 46$	$R^2 = 0.957$	$R_a^2 = 0.953$	$\hat{\sigma} = 1027$	d.f.= 41

By evaluating (5.1) for the different values of the indicator variables, it follows that there is a different regression equation for each of the six (three education and two management) categories as shown in Table 5.2. According to the proposed model, we may say that the indicator variables help to determine the base salary level as a function of education and management status after adjustment for years of experience.

The results of the regression computations for the model given in (5.1) appear in Table 5.3. The proportion of salary variation accounted for by the model is quite high ($R^2 = 0.957$). At this point in the analysis we should investigate the pattern of residuals to check on model specification. We shall postpone that investigation for now and assume that the model is satisfactory so that we can discuss the interpretation of the regression results. Later we shall return to analyze the residuals and find that the model must be altered.

We see that the coefficient of X is 546.16. That is, each additional year of experience is estimated to be worth an annual salary increment of \$546. The other coefficients may be interpreted by looking into Table 5.2. The coefficient of the management indicator variable, δ_1, is estimated to be 6883.50. From Table 5.2 we interpret this amount to be the average incremental value in annual salary associated with a management position. For the education variables, γ_1 measures the salary differential for the H.S. category relative to the advanced degree category and γ_2 measures the differential for the B.S. category relative to the advanced degree category. The difference, $\gamma_2 - \gamma_1$, measures the differential salary for the H.S. category relative to the B.S. category. From the regression results, in terms of salary for computer professionals, we see that an advanced degree is worth $2996 more than a high school diploma, a B.S. is worth $148 more than an advanced degree (this differential is not statistically significant, $t = 0.38$), and a B.S. is worth about $3144 more than a high school diploma. These salary differentials hold for every fixed level of experience.

5.3   INTERACTION VARIABLES

Returning now to the question of model specification, consider Figure 5.1, where the residuals are plotted against X. The plot suggests that there may be three or more specific levels of residuals. Possibly the indicator variables that have been defined are not adequate for explaining the effects of education and management status. Actually, each residual is identified with one of the six education-management combinations. To see this we plot the residuals against Category (a new categorical variable that takes a separate value for each of the six combinations). This graph is, in effect, a plot of residuals versus a potential predictor variable that has not yet been used in the equation. The graph is given in Figure 5.2. It can be seen from the graph that the residuals cluster by size according to their education-management category. The combinations of education and management have not been satisfactorily treated in the model. Within each of the six groups, the residuals are either almost totally positive or totally negative. This behavior implies that the model given in (5.1) does not adequately explain the relationship between salary and experience, education, and management variables. The graph points to some hidden structure in the data that has not been explored.

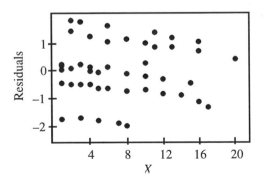

Fig. 5.1 Standardized residuals versus years of experience (X).

The graphs strongly suggest that the effects of education and management status on salary determination are not additive. Note that in the model

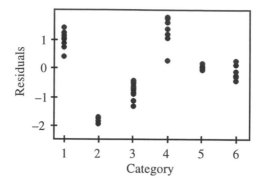

Fig. 5.2   Standardized residuals versus education-management categorical variable.

in (5.1) and its further exposition in Table 5.2, the incremental effects of both variables are determined by additive constants. For example, the effect of a management position is measured as δ_1, independently of the level of educational attainment. The nonadditive effects of these variables can be evaluated by constructing additional variables that are used to measure what may be referred to as *multiplicative* or *interaction effects*. Interaction variables are defined as products of the existing indicator variables $(E_1 \cdot M)$ and $(E_2 \cdot M)$. The inclusion of these two variables on the right-hand side of (5.1) leads to a model that is no longer additive in education and management, but recognizes the multiplicative effect of these two variables.

The expanded model is

$$
\begin{aligned}
S \;=\; & \beta_0 + \beta_1\,X \;+\gamma_1 E_1 + \gamma_2 E_2 + \delta_1 M \\
& +\alpha_1(E_1 \cdot\, M) + \alpha_2(E_2 \cdot\, M) + \varepsilon.
\end{aligned}
\tag{5.2}
$$

The regression results are given in Table 5.4. The residuals from the regression of the expanded model are plotted against X in Figure 5.3. Note that observation 33 is an outlier. Salary is overpredicted by the model. Checking this observation in the listing of the raw data, it appears that this particular person seems to have fallen behind by a couple of hundred dollars in annual salary as compared to other persons with similar characteristics. To be sure that this single observation is not overly affecting the regression estimates, it has been deleted and the regression rerun. The new results are given in Table 5.5.

The regression coefficients are basically unchanged. However, the standard deviation of the residuals has been reduced to $67.28 and the proportion of explained variation has reached 0.9998. The plot of residuals versus X (Figure 5.4) appears to be satisfactory compared with the similar residual plot for the additive model. In addition, the plot of residuals for each education-management category (Figure 5.5) shows that each of these groups has residuals that appear to be symmetrically distributed about zero. Therefore the

Table 5.4 Regression Analysis of Salary Data: Expanded Model.

Variable	Coefficient	s.e.	t-test	p-value
Constant	11203.50	79.07	142.0	< 0.0001
X	496.98	5.57	89.3	< 0.0001
E_1	−1730.69	105.30	−16.4	< 0.0001
E_2	−349.03	97.57	−3.6	0.0009
M	7047.32	102.60	68.7	< 0.0001
$E_1 \cdot M$	−3065.99	149.30	−20.5	< 0.0001
$E_2 \cdot M$	1836.58	131.20	14.0	< 0.0001
$n = 46$	$R^2 = 0.999$	$R_a^2 = 0.999$	$\hat{\sigma} = 173.8$	d.f.= 39

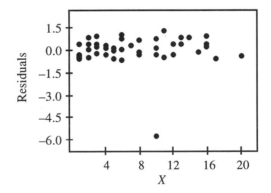

Fig. 5.3 Standardized residuals versus years of experience: Expanded model.

Table 5.5 Regression Analysis of Salary Data: Expanded Model, Observation 33 Deleted.

Variable	Coefficient	s.e.	t-test	p-value
Constant	11199.70	30.54	367.0	< 0.0001
X	498.41	2.15	232.0	< 0.0001
E_1	−1741.28	40.69	−42.8	< 0.0001
E_2	−357.00	37.69	−9.5	< 0.0001
M	7040.49	39.63	178.0	< 0.0001
$E_1 \cdot M$	−3051.72	57.68	−52.9	< 0.0001
$E_2 \cdot M$	1997.62	51.79	38.6	< 0.0001
$n = 45$	$R^2 = 1.0$	$R_a^2 = 1.0$	$\hat{\sigma} = 67.13$	d.f.= 38

introduction of the interaction terms has produced an accurate representation of salary variations. The relationships between salary and experience, education, and management status appear to be adequately described by the model given in (5.2).

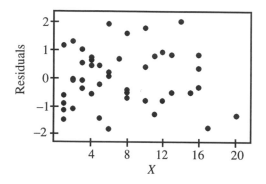

Fig. 5.4 Standardized residuals versus years of experience: Expanded model, observation 33 deleted.

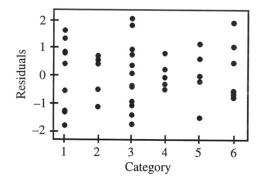

Fig. 5.5 Standardized residuals versus education-management categorical variable: Expanded model, observation 33 deleted.

With the standard error of the residuals estimated to be $67.28, we can believe that we have uncovered the actual and very carefully administered salary formula. Using 95% confidence intervals, each year of experience is estimated to be worth between $494.08 and $502.72. These increments of approximately $500 are added to a starting salary that is specified for each of the six education-management groups. Since the final regression model is not additive, it is rather difficult to directly interpret the coefficients of the indicator variables. To see how the qualitative variables affect salary differentials, we use the coefficients to form estimates of the base salary for each of the six categories. These results are presented in Table 5.6 along with

Table 5.6 Estimates of Base Salary Using the Nonadditive Model in (5.2).

Category	E	M	Coefficients	Estimate of Base Salary[a]	s.e.[a]	95% Confidence Interval
1	1	0	$\beta_0 + \gamma_1$	9459	31	(9398, 9520)
2	1	1	$\beta_0 + \gamma_1 + \delta + \alpha_1$	13448	32	(13385, 13511)
3	2	0	$\beta_0 + \gamma_2$	10843	26	(10792, 10894)
4	2	1	$\beta_0 + \gamma_2 + \delta + \alpha_2$	19880	33	(19815, 19945)
5	3	0	β_0	11200	31	(11139, 11261)
6	3	1	$\beta_0 + \delta$	18240	29	(18183, 18297)

[a] Recorded to the nearest dollar.

standard errors and confidence intervals. The standard errors are computed using Equation (A.10) in the Appendix to Chapter 3.

Using a regression model with indicator variables and interaction terms, it has been possible to account for almost all the variation in salaries of computer professionals selected for this survey. The level of accuracy with which the model explains the data is very rare! We can only conjecture that the methods of salary administration in this company are precisely defined and strictly applied.

In retrospect, we see that an equivalent model may be obtained with a different set of indicator variables and regression parameters. One could define five variables, each taking on the values of 1 or 0, corresponding to five of the six education-management categories. The numerical estimates of base salary and the standard errors of Table 5.6 would be the same. The advantage to proceeding as we have is that it allows us to separate the effects of the three sets of predictor variables, (1) education, (2) management, and (3) education-management interaction. Recall that interaction terms were included only after we found that an additive model did not satisfactorily explain salary variations. In general, we start with simple models and proceed sequentially to more complex models if necessary. We shall always hope to retain the simplest model that has an acceptable residual structure.

5.4 SYSTEMS OF REGRESSION EQUATIONS: COMPARING TWO GROUPS

A collection of data may consist of two or more distinct subsets, each of which may require a separate regression equation. Serious bias may be incurred if one regression relationship is used to represent the pooled data set. An analysis of this problem can be accomplished using indicator variables. An analysis of separate regression equations for subsets of the data may be applied to *cross-sectional* or *time series* data. The example discussed below treats cross-sectional data. Applications to time series data are discussed in Section 5.5.

The model for the two groups can be different in all aspects or in only some aspects. In this section we discuss three distinct cases:

1. Each group has a separate regression model.

2. The models have the same intercept but different slopes.

3. The models have the same slope but different intercepts.

We illustrate these cases below when we have only one quantitative predictor variable. These ideas can be extended straightforwardly to the cases where there are more than one quantitative predictor variable.

5.4.1   Models With Different Slopes and Different Intercepts

We illustrate this case with an important problem concerning equal opportunity in employment. Many large corporations and government agencies administer a preemployment test in an attempt to screen job applicants. The test is supposed to measure an applicant's aptitude for the job and the results are used as part of the information for making a hiring decision. The federal government has ruled[3] that these tests (1) must measure abilities that are directly related to the job under consideration and (2) must not discriminate on the basis of race or national origin. Operational definitions of requirements (1) and (2) are rather elusive. We shall not try to resolve these operational problems. We shall take one approach involving race represented as two groups, white and minority. The hypothesis that there are separate regressions relating test scores to job performance for the two groups will be examined. The implications of this hypothesis for discrimination in hiring are discussed.

Let Y represent job performance and let X be the score on the preemployment test. We want to compare

$$\begin{aligned}
\text{Model 1 (Pooled)} : &\quad y_{ij} = \beta_0 + \beta_1 \, x_{ij} + \varepsilon_{ij}, \quad j = 1, 2; \quad i = 1, 2, \ldots, n_j, \\
\text{Model 2 (Minority)} : &\quad y_{i1} = \beta_{01} + \beta_{11} x_{i1} + \varepsilon_{i1}, \qquad\qquad\qquad\qquad\qquad (5.3) \\
\text{Model 2 (White)} : &\quad y_{i2} = \beta_{02} + \beta_{12} x_{i2} + \varepsilon_{i2}.
\end{aligned}$$

Figure 5.6 depicts the two models. In model 1, race distinction is ignored, the data are pooled, and there is one regression line. In model 2 there is a separate regression relationship for the two subgroups, each with distinct regression coefficients. We shall assume that the variances of the residual terms are the same in each subgroup.

Before analyzing the data, let us briefly consider the types of errors that could be present in interpreting and applying the results. If Y_0, as seen on the

---

[3]Tower amendment to Title VII, Civil Rights Act of 1964.

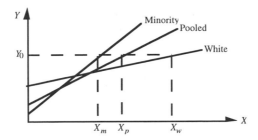

Fig. 5.6 Requirements for employment on pretest.

graph, has been set as the minimum required level of performance, then using Model 1, an acceptable score on the test is one that exceeds X_p. However, if Model 2 is in fact correct, the appropriate test score for whites is X_w and for minorities is X_m. Using X_p in place of X_m and X_w represents a relaxation of the pretest requirement for whites and a tightening of that requirement for minorities. Since inequity can result in the selection procedure if the wrong model is used to set cutoff values, it is necessary to examine the data carefully. It must be determined whether there are two distinct relationships or whether the relationship is the same for both groups and a single equation estimated from the pooled data is adequate. Note that whether Model 1 or Model 2 is chosen, the values X_m, X_w, and X_p are estimates subject to sampling errors and should only be used in conjunction with appropriate confidence intervals. (Construction of confidence intervals is discussed in the following paragraphs.)

Data were collected for this analysis using a special employment program. Twenty applicants were hired on a trial basis for six weeks. One week was spent in a training class. The remaining five weeks were spent on the job. The participants were selected from a pool of applicants by a method that was not related to the preemployment test scores. A test was given at the end of the training period and a work performance evaluation was developed at the end of the six-week period. These two scores were combined to form an index of job performance. (Those employees with unsatisfactory performance at the end of the six-week period were dropped.) The data appear in Table 5.7 and can be obtained from the book's Web site. We refer to this data set as the *Preemployment Testing* data.

Formally, we want to test the null hypothesis $H_0 : \beta_{11} = \beta_{12}, \beta_{01} = \beta_{02}$ against the alternative that there are substantial differences in these parameters. The test can be performed using indicator variables. Let z_{ij} be defined to take the value 1 if $j = 1$ and to take the value zero if $j - 2$. That is, Z is a new variable that has the value 1 for a minority applicant and the value zero for a white applicant. We consider the two models,

$$Model\ 1: \quad y_{ij} = \beta_0 + \beta_1 x_{ij} + \varepsilon_{ij}$$
$$Model\ 3: \quad y_{ij} = \beta_0 + \beta_1 x_{ij} + \gamma z_{ij} + \delta(z_{ij} \cdot x_{ij}) + \varepsilon_{ij}. \quad (5.4)$$

Table 5.7 Data on Preemployment Testing Program.

Row	TEST	RACE	JPERF	Row	TEST	RACE	JPERF
1	0.28	1	1.83	11	2.36	0	3.25
2	0.97	1	4.59	12	2.11	0	5.30
3	1.25	1	2.97	13	0.45	0	1.39
4	2.46	1	8.14	14	1.76	0	4.69
5	2.51	1	8.00	15	2.09	0	6.56
6	1.17	1	3.30	16	1.50	0	3.00
7	1.78	1	7.53	17	1.25	0	5.85
8	1.21	1	2.03	18	0.72	0	1.90
9	1.63	1	5.00	19	0.42	0	3.85
10	1.98	1	8.04	20	1.53	0	2.95

The variable $(z_{ij} \cdot x_{ij})$ represents the interaction between the group (race) variable, Z and the preemployment test X. Note that Model 3 is equivalent to Model 2. This can be seen if we observe that for the minority group, $x_{ij} = x_{i1}$ and $z_{ij} = 1$; hence Model 3 becomes

$$
\begin{aligned}
y_{i1} &= \beta_0 + \beta_1 x_{i1} + \gamma + \delta x_{i1} + \varepsilon_{i1} \\
&= (\beta_0 + \gamma) + (\beta_1 + \delta) x_{i1} + \varepsilon_{i1} \\
&= \beta_{01} + \beta_{11} x_{i1} + \varepsilon_{i1},
\end{aligned}
$$

which is the same as Model 2 for minority with $\beta_{01} = \beta_0 + \gamma$ and $\beta_{11} = \beta_1 + \delta$. Similarly, for the white group, we have $x_{ij} = x_{i2}$, $z_{ij} = 0$, and Model 3 becomes

$$
y_{i2} = \beta_0 + \beta_1 x_{i2} + \varepsilon_{i2},
$$

which is the same as Model 2 for white with $\beta_{02} = \beta_0$ and $\beta_{12} = \beta_1$. Therefore, a comparison between Models 1 and 2 is equivalent to a comparison between Models 1 and 3. Note that Model 3 can be viewed as a full model (FM) and Model 1 as a restricted model (RM) because Model 1 is obtained from Model 3 by setting $\gamma = \delta = 0$. Thus, our null hypothesis H_0 now becomes $H_0 : \gamma = \delta = 0$. The hypothesis is tested by constructing an F-test for the comparison of two models as described in Chapter 3. In this case, the test statistics is

$$
F = \frac{[SSE(RM) - SSE(FM)]/2}{SSE(FM)/16},
$$

which has 2 and 16 degrees of freedom. (Why?) Proceeding with the analysis of the data, the regression results for Model 1 and Model 3 are given in Tables 5.8 and 5.9. The plots of residuals against the predictor variable (Figures 5.7 and 5.8) look acceptable in both cases. The one residual at the lower right in Model 1 may require further investigation.

Table 5.8   Regression Results, Preemployment Testing Data: Model 1.

Variable	Coefficient	s.e.	t-test	p-value
Constant	1.03	0.87	1.19	0.2486
TEST (X)	2.36	0.54	4.39	0.0004
$n = 20$	$R^2 = 0.52$	$R_a^2 = 0.49$	$\hat{\sigma} = 1.59$	d.f.$= 18$

Table 5.9   Regression Results, Preemployment Testing Data: Model 3.

Variable	Coefficient	s.e.	t-test	p-value
Constant	2.01	1.05	1.91	0.0736
TEST (X)	1.31	0.67	1.96	0.0677
RACE (Z)	−1.91	1.54	−1.24	0.2321
(RACE · TEST) $(X \cdot Z)$	2.00	0.95	2.09	0.0527
$n = 20$	$R^2 = 0.664$	$R_a^2 = 0.601$	$\hat{\sigma} = 1.41$	d.f.$= 16$

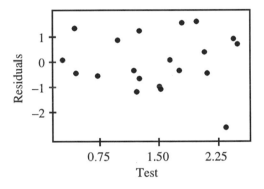

Fig. 5.7   Standardized residuals versus test score: Model 1.

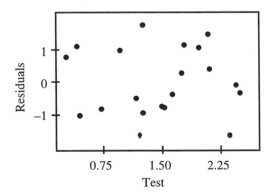

Fig. 5.8   Standardized residuals versus test score: Model 3.

Table 5.10 Separate Regression Results.

Sample	$\hat{\beta}_0$	$\hat{\beta}_1$	t_1	R^2	$\hat{\sigma}$	d.f.
Minority	0.10	3.31	5.31	0.78	1.29	8
White	2.01	1.31	1.82	0.29	1.51	8

To evaluate the formal hypothesis we compute the F-ratio specified previously, which is equal to

$$F = \frac{(45.51 - 31.81)/2}{31.81/16} = 3.4$$

and is significant at a level slightly above 5%. Therefore, on the basis of this test we would conclude that the relationship is probably different for the two groups. Specifically, for minorities we have

$$Y_1 = 0.10 + 3.31X_1$$

and for whites we have

$$Y_2 = 2.01 + 1.32X_2.$$

The results are very similar to those that were described in Figure 5.5 when the problem of bias was discussed. The straight line representing the relationship for minorities has a larger slope and a smaller intercept than the line for whites. If a pooled model were used, the types of biases discussed in relation to Figure 5.6 would occur.

Although the formal procedure using indicator variables has led to the plausible conclusion that the relationships are different for the two groups, the data for the individual groups have not been looked at carefully. Recall that it was assumed that the variances were identical in the two groups. This assumption was required so that the only distinguishing characteristic between the two samples was the pair of regression coefficients. In Figure 5.9 a plot of residuals versus the indicator variable is presented. There does not appear to be a difference between the two sets of residuals. We shall now look more closely at each group. The regression coefficients for each sample taken separately are presented in Table 5.10. The residuals are shown in Figures 5.10 and 5.11. The regression coefficients are, of course, the values obtained from Model 3. The standard errors of the residuals are 1.29 and 1.51 for the minority and white samples, respectively. The residual plots against the test score are acceptable in both cases. An interesting observation that was not available in the earlier analysis is that the preemployment test accounts for a major portion of the variation in the minority sample, but the test is only marginally useful in the white sample.

Our previous conclusion is still valid. The two regression equations are different. Not only are the regression coefficients different, but the residual

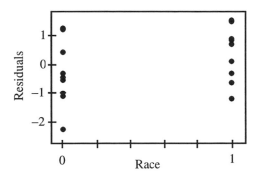

Fig. 5.9 Standardized residuals versus race: Model 1.

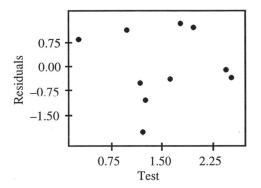

Fig. 5.10 Standardized residuals versus test: Model 1, minority only.

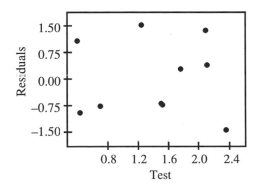

Fig. 5.11 Standardized residuals versus test: Model 1, white only.

mean squares also show slight differences. Of more importance, the values of R^2 are greatly different. For the white sample, $R^2 = 0.29$ is so small ($t = 1.82; 2.306$ is required for significance) that the preemployment test score is not deemed an adequate predictor of job success. This finding has bearing on our original objective since it should be a prerequisite for comparing regressions in two samples that the relationships be valid in each of the samples when taken alone. Concerning the validity of the preemployment test, we conclude that if applied as the law prescribes, with indifference to race, it will give biased results for both racial groups. Moreover, based on these findings, we may be justified in saying that the test is of no value for screening white applicants.

We close the discussion with a note about determining the appropriate cutoff test score if the test were used. Consider the results for the minority sample. If Y_m is designated as the minimum acceptable job performance value to be considered successful, then from the regression equation (also see Figure 5.6)

$$X_m = \frac{Y_m - \hat{\beta}_0}{\hat{\beta}_1} \, ,$$

where $\hat{\beta}_0$ and $\hat{\beta}_1$ are the estimated regression coefficients. X_m is an estimate of the minimum acceptable test score required to attain Y_m. Since X_m is defined in terms of quantities with sampling variation, X_m is also subject to sampling variation. The variation is most easily summarized by constructing a confidence interval for X_m. An approximate 95% level confidence interval takes the form (Scheffé, 1959, p. 52)

$$X_m \pm \frac{t_{(n-2,\alpha/2)}(\hat{\sigma}/n)}{\hat{\beta}_1} \, ,$$

where $t_{(n-2,\alpha/2)}$ is the appropriate percentile point of the t-distribution and $\hat{\sigma}^2$ is the least squares estimate of σ^2. If Y_m is set at 4, then $X_m = (4 - 0.10)/3.31 = 1.18$ and a 95% confidence interval for the test cutoff score is $(1.09, 1.27)$.

5.4.2   Models With Same Slope and Different Intercepts

In the previous subsection we dealt with the case where the two groups have distinct models with different sets of coefficients as given by Models 1 and 2 in (5.3) and as depicted in Figure 5.6. Suppose now that there is a reason to believe that the two groups have the same slope, β_1, and we wish to test the hypothesis that the two groups also have the same intercept, that is, $H_0 : \beta_{01} = \beta_{02}$. In this case we compare

$$
\begin{aligned}
&\text{Model 1 (Pooled)}: \quad &&y_{ij} = \beta_0 + \beta_1 x_{ij} + \varepsilon_{ij}, \quad j = 1, 2; i = 1, 2, \ldots, n_j, \\
&\text{Model 2 (Minority)}: \quad &&y_{i1} = \beta_{01} + \beta_1 x_{i1} + \varepsilon_{i1}, \quad\quad\quad\quad\quad\quad\quad (5.5) \\
&\text{Model 2 (White)}: \quad &&y_{i2} = \beta_{02} + \beta_1 x_{i2} + \varepsilon_{i2}.
\end{aligned}
$$

Notice that the two models have the same value of the slope β_1 but different values of the intercepts β_{01} and β_{02}. Using the indicator variable Z defined earlier, we can write Model 2 as

$$Model\ 3 : y_{ij} = \beta_0 + \beta_1 x_{ij} + \gamma z_{ij} + \varepsilon_{ij}. \tag{5.6}$$

Note the absence of the interaction variable $(z_{ij} \cdot x_{ij})$ from Model 3 in (5.6). If it is present, as it is in (5.4), the two groups would have two models with different slopes and different intercepts.

The equivalence of Models 2 and 3 can be seen by noting that for the minority group, where $x_{ij} = x_{i1}$ and $z_{ij} = 1$, Model 3 becomes

$$
\begin{aligned}
y_{i1} &= \beta_0 + \beta_1 x_{i1} + \gamma + \varepsilon_{i1} \\
&= (\beta_0 + \gamma) + \beta_1 x_{i1} + \varepsilon_{i1} \\
&= \beta_{01} + \beta_1 x_{i1} + \varepsilon_{i1},
\end{aligned}
$$

which is the same as Model 2 for minority with $\beta_{01} = \beta_0 + \gamma$. Similarly, Model 3 for the white group becomes

$$y_{i2} = \beta_0 + \beta_1 x_{i2} + \varepsilon_{i2}.$$

Thus, Model 2 (or equivalently, Model 3) represents two parallel lines[4] (same slope) with intercepts $\beta_0 + \gamma$ and β_0. Therefore, our null hypothesis implies a restriction on γ in Model 3, namely, $H_0 : \gamma = 0$. To test this hypothesis, we use the F-test

$$F = \frac{[SSE(RM) - SSE(FM)]/1}{SSE(FM)/17},$$

which has 1 and 17 degrees of freedom. Equivalently, we can use the t-test for testing $\gamma = 0$ in Model 3, which is

$$t = \frac{\hat{\gamma}}{s.e.(\hat{\gamma})},$$

which has 17 degrees of freedom. Again, the validation of the assumptions of Model 3 should be done before any conclusions are drawn from these tests. For the current example, we leave the computations of the above tests and the conclusions based on them, as an exercise for the reader.

5.4.3 Models With Same Intercept and Different Slopes

Now we deal with the third case where the two groups have the same intercept, β_0, and we wish to test the hypothesis that the two groups also have the same

---

[4]In the general case where the model contains $X_1, X_2, \ldots, X_p$ plus one indicator variable Z, Model 3 represents two parallel (hyper-) planes that differ only in the intercept.

slope, that is, $H_0 : \beta_{11} = \beta_{12}$. In this case we compare

Model 1 (Pooled) :	$y_{ij} = \beta_0 + \beta_1 \, x_{ij} + \varepsilon_{ij}, \quad j = 1, 2; \quad i = 1, 2, \ldots, n_j,$	
Model 2 (Minority) :	$y_{i1} = \beta_0 + \beta_{11} x_{i1} + \varepsilon_{i1},$	(5.7)
Model 2 (White) :	$y_{i2} = \beta_0 + \beta_{12} x_{i2} + \varepsilon_{i2}.$	

Note that the two models have the same value of the intercept β_0 but different values of the slopes β_{11} and β_{12}. Using the indicator variable Z defined earlier, we can write Model 2 as

$$Model\ 3 : y_{ij} = \beta_0 + \beta_1 x_{ij} + \delta(z_{ij} \cdot x_{ij}) + \varepsilon_{ij}. \tag{5.8}$$

Observe the presence of the interaction variable $(z_{ij} \cdot x_{ij})$ but the absence of the individual contribution of the variable Z. The equivalence of Models 2 and 3 can be seen by observing that for the minority group, where $x_{ij} = x_{i1}$ and $z_{ij} = 1$, Model 3 becomes

$$
\begin{aligned}
y_{i1} &= \beta_0 + \beta_1 x_{i1} + \delta x_{i1} + \varepsilon_{i1} \\
&= \beta_0 + (\beta_1 + \delta) x_{i1} + \varepsilon_{i1} \\
&= \beta_0 + \beta_{11} x_{i1} + \varepsilon_{i1},
\end{aligned}
$$

which is the same as Model 2 for minority with $\beta_{11} = \beta_1 + \delta$. Similarly, Model 3 for the white group becomes

$$y_{i2} = \beta_0 + \beta_{12} x_{i2} + \varepsilon_{i2}.$$

Therefore, our null hypothesis implies a restriction on δ in Model 3, namely, $H_0 : \delta = 0$. To test this hypothesis, we use the F-test

$$F = \frac{[SSE(RM) - SSE(FM)]/1}{SSE(FM)/17},$$

which has 1 and 17 degrees of freedom. Equivalently, we can use the t-test for testing $\delta = 0$ in Model 3, which is

$$t = \frac{\hat{\delta}}{s.e.(\hat{\delta})},$$

which has 17 degrees of freedom. Validation of the assumptions of Model 3, the computations of the above tests, and the conclusions based on them are left as an exercise for the reader.

5.5   OTHER APPLICATIONS OF INDICATOR VARIABLES

Applications of indicator variables such as those described in Section 5.4 can be extended to cover a variety of problems (see, e.g., Fox (1984), and Kmenta

(1986) for a variety of applications). Suppose, for example, that we wish to compare the means of $k \geq 2$ populations or groups. The techniques commonly used here is known as the *analysis of variance* (ANOVA). A random sample of size n_j is taken from the jth population, $j = 1, \ldots, k$. We have a total of $n = n_1 + \ldots + n_k$ observations on the response variable. Let y_{ij} be the ith response in the jth sample. Then y_{ij} can be modeled as

$$y_{ij} = \mu_0 + \mu_1 x_{i1} + \ldots + \mu_p x_{ip} + \varepsilon_{ij}. \tag{5.9}$$

In this model there are $p = k - 1$ indicator predictor variables $x_{i1}, \ldots, x_{ip}$. Each variable x_{ij} is 1 if the corresponding response is from population j, and zero otherwise. The population that is left out is usually known as the *control* group. All indicator variables for the control group are equal to zero. Thus, for the control group, (5.9) becomes

$$y_{ij} = \mu_0 + \varepsilon_{ij}. \tag{5.10}$$

In both (5.9) and (5.10), ε_{ij} are random errors assumed to be independent normal variables with zero means and constant variance σ^2. The constant μ_0 represents the mean of the control group and the regression coefficient μ_j can be interpreted as the difference between the means of the control and jth groups. If $\mu_j = 0$, then the means of the control and jth groups are equal. The null hypothesis $H_0 : \mu_1 = \ldots = \mu_p = 0$ that all groups have the same mean can be represented by the model in (5.10). The alternate hypothesis that at least one of the μ_j's is different from zero can be represented by the model in (5.9). The models in (5.9) and (5.10) can be viewed as full and reduced models, respectively. Hence H_0 can be tested using the F-test given in (3.33). Thus, the use of indicator variables allowed us to express ANOVA techniques as a special case of regression analysis. Both the number of quantitative predictor variables and the number of distinct groups represented in the data by indicator variables may be increased.

Note that the examples discussed above are based on cross-sectional data. Indicator variables can also be utilized with time series data. In addition, there are some models of growth processes where an indicator variable is used as the dependent variable. These models, known as *logistic regression models*, are discussed in Chapter 12.

In Sections 5.6 and 5.7 we discuss the use of indicator variables with time series data. In particular, notions of *seasonality* and *stability* of parameters over time are discussed. These problems are formulated and the data are provided. The analyses are left to the reader.

5.6 SEASONALITY

The data set we use as an example here, referred to as the Ski Sales data, is shown in Table 5.11 and can be obtained from the book's Web Site. The data consist of two variables: the sales, S, in millions for a firm that manufactures skis and related equipment for the years 1964–1973, and personal disposable income, PDI.[5] Each of these variables is measured quarterly. We use these data in Chapter 8 to illustrate the problem of *correlated errors*. The model is an equation that relates S to PDI,

$$S_t = \beta_0 + \beta_1 \cdot PDI_t + \varepsilon_t,$$

where S_t is sales in millions in the tth period and PDI_t is the corresponding personal disposable income. Our approach here is to assume the existence of a seasonal effect on sales that is determined on a quarterly basis. To measure this effect we may define indicator variables to characterize the seasonality. Since we have four quarters, we define three indicator variables, Z_1, Z_2, and Z_3, where

$$z_{t1} = \begin{cases} 1, & \text{if the } t\text{th period is a first quarter,} \\ 0, & \text{otherwise,} \end{cases}$$

$$z_{t2} = \begin{cases} 1, & \text{if the } t\text{th period is a second quarter,} \\ 0, & \text{otherwise,} \end{cases}$$

$$z_{t3} = \begin{cases} 1, & \text{if the } t\text{th period is a third quarter,} \\ 0, & \text{otherwise.} \end{cases}$$

The analysis and interpretation of this data set are left to the reader. The authors have analyzed these data and found that there are actually only two seasons. (See the discussion of these sales data in Chapter 8 for an analysis using only one indicator variable, two seasons.) See Kmenta (1986) for further discussion on using indicator variables for analyzing seasonality.

5.7   STABILITY OF REGRESSION PARAMETERS OVER TIME

Indicator variables may also be used to analyze the stability of regression coefficients over time or to test for structural change. We consider an extension of the system of regressions problem when data are available on a cross-section of observations and over time. Our objective is to analyze the constancy of the relationships over time. The methods described here are suitable for intertemporal and interspatial comparisons. To outline the method we use the Education Expenditure data shown in Tables 5.12–5.14. The measured variables for the 50 states are:

---

[5] Aggregate measure of purchasing potential.

Table 5.11   Disposable Income and Ski Sales for Years 1964–1973.

Row	Date	Sales	PDI	Row	Date	Sales	PDI
1	Q1/64	37.0	109	21	Q1/69	44.9	153
2	Q2/64	33.5	115	22	Q2/69	41.6	156
3	Q3/64	30.8	113	23	Q3/69	44.0	160
4	Q4/64	37.9	116	24	Q4/69	48.1	163
5	Q1/65	37.4	118	25	Q1/70	49.7	166
6	Q2/65	31.6	120	26	Q2/70	43.9	171
7	Q3/65	34.0	122	27	Q3/70	41.6	174
8	Q4/65	38.1	124	28	Q4/70	51.0	175
9	Q1/66	40.0	126	29	Q1/71	52.0	180
10	Q2/66	35.0	128	30	Q2/71	46.2	184
11	Q3/66	34.9	130	31	Q3/71	47.1	187
12	Q4/66	40.2	132	32	Q4/71	52.7	189
13	Q1/67	41.9	133	33	Q1/72	52.2	191
14	Q2/67	34.7	135	34	Q2/72	47.0	193
15	Q3/67	38.8	138	35	Q3/72	47.8	194
16	Q4/67	43.7	140	36	Q4/72	52.8	196
17	Q1/68	44.2	143	37	Q1/73	54.1	199
18	Q2/68	40.4	147	38	Q2/73	49.5	201
19	Q3/68	38.4	148	39	Q3/73	49.5	202
20	Q4/68	45.4	151	40	Q4/73	54.3	204

Y Per capita expenditure on public education

X_1 Per capita personal income

X_2 Number of residents per thousand under 18 years of age

X_3 Number of people per thousand residing in urban areas

The variable Region is a categorical variable representing geographical regions (1 = Northeast, 2 = North Central, 3 = South, 4 = West). This data set is used in Chapter 7 to demonstrate methods of dealing with heteroscedasticity in multiple regression and to analyze the effects of regional characteristics on the regression relationships. Here we focus on the stability of the expenditure relationship with respect to time.

Data have been developed on the four variables described above for each state in 1960, 1970, and 1975. Assuming that the relationship can be identically specified in each of the three years,[6] the analysis of stability can be carried out by evaluating the variation in the estimated regression coefficients over time. Working with the pooled data set of 150 observations (50 states

---

[6] *Specification* as used here means that the same variables appear in each equation. Any transformations that are used apply to each equation. The assumption concerning identical specification should be empirically validated.

each in 3 years) we define two indicator variables, T_1 and T_2, where

$$T_{i1} = \begin{cases} 1, & \text{if the } i\text{th observation was from 1960,} \\ 0, & \text{otherwise,} \end{cases}$$

$$T_{i2} = \begin{cases} 1, & \text{if the } i\text{th observation was from 1970,} \\ 0, & \text{otherwise.} \end{cases}$$

Using Y to represent per capita expenditure on schools, the model takes the form

$$\begin{aligned} Y = {} & \beta_0 + \beta_1 X_1 + \beta_2 X_2 + \beta_3 X_3 + \gamma_1 T_1 + \gamma_2 T_2 + \delta_1 T_1 \cdot X_1 \\ & + \delta_2 T_1 \cdot X_2 + \delta_3 T_1 \cdot X_3 + \alpha_1 T_2 \cdot X_1 + \alpha_2 T_2 \cdot X_2 \\ & + \alpha_3 T_2 \cdot X_3 + \varepsilon. \end{aligned}$$

From the definitions of T_1 and T_2, the above model is equivalent to

$$\begin{aligned} \text{For 1960}: \quad & Y = (\beta_0 + \gamma_1) + (\beta_1 + \delta_1)X_1 + (\beta_2 + \delta_2)X_2 \\ & \qquad + (\beta_3 + \delta_3)X_3 + \varepsilon, \\ \text{For 1970}: \quad & Y = (\beta_0 + \gamma_2) + (\beta_1 + \alpha_1)X_1 + (\beta_2 + \alpha_2)X_2 \\ & \qquad + (\beta_3 + \alpha_3)X_3 + \varepsilon, \\ \text{For 1975}: \quad & Y = \beta_0 + \beta_1 X_1 + \beta_2 X_2 + \beta_3 X_3 + \varepsilon. \end{aligned}$$

As noted earlier, this method of analysis necessarily implies that the variability about the regression function is assumed to be equal for all three years. One formal hypothesis of interest is

$$H_0 : \gamma_1 = \gamma_2 = \delta_1 = \delta_2 = \delta_3 = \alpha_1 = \alpha_2 = \alpha_3 = 0,$$

which implies that the regression system has remained unchanged throughout the period of investigation (1960–1975).

The data for this example, which we refer to as the Education Expenditures data, appear in Tables 5.12, 5.13, and 5.14 and can be obtained from the book's Web Site. The reader is invited to perform the analysis described above as an exercise.

EXERCISES

5.1 Using the model defined in (5.6):
  (a) Check to see if the usual least squares assumptions hold.
  (b) Test $H_0 : \gamma = 0$ using the F-test.
  (c) Test $H_0 : \gamma = 0$ using the t-test.
  (d) Verify the equivalence of the two tests above.

5.2 Using the model defined in (5.8):
  (a) Check to see if the usual least squares assumptions hold.
  (b) Test $H_0 : \delta = 0$ using the F-test.

Table 5.12 Education Expenditures Data (1960).

Row	STATE	Y	X_1	X_2	X_3	Region
1	ME	61	1704	388	399	1
2	NH	68	1885	372	598	1
3	VT	72	1745	397	370	1
4	MA	72	2394	358	868	1
5	RI	62	1966	357	899	1
6	CT	91	2817	362	690	1
7	NY	104	2685	341	728	1
8	NJ	99	2521	353	826	1
9	PA	70	2127	352	656	1
10	OH	82	2184	387	674	2
11	IN	84	1990	392	568	2
12	IL	84	2435	366	759	2
13	MI	104	2099	403	650	2
14	WI	84	1936	393	621	2
15	MN	103	1916	402	610	2
16	IA	86	1863	385	522	2
17	MO	69	2037	364	613	2
18	ND	94	1697	429	351	2
19	SD	79	1644	411	390	2
20	NB	80	1894	379	520	2
21	KS	98	2001	380	564	2
22	DE	124	2760	388	326	3
23	MD	92	2221	393	562	3
24	VA	67	1674	402	487	3
25	WV	66	1509	405	358	3
26	NC	65	1384	423	362	3
27	SC	57	1218	453	343	3
28	GA	60	1487	420	498	3
29	FL	74	1876	334	628	3
30	KY	49	1397	594	377	3
31	TN	60	1439	346	457	3
32	AL	59	1359	637	517	3
33	MS	68	1053	448	362	3
34	AR	56	1225	403	416	3
35	LA	72	1576	433	562	3
36	OK	80	1740	378	610	3
37	TX	79	1814	409	727	3
38	MT	95	1920	412	463	4
39	ID	79	1701	418	414	4
40	WY	142	2088	415	568	4
41	CO	108	2047	399	621	4
42	NM	94	1838	458	618	4
43	AZ	107	1932	425	699	4
44	UT	109	1753	494	665	4
45	NV	114	2569	372	663	4
46	WA	112	2160	386	584	4
47	OR	105	2006	382	534	4
48	CA	129	2557	373	717	4
49	AK	107	1900	434	379	4
50	HI	77	1852	431	693	4

Table 5.13   Education Expenditures Data (1970).

Row	STATE	Y	X_1	X_2	X_3	Region
1	ME	189	2828	351	508	1
2	NH	169	3259	346	564	1
3	VT	230	3072	348	322	1
4	MA	168	3835	335	846	1
5	RI	180	3549	327	871	1
6	CT	193	4256	341	774	1
7	NY	261	4151	326	856	1
8	NJ	214	3954	333	889	1
9	PA	201	3419	326	715	1
10	OH	172	3509	354	753	2
11	IN	194	3412	359	649	2
12	IL	189	3981	349	830	2
13	MI	233	3675	369	738	2
14	WI	209	3363	361	659	2
15	MN	262	3341	365	664	2
16	IA	234	3265	344	572	2
17	MO	177	3257	336	701	2
18	ND	177	2730	369	443	2
19	SD	187	2876	369	446	2
20	NB	148	3239	350	615	2
21	KS	196	3303	340	661	2
22	DE	248	3795	376	722	3
23	MD	247	3742	364	766	3
24	VA	180	3068	353	631	3
25	WV	149	2470	329	390	3
26	NC	155	2664	354	450	3
27	SC	149	2380	377	476	3
28	GA	156	2781	371	603	3
29	FL	191	3191	336	805	3
30	KY	140	2645	349	523	3
31	TN	137	2579	343	588	3
32	AL	112	2337	362	584	3
33	MS	130	2081	385	445	3
34	AR	134	2322	352	500	3
35	LA	162	2634	390	661	3
36	OK	135	2880	330	680	3
37	TX	155	3029	369	797	3
38	MT	238	2942	369	534	4
39	ID	170	2668	368	541	4
40	WY	238	3190	366	605	4
41	CO	192	3340	358	785	4
42	NM	227	2651	421	698	4
43	AZ	207	3027	387	796	4
44	UT	201	2790	412	804	4
45	NV	225	3957	385	809	4
46	WA	215	3688	342	726	4
47	OR	233	3317	333	671	4
48	CA	273	3968	348	909	4
49	AK	372	4146	440	484	4
50	HI	212	3513	383	831	4

Table 5.14   Education Expenditures Data (1975).

Row	STATE	Y	X_1	X_2	X_3	Region
1	ME	235	3944	325	508	1
2	NH	231	4578	323	564	1
3	VT	270	4011	328	322	1
4	MA	261	5233	305	846	1
5	RI	300	4780	303	871	1
6	CT	317	5889	307	774	1
7	NY	387	5663	301	856	1
8	NJ	285	5759	310	889	1
9	PA	300	4894	300	715	1
10	OH	221	5012	324	753	2
11	IN	264	4908	329	649	2
12	IL	308	5753	320	830	2
13	MI	379	5439	337	738	2
14	WI	342	4634	328	659	2
15	MN	378	4921	330	664	2
16	IA	232	4869	318	572	2
17	MO	231	4672	309	701	2
18	ND	246	4782	333	443	2
19	SD	230	4296	330	446	2
20	NB	268	4827	318	615	2
21	KS	337	5057	304	661	2
22	DE	344	5540	328	722	3
23	MD	330	5331	323	766	3
24	VA	261	4715	317	631	3
25	WV	214	3828	310	390	3
26	NC	245	4120	321	450	3
27	SC	233	3817	342	476	3
28	GA	250	4243	339	603	3
29	FL	243	4647	287	805	3
30	KY	216	3967	325	523	3
31	TN	212	3946	315	588	3
32	AL	208	3724	332	584	3
33	MS	215	3448	358	445	3
34	AR	221	3680	320	500	3
35	LA	244	3825	355	661	3
36	OK	234	4189	306	680	3
37	TX	269	4336	335	797	3
38	MT	302	4418	335	534	4
39	ID	268	4323	344	541	4
40	WY	323	4813	331	605	4
41	CO	304	5046	324	785	4
42	NM	317	3764	366	698	4
43	AZ	332	4504	340	796	4
44	UT	315	4005	378	804	4
45	NV	291	5560	330	809	4
46	WA	312	4989	313	726	4
47	OR	316	4697	305	671	4
48	CA	332	5438	307	909	4
49	AK	546	5613	386	484	4
50	HI	311	5309	333	831	4

Table 5.15   Corn Yields by Fertilizer Group.

Fertilizer 1	Fertilizer 2	Fertilizer 3	Control Group
31	27	36	33
34	27	37	27
34	25	37	35
34	34	34	25
43	21	37	29
35	36	28	20
38	34	33	25
36	30	29	40
36	32	36	35
45	33	42	29

(c)   Test $H_0 : \delta = 0$ using the t-test.

(d)   Verify the equivalence of the two tests above.

5.3   Perform a thorough analysis of the Ski Sales data in Table 5.11 using the ideas presented in Section 5.6.

5.4   Perform a thorough analysis of the Education Expenditures data in Tables 5.12, 5.13, and 5.14 using the ideas presented in Section 5.7.

5.5   Three types of fertilizer are to be tested to see which one yields more corn crop. Forty similar plots of land were available for testing purposes. The 40 plots are divided at random into four groups, ten plots in each group. Fertilizer 1 was applied to each of the ten corn plots in Group 1. Similarly, Fertilizers 2 and 3 were applied to the plots in Groups 2 and 3, respectively. The corn plants in Group 4 were not given any fertilizer; it will serve as the control group. Table 5.15 gives the corn yield y_{ij} for each of the forty plots.

(a)   Create three indicator variables F_1, F_2, F_3, one for each of the three fertilizer groups.

(b)   Fit the model $y_{ij} = \mu_0 + \mu_1 F_{i1} + \mu_2 F_{i2} + \mu_3 F_{i3} + \varepsilon_{ij}$.

(c)   Test the hypothesis that, on the average, none of the three types of fertilizer has an effect on corn crops. Specify the hypothesis to be tested, the test used, and your conclusions at the 5% significance level.

(d)   Test the hypothesis that, on the average, the three types of fertilizer have equal effects on corn crop but different from that of the control group. Specify the hypothesis to be tested, the test used, and your conclusions at the 5% significance level.

(e)   Which of the three fertilizers has the greatest effects on corn yield?

5.6   In a statistics course personal information was collected on all the students for class analysis. Data on age (in years), height (in inches), and weight (in pounds) of the students are given in Table 5.16 and can be obtained from the book's Web Site. The sex of each student is also

Table 5.16 Class Data on Age (in Years), Height (in Inches), Weight (in Pounds), and Sex (1 = Female, 0 = Male).

Age	Height	Weight	Sex	Age	Height	Weight	Sex
19	61	180	0	19	65	135	1
19	70	160	0	19	70	120	0
19	70	135	0	21	69	142	0
19	71	195	0	20	63	108	1
19	64	130	1	19	63	118	1
19	64	120	1	20	72	135	0
21	69	135	1	19	73	169	0
19	67	125	0	19	69	145	0
19	62	120	1	27	69	130	1
20	66	145	0	18	64	135	0
19	65	155	0	20	61	115	1
19	69	135	1	19	68	140	0
19	66	140	0	21	70	152	0
19	63	120	1	19	64	118	1
19	69	140	0	19	62	112	1
18	66	113	1	19	64	100	1
18	68	180	0	20	67	135	1
19	72	175	0	20	63	110	1
19	70	169	0	20	68	135	0
19	74	210	0	18	63	115	1
20	66	104	1	19	68	145	0
20	64	105	1	19	65	115	1
20	65	125	1	19	63	128	1
20	71	120	1	20	68	140	1
19	69	119	1	19	69	130	0
20	64	140	1	19	69	165	0
20	67	185	1	19	69	130	0
19	60	110	1	20	70	180	0
20	66	120	1	28	65	110	1
19	71	175	0	19	55	155	0

noted and coded as 1 for women and 0 for men. We want to study the relationship between the height and weight of students. Weight is taken as the response variable, and the height as the predictor variable.

(a) Do you agree or do you think the roles of the variables should be reversed?

(b) Is a single equation adequate to describe the relationship between height and weight for the two groups of students? Examine the standardized residual plot from the model fitted to the pooled data, distinguishing between the male and female students.

(c) Find the best model that describes the relationship between the weight and the height of students. Use interaction variables, and the methodology described in this chapter.

(d) Do you think we should include age as a variable to predict weight? Give an intuitive justification for your answer.

Table 5.17   Presidential Election Data (1916–1996).

Year	V	I	D	W	G	P	N
1916	0.5168	1	1	0	2.229	4.252	3
1920	0.3612	1	0	1	−11.463	16.535	5
1924	0.4176	−1	−1	0	−3.872	5.161	10
1928	0.4118	−1	0	0	4.623	0.183	7
1932	0.5916	−1	−1	0	−14.901	7.069	4
1936	0.6246	1	1	0	11.921	2.362	9
1940	0.5500	1	1	0	3.708	0.028	8
1944	0.5377	1	1	1	4.119	5.678	14
1948	0.5237	1	1	1	1.849	8.722	5
1952	0.4460	1	0	0	0.627	2.288	6
1956	0.4224	−1	−1	0	−1.527	1.936	5
1960	0.5009	−1	0	0	0.114	1.932	5
1964	0.6134	1	1	0	5.054	1.247	10
1968	0.4960	1	0	0	4.836	3.215	7
1972	0.3821	−1	−1	0	6.278	4.766	4
1976	0.5105	−1	0	0	3.663	7.657	4
1980	0.4470	1	1	0	−3.789	8.093	5
1984	0.4083	−1	−1	0	5.387	5.403	7
1988	0.4610	−1	0	0	2.068	3.272	6
1992	0.5345	−1	−1	0	2.293	3.692	1
1996	0.5474	1	1	0	2.918	2.268	3

Table 5.18   Variables for the Presidential Election Data (1916–1996) in Table 5.17.

Variable	Definition
YEAR	Election year
V	Democratic share of the two-party presidential vote
I	Indicator variable (1 if there is a Democratic incumbent at the time of the election and −1 if there is a Republican incumbent)
D	Indicator variable (1 if a Democratic incumbent is running for election, −1 if a Republican incumbent is running for election, and 0 otherwise)
W	Indicator variable (1 for the elections of 1920, 1944, and 1948, and 0 otherwise)
G	Growth rate of real per capita GDP in the first three quarters of the election year
P	Absolute value of the growth rate of the GDP deflator in the first 15 quarters of the administration
N	Number of quarters in the first 15 quarters of the administration in which the growth rate of real per capita GDP is greater than 3.2%

5.7 Presidential Election Data (1916–1996): The data in Table 5.17 were kindly provided by Professor Ray Fair of Yale University, who has found that the proportion of votes obtained by a presidential candidate in a United States presidential election can be predicted accurately by three macroeconomic variables, incumbency, and a variable which indicates whether the election was held during or just after a war. The variables considered are given in Table 5.18. All growth rates are annual rates in percentage points. Consider fitting the initial model

$$
\begin{aligned}
V \;=\; & \beta_0 + \beta_1 \cdot I + \beta_2 \cdot D + \beta_3 \cdot W + \beta_4 \cdot (G \cdot I) \\
& + \beta_5 \cdot P + \beta_6 \cdot N + \varepsilon
\end{aligned}
\tag{5.11}
$$

to the data.

(a) Do we need to keep the variable I in the above model?

(b) Do we need to keep the interaction variable $(G \cdot I)$ in the above model?

(c) Examine different models to produce the model or models that might be expected to perform best in predicting future presidential elections. Include interaction terms if needed.

6

Transformation of Variables

6.1 INTRODUCTION

Data do not always come in a form that is immediately suitable for analysis. We often have to transform the variables before carrying out the analysis. Transformations are applied to accomplish certain objectives such as to ensure linearity, to achieve normality, or to stabilize the variance. It often becomes necessary to fit a linear regression model to the transformed rather than the original variables. This is common practice. In this chapter, we discuss the situations where it is necessary to transform the data, the possible choices of transformation, and the analysis of transformed data.

We illustrate transformation mainly using simple regression. In multiple regression where there are several predictors, some may require transformation and others may not. Although the same technique can be applied to multiple regression, transformation in multiple regression requires more effort and care.

The necessity for transforming the data arises because the original variables, or the model in terms of the original variables, violates one or more of the standard regression assumptions. The most commonly violated assumptions are those concerning the linearity of the model and the constancy of the error variance. As mentioned in Chapters 2 and 3, a regression model is linear when the parameters present in the model occur linearly even if the predictor variables occur nonlinearly. For example, each of the four following models is linear:

$$Y = \beta_0 + \beta_1 X + \varepsilon,$$
$$Y = \beta_0 + \beta_1 X + \beta_2 X^2 + \varepsilon,$$

$$Y = \beta_0 + \beta_1 \log X + \varepsilon,$$
$$Y = \beta_0 + \beta_1 \sqrt{X} + \varepsilon,$$

because the model parameters $\beta_0, \beta_1, \beta_2$ enter linearly. On the other hand,

$$Y = \beta_0 + e^{\beta_1 X} + \varepsilon$$

is a nonlinear model because the parameter β_1 does not enter the model linearly. To satisfy the assumptions of the standard regression model, instead of working with the original variables, we sometimes work with transformed variables. Transformations may be necessary for several reasons.

1. Theoretical considerations may specify that the relationship between two variables is nonlinear. An appropriate transformation of the variables can make the relationship between the transformed variables linear. Consider an example from learning theory (experimental psychology). A learning model that is widely used states that the time taken to perform a task on the ith occasion (T_i) is

$$T_i = \alpha\beta^i, \quad \alpha > 0, \quad 0 < \beta < 1. \tag{6.1}$$

The relationship between (T_i) and i as given in (6.1) is nonlinear, and we cannot directly apply techniques of linear regression. On the other hand, if we take logarithms of both sides, we get

$$\log T_i = \log \alpha + i \log \beta, \tag{6.2}$$

showing that $\log T_i$ and i are linearly related. The transformation enables us to use standard regression methods. Although the relationship between the original variables was nonlinear, the relationship between transformed variables is linear. A transformation is used to achieve the linearity of the fitted model.

2. The response variable Y, which is analyzed, may have a probability distribution whose variance is related to the mean. If the mean is related to the value of the predictor variable X, then the variance of Y will change with X, and will not be constant. The distribution of Y will usually also be non-normal under these conditions. Non-normality invalidates the standard tests of significance (although not in a major way with large samples) since they are based on the normality assumption. The unequal variance of the error terms will produce estimates that are unbiased, but are no longer best in the sense of having the smallest variance. In these situations we often transform the data so as to ensure normality and constancy of error variance. In practice, the transformations are chosen to ensure the constancy of variance (*variance-stabilizing transformations*). It is a fortunate coincidence that the variance-stabilizing transformations are also good normalizing transforms.

Table 6.1 Linearizable Simple Regression Functions with Corresponding Transformations.

Function	Transformation	Linear Form	Graph
$Y = \alpha X^{\beta}$	$Y' = \log Y, X' = \log X$	$Y' = \log \alpha + \beta X'$	Figure 6.1
$Y = \alpha e^{\beta X}$	$Y' = \ln Y$	$Y' = \ln \alpha + \beta X$	Figure 6.2
$Y = \alpha + \beta \log X$	$X' = \log X$	$Y = \alpha + \beta X'$	Figure 6.3
$Y = \frac{X}{\alpha X - \beta}$	$Y' = \frac{1}{Y}, X' = \frac{1}{X}$	$Y' = \alpha - \beta X'$	Figure 6.4(a)
$Y = \frac{e^{\alpha + \beta X}}{1 + e^{\alpha + \beta X}}$	$Y' = \ln \frac{Y}{1-Y}$	$Y' = \alpha + \beta X$	Figure 6.4(b)

In Chapter 6 we describe an application using the last transformation.

3. There are neither prior theoretical nor probabilistic reasons to suspect that a transformation is required. The evidence comes from examining the residuals from the fit of a linear regression model in which the original variables are used.

Each of these cases where transformation is needed is illustrated in the following sections.

6.2   TRANSFORMATIONS TO ACHIEVE LINEARITY

One of the standard assumptions made in regression analysis is that the model which describes the data is linear. From theoretical considerations, or from an examination of scatter plot of Y against each predictor X_j, the relationship between Y and X_j may appear to be nonlinear. There are, however, several simple nonlinear regression models which by appropriate transformations can be made linear. We list some of these linearizable curves in Table 6.1. The corresponding graphs are given in Figures 6.1 to 6.4.

When curvature is observed in the scatter plot of Y against X, a linearizable curve from one of those given in Figures 6.1 to 6.4 may be chosen to represent the data. There are, however, many simple nonlinear models that cannot be linearized. Consider for example, $Y = \alpha + \beta \delta^X$, a modified exponential curve, or

$$Y = \alpha_1 e^{\theta_1 X} + \alpha_2 e^{\theta_2 X},$$

which is the sum of two exponential functions. The strictly nonlinear models (i.e., those not linearizable by variable transformation) require very different methods for fitting. We do not describe them in this book but refer the interested reader to Bates and Watts (1988) and Seber and Wild (1989), and Ratkowsky (1990).

In the following example, theoretical considerations lead to a model that is nonlinear. The model is, however, linearizable and we indicate the appropriate analysis.

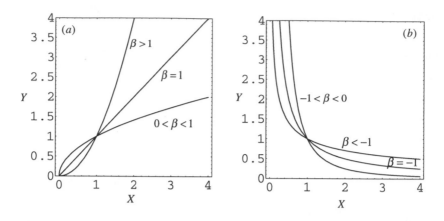

Fig. 6.1   Graphs of the linearizable function $Y = \alpha\, X^\beta$.

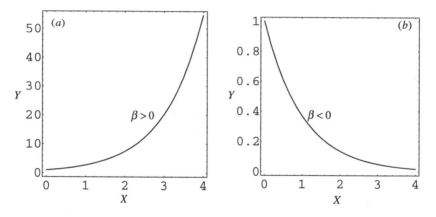

Fig. 6.2   Graphs of the linearizable function $Y = \alpha\, e^{\beta X}$.

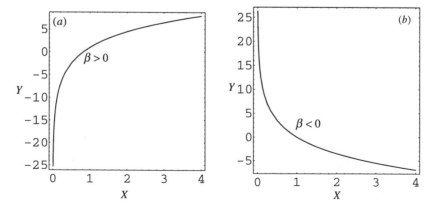

Fig. 6.3   Graphs of the linearizable function $Y = \alpha + \beta \log X$.

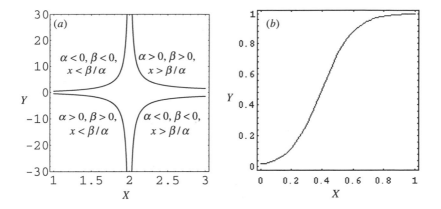

Fig. 6.4 Graphs of the linearizable functions: (a) $Y = X/(\alpha X - \beta)$, and (b) $Y = (e^{\alpha + \beta X})/(1 + e^{\alpha + \beta X})$.

Table 6.2 Number of Surviving Bacteria (Units of 100).

t	n_t	t	n_t	t	n_t
1	355	6	106	11	36
2	211	7	104	12	32
3	197	8	60	13	21
4	166	9	56	14	19
5	142	10	38	15	15

6.3   BACTERIA DEATHS DUE TO X-RAY RADIATION

The data given in Table 6.2 represent the number of surviving bacteria (in hundreds) as estimated by plate counts in an experiment with marine bacterium following exposure to 200-kilovolt X-rays for periods ranging from $t = 1$ to 15 intervals of 6 minutes. The data can also be found in the book's Web Site.[1] The response variable n_t represents the number surviving after exposure time t. The experiment was carried out to test the single-hit hypothesis of X-ray action under constant field of radiation. According to this theory, there is a single vital center in each bacterium, and this must be hit by a ray before the bacteria is inactivated or killed. The particular bacterium studied does not form clumps or chains, so the number of bacterium can be estimated directly from plate counts.

If the theory is applicable, then n_t and t should be related by

$$n_t = n_0 e^{\beta_1 t}, \quad t \geq 0, \tag{6.3}$$

[1]http://www.ilr.cornell.edu/~hadi/RABE

where n_0 and β_1 are parameters. These parameters have simple physical interpretations; n_0 is the number of bacteria at the start of the experiment, and β_1 is the destruction (decay) rate. Taking logarithms of both sides of (6.3), we get

$$\ln n_t = \ln n_0 + \beta_1 t = \beta_0 + \beta_1 t, \tag{6.4}$$

where $\beta_0 = \ln n_0$ and we have $\ln n_t$ as a linear function of t. If we introduce ε_t as the random error, our model becomes

$$\ln n_t = \beta_0 + \beta_1 t + \varepsilon_t \tag{6.5}$$

and we can now apply standard least squares methods.

To get the error ε_t in the transformed model (6.5) to be additive, the error must occur in the multiplicative form in the original model (6.3). The correct representation of the model should be

$$n_t = n_0 e^{\beta_1 t} \varepsilon_t', \tag{6.6}$$

where ε_t' is the multiplicative random error. By comparing (6.5) and (6.6), it is seen that $\varepsilon_t = \ln \varepsilon_t'$. For standard least squares analysis ε_t should be normally distributed, which in turn implies that ε_t', has a log-normal distribution.[2] In practice, after fitting the transformed model we look at the residuals from the fitted model to see if the model assumptions hold. No attempt is usually made to investigate the random component, ε_t', of the original model.

6.3.1   Inadequacy of a Linear Model

The first step in the analysis is to plot the raw data n_t versus t. The plot, shown in Figure 6.5, suggests a nonlinear relationship between n_t and t. However, we proceed by fitting the simple linear model and investigate the consequences of misspecification. The model is

$$n_t = \beta_0 + \beta_1 t + \varepsilon_t, \tag{6.7}$$

where β_0 and β_1 are constants; ε_t's are the random errors, with zero means and equal variances, and are uncorrelated with each other. Estimates of β_0, β_1, their standard errors, and the square of the correlation coefficient are given in Table 6.3. Despite the fact that the regression coefficient for the time variable is significant and we have a high value of R^2, the linear model is not appropriate. The plot of n_t against t shows departure from linearity for high values of t (Figure 6.5). We see this even more clearly if we look at a plot of the standardized residuals against time (Figure 6.6). The distribution of residuals has a distinct pattern. The residuals for $t = 2$ through 11 are all

---

[2]The random variable Y is said to have a log-normal distribution if $\ln Y$ has a normal distribution.

Table 6.3 Estimated Regression Coefficients From Model (6.7).

Variable	Coefficient	s.e.	t-test	p-value
Constant	259.58	22.73	11.42	< 0.0001
TIME (t)	−19.46	2.50	−7.79	< 0.0001
	$n = 15$	$R^2 = 0.823$	$\hat{\sigma} = 41.83$	$d.f. = 13$

negative, for $t = 12$ through 15 are all positive, whereas the residual for $t = 1$ appears to be an outlier. This systematic pattern of deviation confirms that the linear model in (6.7) does not fit the data.

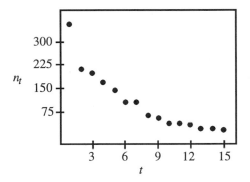

Fig. 6.5 Plot of n_t against time t.

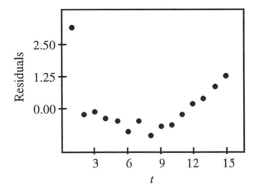

Fig. 6.6 Plot of the standardized residuals from (6.7) against time t.

Table 6.4   Estimated Regression Coefficients When $\ln n_t$ Is Regressed on Time t.

Variable	Coefficient	s.e.	t-test	p-value
Constant	5.973	0.0598	99.9	< 0.0001
TIME (t)	-0.218	0.0066	-33.2	< 0.0001
	$n = 15$	$R^2 = 0.988$	$\hat{\sigma} = 0.11$	$d.f. = 13$

6.3.2   Logarithmic Transformation for Achieving Linearity

The relation between n_t and t appears distinctly nonlinear and we will work with the transformed variable $\ln n_t$, which is suggested from theoretical considerations as well as by Figure 6.7. The plot of $\ln n_t$ against t appears linear, indicating that the logarithmic transformation is appropriate. The results of fitting (6.5) appear in Table 6.4. The coefficients are highly significant, the standard errors are reasonable, and nearly 99% of the variation in the data is explained by the model. The standardized residuals are plotted against t in Figure 6.8. There are no systematic patterns to the distribution of the residuals and the plot is satisfactory. The single-hit hypothesis of X-ray action, which postulates that $\ln n_t$ should be linearly related to t, is confirmed by the data.

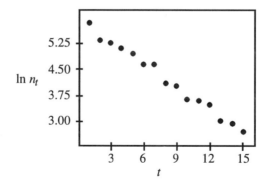

Fig. 6.7   Plot of $\ln n_t$ against time t.

While working with transformed variables, careful attention must be paid to the estimates of the parameters of the model. In our example the point estimate of β_1 is -0.218 and the 95% confidence interval for the same parameter is $(-0.232, -0.204)$. The estimate of the constant term in the equation is the best linear unbiased estimate of $\ln n_0$. If $\hat{\beta}_0$ denotes the estimate, $e^{\hat{\beta}_0}$ may be used as an estimate of n_0. With $\hat{\beta}_0 = 5.973$, the estimate of n_0 is $e^{\hat{\beta}_0} = 392.68$. This estimate is not an unbiased estimate of n_0; that is, the true size of the bacteria population at the start of the experiment was prob-

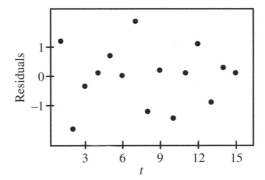

Fig. 6.8 Plot of the standardized residuals against time t after transformation.

somewhat smaller than 392.68. A correction can be made to reduce the bias in the estimate of n_0. The estimate $exp[\hat{\beta}_0 - \frac{1}{2}\text{Var}(\hat{\beta}_0)]$ is nearly unbiased of n_0. In our present example, the modified estimate of n_0 is 381.11. Note that the bias in estimating n_0 has no effect on the test of the theory or the estimation of the decay rate.

In general, if nonlinearity is present, it will show up in a plot of the data. If the plot corresponds approximately to one of the graphs given in Figures 6.1 to 6.4, one of those curves can be fitted after transforming the data. The adequacy of the transformed model can then be investigated by methods outlined in Chapter 4.

6.4 TRANSFORMATIONS TO STABILIZE VARIANCE

We have discussed in the preceding section the use of transformations to achieve linearity of the regression function. Transformations are also used to stabilize the error variance, that is, to make the error variance constant for all the observations. The constancy of error variance is one of the standard assumptions of least squares theory. It is often referred to as the assumption of *homoscedasticity*. When the error variance is not constant over all the observations, the error is said to be *heteroscedastic*. *Heteroscedasticity* is usually detected by suitable graphs of the residuals such as the scatter plot of the standardized residuals against the fitted values or against each of the predictor variables. A plot with the characteristics of Figure 6.9 typifies the situation. The residuals tend to have a funnel-shaped distribution, either fanning out or closing in with the values of X.

If heteroscedasticity is present, and no corrective action is taken application of OLS to the raw data will result in estimated coefficients which lack precision in a theoretical sense. The estimated standard errors of the regression coefficients are often understated, giving a false sense of accuracy.

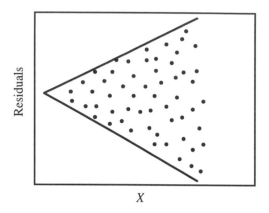

Fig. 6.9   An example of heteroscedastic residuals.

Heteroscedasticity can be removed by means of a suitable transformation. We describe an approach for (a) detecting heteroscedasticity and its effects on the analysis, and (b) removing heteroscedasticity from the data analyzed using transformations.

The response variable Y, in a regression problem, may follow a probability distribution whose variance is a function of the mean of that distribution. One property of the normal distribution, that many other probability distributions do not have, is that its mean and variance are independent in the sense that one is not a function of the other. The binomial and Poisson are but two examples of common probability distributions that have this characteristic. We know, for example, that a variable that is distributed binomially with parameters n and π has mean $n\pi$ and variance $n\pi(1 - \pi)$. It is also known that the mean and variance of a Poisson random variable are equal. When the relationship between the mean and variance of a random variable is known, it is possible to find a simple transformation of the variable, which makes the variance approximately constant (stabilizes the variance). We list in Table 6.5, for convenience and easy reference, transformations that stabilize the variance for some random variables with commonly occurring probability distributions whose variances are functions of their means. The transformations listed in Table 6.5 not only stabilize the variance, but also have the effect of making the distribution of the transformed variable closer to the normal distribution. Consequently, these transformations serve the dual purpose of normalizing the variable as well as making the variance functionally independent of the mean.

As an illustration, consider the following situation: Let Y be the number of accidents and X the speed of operating a lathe in a machine shop. We want to study the relationship between the number of accidents Y and the speed of lathe operation X. Suppose that a linear relationship is postulated

Table 6.5 Transformations to Stabilize Variance.

Probability Distribution of Y	$Var(Y)$ in Terms of Its Mean μ	Transformation	Resulting Variance
Poisson[a]	μ	$\sqrt{Y}$ or $(\sqrt{Y} + \sqrt{Y+1})$	0.25
Binomial[b]	$\mu(1-\mu)/n$	$\sin^{-1}\sqrt{Y}$ (degrees)	$821/n$
		$\sin^{-1}\sqrt{Y}$ (radians)	$0.25/n$
Negative Binomial[c]	$\mu + \lambda^2\mu^2$	$\lambda^{-1}\sinh^{-1}(\lambda\sqrt{Y})$ or	
		$\lambda^{-1}\sinh^{-1}(\lambda\sqrt{Y} + 0.5)$	0.25

[a] For small values of Y, $\sqrt{Y + 0.5}$ is sometimes recommended.
[b] n is an index describing the sample size; for $Y = r/n$ a slightly better transformation is $sin^{-1}\sqrt{(r+3/8)/(n+3/4)}$.
[c] Note that the parameter $\lambda = 1/\sqrt{r}$.

between Y and X and is given by

$$Y = \beta_0 + \beta_1 X + \varepsilon,$$

where ε is the random error. The mean of Y is seen to increase with X. It is known from empirical observation that rare events (events with small probabilities of occurrence) often have a Poisson distribution. Let us assume that Y has a Poisson distribution. Since the mean and variance of Y are the same,[3] it follows that the variance of Y is a function of X, and consequently the assumption of homoscedasticity will not hold. From Table 6.5 we see that the square root of a Poisson variable ($\sqrt{Y}$) has a variance independent of the mean and is approximately equal to 0.25. To ensure homoscedasticity we, therefore, regress $\sqrt{Y}$ on X. Here the transformation is chosen to stabilize the variance, the specific form being suggested by the assumed probability distribution of the response variable. An analysis of data employing transformations suggested by probabilistic considerations is demonstrated in the following example.

Injury Incidents in Airlines

The number of injury incidents and the proportion of total flights from New York for nine ($n = 9$) major United States, airlines for a single year is given in Table 6.6 and plotted in Figure 6.10. Let f_i and y_i denote the total flights and the number of injury incidents for the ith airline that year. Then the

---

[3]The probability mass function of a Poisson random variable Y is $Pr(Y = y) = e^{-\lambda}\lambda^y/y!$; $y = 0, 1, \ldots$, where λ is a parameter. The mean and variance of a Poisson random variable are equal to λ.

Table 6.6   Number of Injury Incidents Y and Proportion of Total Flights N.

Row	Y	N	Row	Y	N	Row	Y	N
1	11	0.0950	4	19	0.2078	7	3	0.1292
2	7	0.1920	5	9	0.1382	8	1	0.0503
3	7	0.0750	6	4	0.0540	9	3	0.0629

proportion of total flights n_i made by the ith airline is

$$n_i = \frac{f_i}{\sum f_i} .$$

If all the airlines are equally safe, the injury incidents can be explained by the model

$$y_i = \beta_0 + \beta_1 n_i + \varepsilon_i,$$

where β_0 and β_1 are constants and ε_i is the random error.

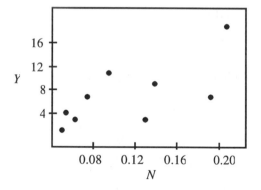

Fig. 6.10   Plot of Y against N.

The results of fitting the model are given in Table 6.7. The plot of residuals against n_i is given in Figure 6.11. The residuals are seen to increase with n_i in Figure 6.11 and, consequently, the assumption of homoscedasticity seems to be violated. This is not surprising, since the injury incidents may behave as a Poisson variable which has a variance proportional to its mean. To ensure the assumption of homoscedasticity, we make the square root transformation. Instead of working with Y we work with $\sqrt{Y}$, a variate which has an approximate variance of 0.25, and is more normally distributed than the original variable.

Consequently, the model we fit is

$$\sqrt{y_i} = \beta_0' + \beta_1' n_i + \varepsilon_i. \tag{6.8}$$

The result of fitting (6.8) is given in Table 6.8. The residuals from (6.8) when plotted against n_i are shown in Figure 6.12. The residuals for the transformed

Table 6.7 Estimated Regression Coefficients (When Y is Regressed on N).

Variable	Coefficient	s.e.	t-test	p-value
Constant	−0.14	3.14	−0.045	0.9657
N	64.98	25.20	2.580	0.0365
	$n = 9$	$R^2 = 0.487$	$\hat{\sigma} = 4.201$	$d.f. = 7$

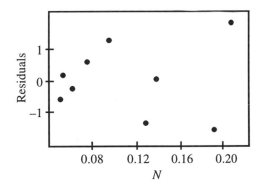

Fig. 6.11 Plot of the standardized residuals versus N.

Table 6.8 Estimated Regression Coefficients When $\sqrt{y_i}$ Is Regressed on n_i.

Variable	Coefficient	s.e.	t-test	p-value
Constant	1.169	0.578	2.02	0.0829
N	11.856	4.638	2.56	0.0378
	$n = 9$	$R^2 = 0.483$	$\hat{\sigma} = 0.773$	$d.f. = 7$

model do not seem to increase with n_i. This suggests that for the transformed model the homoscedastic assumption is not violated. The analysis of the model in terms of $\sqrt{y_i}$ and n_i can now proceed using standard techniques. The regression is significant here (as judged by the t statistic) but is not very strong. Only 48% of the total variability of the injury incidents of the airlines is explained by the variation in their number of flights. It appears that for a better explanation of injury incidents other factors have to be considered.

In the preceding example the nature of the response variable (injury incidents) suggested that the error variance was not constant about the fitted line. The square root transformation was considered based on the well-established empirical fact that the occurrence of accidents tend to follow the Poisson probability distribution. For Poisson variables, the square root is the appropriate transformation (Table 6.5). There are situations, however, when the

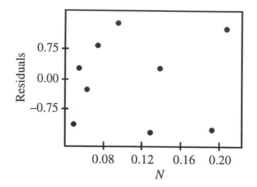

Fig. 6.12 Plot of the standardized residuals from the regression of $\sqrt{y_i}$ on n_i.

error variance is not constant and there is no a priori reason to suspect that this would be the case. Empirical analysis will reveal the problem, and by making an appropriate transformation this effect can be eliminated. If the unequal error variance is not detected and eliminated, the resulting estimates will have large standard errors, but will be unbiased. This will have the effect of producing wide confidence intervals for the parameters, and tests with low sensitivity. We illustrate the method of analysis for a model with this type of heteroscedasticity in the next example.

6.5   DETECTION OF HETEROSCEDASTIC ERRORS

In a study of 27 industrial establishments of varying size, the number of supervised workers (X) and the number of supervisors (Y) were recorded (Table 6.9). The data can also be found in the book's Web site. It was decided to study the relationship between the two variables, and as a start a linear model

$$y_i = \beta_0 + \beta_1 x_i + \varepsilon_i \tag{6.9}$$

was postulated. A plot of Y versus X suggests a simple linear model as a starting point (Figure 6.13). The results of fitting the linear model are given in Table 6.10.

The plot of the standardized residuals versus X (Figure 6.14) shows that the residual variance tends to increase with X. The residuals tend to lie in a band that diverges as one moves along the X axis. In general, if the band within which the residuals lie diverges (i.e., becomes wider) as X increases, the error variance is also increasing with X. On the other hand, if the band converges (i.e., becomes narrower), the error variance decreases with X. If the band that contains the residual plots consists of two lines parallel to the X axis, there is no evidence of heteroscedasticity. A plot of the standardized residuals against the predictor variable points up the presence of heteroscedastic errors.

Table 6.9 Number of Supervised Workers and Supervisors in 27 Industrial Establishments.

Row	X	Y	Row	X	Y	Row	X	Y
1	294	30	10	697	78	19	700	106
2	247	32	11	688	80	20	850	128
3	267	37	12	630	84	21	980	130
4	358	44	13	709	88	22	1025	160
5	423	47	14	627	97	23	1021	97
6	311	49	15	615	100	24	1200	180
7	450	56	16	999	109	25	1250	112
8	534	62	17	1022	114	26	1500	210
9	438	68	18	1015	117	27	1650	135

Table 6.10 Estimated Regression Coefficients When Number of Supervisors (Y) Is Regressed on the Number Supervised (X).

Variable	Coefficient	s.e.	t-test	p-value
Constant	14.448	9.562	1.51	0.1350
X	0.105	0.011	9.30	< 0.0001
	$n = 27$	$R^2 = 0.776$	$\hat{\sigma} = 21.73$	$d.f. = 25$

As can be seen in Figure 6.14, in our present example the residuals tend to increase with X.

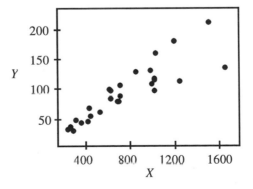

Fig. 6.13 Number of supervisors (Y) versus number supervised (X).

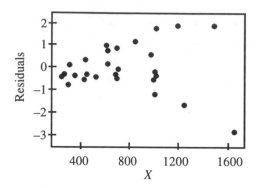

Fig. 6.14 Plot of the standardized residuals against X when number of supervisors (Y) is regressed on the number supervised (X).

6.6   REMOVAL OF HETEROSCEDASTICITY

In many industrial, economic, and biological applications, when unequal error variances are encountered, it is often found that the standard deviation of residuals tends to increase as the predictor variable increases. Based on this empirical observation, we will hypothesize in the present example that the standard deviation of the residuals is proportional to X (some indication of this is available from the plot of the residuals in Figure 6.14):

$$Var(\varepsilon_i) = k^2 x_i^2, \quad k > 0. \tag{6.10}$$

Dividing both sides of (6.9) by x_i, we obtain

$$\frac{y_i}{x_i} = \frac{\beta_0}{x_i} + \beta_1 + \frac{\varepsilon_i}{x_i}. \tag{6.11}$$

Now, define a new set of variables and coefficients,

$$Y' = \frac{Y}{X}, \ X' = \frac{1}{X}, \ \beta_0' = \beta_1, \ \beta_1' = \beta_0, \ \varepsilon' = \frac{\varepsilon}{X}.$$

In terms of the new variables (6.11) reduces to

$$y_i' = \beta_0' + \beta_1' x_i' + \varepsilon_i'. \tag{6.12}$$

Note that for the transformed model, $Var(\varepsilon_i')$ is constant and equals k^2. If our assumption about the error term as given in (6.10) holds, to fit the model properly we must work with the transformed variables: Y/X and $1/X$ as response and predictor variables, respectively. If the fitted model for the transformed data is $\hat{\beta}_0' + \hat{\beta}_1'/X$, the fitted model in terms of the original variables is

$$\hat{Y} = \hat{\beta}_1' + \hat{\beta}_0' X. \tag{6.13}$$

Table 6.11   Estimated Regression Coefficients of the Original Equation When Fitted by the Transformed Variables Y/X and $1/X$.

Variable	Coefficient	s.e.	t-test	p-value
Constant	0.121	0.009	13.44	< 0.0001
$1/X$	3.803	4.570	0.832	0.4131
	$n = 27$	$R^2 = 0.758$	$\hat{\sigma} = 22.577$	$d.f. = 25$

The constant in the transformed model is the regression coefficient of X in the original model, and vice versa. This can be seen from comparing (6.11) and (6.12).

The residuals obtained after fitting the transformed model are plotted against the predictor variable in Figure 6.15. It is seen that the residuals are randomly distributed and lie roughly within a band parallel to the horizontal axis. There is no marked evidence of heteroscedasticity in the transformed model. The distribution of residuals shows no distinct pattern and we conclude that the transformed model is adequate. Our assumption about the error term appears to be correct; the transformed model has homoscedastic errors and the standard assumptions of least squares theory hold. The result of fitting Y/X and $1/X$ leads to estimates of β_0' and β_1' which can be used for the original model.

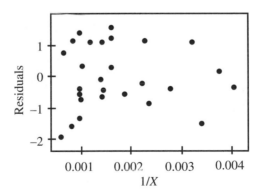

Fig. 6.15   Plot of the standardized residuals against $1/X$ when Y/X is regressed on $1/X$.

The equation for the transformed variables is $Y/X = 0.121 + 3.803/X$. In terms of the original variables, we have $\hat{Y} = 3.803 + 0.121X$. The results are summarized in Table 6.11. By comparing Tables 6.10 and 6.11 we see the reduction in standard errors that is accomplished by working with transformed variables. The variance of the estimate of the slope is reduced by 33%.

6.7   WEIGHTED LEAST SQUARES

Linear regression models with heteroscedastic errors can also be fitted by a method called the *weighted least squares* (WLS), where parameter estimates are obtained by minimizing a weighted sum of squares of residuals where the weights are inversely proportional to the variance of the errors. This is in contrast to ordinary least squares (OLS), where the parameter estimates are obtained by minimizing equally weighted sum of squares of residuals. In the preceding example, the WLS estimates are obtained by minimizing

$$\sum \frac{1}{x_i^2}(y_i - \beta_0 - \beta_1 x_i)^2 \tag{6.14}$$

as opposed to minimizing

$$\sum (y_i - \beta_0 - \beta_1 x_i)^2. \tag{6.15}$$

It can be shown that WLS is equivalent to performing OLS on the transformed variables Y/X and $1/X$. We leave this as an exercise for the reader.

Weighted least squares as an estimation method is discussed in more detail in Chapter 7.

6.8   LOGARITHMIC TRANSFORMATION OF DATA

The logarithmic transformation is one of the most widely used transformations in regression analysis. Instead of working directly with the data, the statistical analysis is carried out on the logarithms of the data. This transformation is particularly useful when the variable analyzed has a large standard deviation compared to its mean. Working with the data on a log scale often has the effect of dampening variability and reducing asymmetry. This transformation is also effective in removing heteroscedasticity. We illustrate this point by using the industrial data given in Table 6.9, where heteroscedasticity has already been detected. Besides illustrating the use of log (logarithmic) transformation to remove heteroscedasticity, we also show in this example that for a given body of data there may exist several adequate descriptions (models).

Instead of fitting the model given in (6.9), we now fit the model

$$\ln y_i = \beta_0 + \beta_1 x_i + \varepsilon_i, \tag{6.16}$$

(i.e., instead of regressing Y on X, we regress $\ln Y$ on X). The corresponding scatter plot is given in Figure 6.16. The results of fitting (6.16) are given in Table 6.12. The coefficients are significant, and the value of R^2 (0.77) is comparable to that obtained from fitting the model given in (6.9).

The plot of the residuals against X is shown in Figure 6.17. The plot is quite revealing. Heteroscedasticity has been removed, but the plot shows

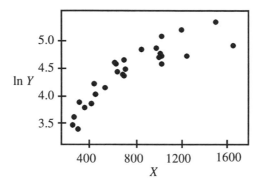

Fig. 6.16   Scatter plot of ln Y versus X.

Table 6.12   Estimated Regression Coefficients When ln Y Is Regressed on X.

Variable	Coefficient	s.e.	t-test	p-value
Constant	3.5150	0.1110	31.65	< 0.0001
X	0.0012	0.0001	9.15	< 0.0001
	$n = 27$	$R^2 = 0.77$	$\hat{\sigma} = 0.252$	$d.f. = 25$

distinct nonlinearity. The residuals display a quadratic effect, suggesting that a more appropriate model for the data may be

$$\ln y_i = \beta_0 + \beta_1 x_i + \beta_2 x_i^2 + \varepsilon_i. \tag{6.17}$$

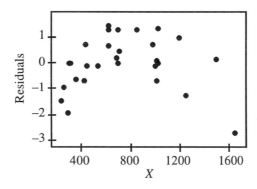

Fig. 6.17   Plot of the standardized residuals against X when ln Y is regressed on X.

Equation (6.17) is a multiple regression model because it has two predictor variables, X and X^2. As discussed in Chapter 4, residual plots can also be used in the detection of model deficiencies in multiple regression. To show the

Table 6.13 Estimated Regression Coefficients When $\ln Y$ is Regressed on X and X^2.

Variable	Coefficient	s.e.	t-test	p-value
Constant	2.8516	0.1566	18.2	< 0.0001
X	3.11267E$-$3	0.0004	7.80	< 0.0001
X^2	$-$1.10226E$-$6	0.220E$-$6	$-$4.93	< 0.0001
	$n = 27$	$R^2 = 0.886$	$\hat{\sigma} = 0.1817$	d.f. = 24

effectiveness of residual plots in detecting model deficiencies and their ability to suggest possible corrections, we present the results of fitting model (6.17) in Table 6.13. Plots of the standardized residuals against the fitted values and against each of the predictor variables X and X^2 are presented in Figures 6.18–6.20, respectively.[4]

Residuals from the model containing a quadratic term appear satisfactory. There is no appearance of heteroscedasticity or nonlinearity in the residuals. We now have two equally acceptable models for the same data. The model given in Table 6.13 may be slightly preferred because of the higher value of R^2. The model given in Table 6.11 is, however, easier to interpret since it is based on the original variables.

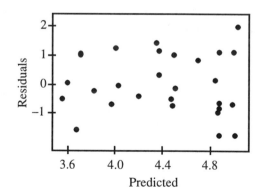

Fig. 6.18 Plot of standardized residuals against the fitted values when $\ln Y$ is regressed on X and X^2.

---

[4]Recall from our discussion in Chapter 4 that in simple regression the plots of residuals against fitted values and against the predictor variable X_1 are identical; hence one needs to examine only one of the two plots but not both. In multiple regression the plot of residuals against the fitted values is distinct from the plots of residuals against each of the predictors.

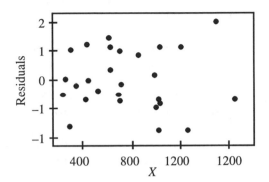

Fig. 6.19 Plot of standardized residuals against X when $\ln Y$ is regressed on X and X^2.

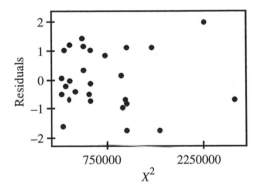

Fig. 6.20 Plot of standardized residuals against X^2 when $\ln Y$ is regressed on X and X^2.

6.9   POWER TRANSFORMATION

In the previous section we used several types of transformations (such as the reciprocal transformation, $1/Y$, the square root transformation, $\sqrt{Y}$, and the logarithmic transformation, lnY). These transformation have been chosen based on theoretical or empirical evidence to obtain linearity of the model, to achieve normality, and/or to stabilize the error variance. These transformation can be thought of as a general case of power transformation. In power transformation, we raise the response variable Y and/or some of the predictor variables to a power. For example, instead of using Y we use Y^λ, where λ is an exponent to be chosen by the data analyst based on either theoretical or empirical evidence. When $\lambda = -1$ we obtain the reciprocal transformation, $\lambda = 0.5$ gives the square root transformation, and when $\lambda = 0$ we obtain the logarithmic transformation.[5] Values of $\lambda = 1$ implies no transformation is needed.

If λ cannot be determined by theoretical considerations, the data can be used to determine the appropriate value of λ. This can be done using numerical methods. In practice, several values of λ are tried and the best value is chosen. Values of λ commonly tried are: 2, 1.5, 1.0, 0.5, 0, −0.5, −1, −1.5, −2. These values of lambda are chosen because they are easy to interpret. They are known as a *ladder of transformation*. This is illustrated in the following example.

Example: The Brain Data

The data set shown in Table 6.14 represent a sample taken from a larger data set. The data can also be found in the book's Web site. The original sources of the data is Jerison (1973). It has also been analyzed by Rousseeuw and Leroy (1987). The average brain weight (in grams), Y, and the average body weight (in kilograms), X, are measured for 28 animals. One purpose of the data is to determine whether a larger brain is required to govern a heavier body. Another purpose is to see whether the ratio of the brain weight to the body weight can be used as a measure of intelligence. The scatter plot of the data (Figure 6.21) does not show an obvious relationship. This is mainly due to the presence of very large animals (e.g., two elephants and three dinosaurs). Let us apply the power transformation to both Y and X. The scatter plots of Y^λ versus X^λ for several values of λ in the ladder of transformation are given in Figure 6.22. It can be seen that the values of $\lambda = 0$ (corresponding to the log transformation) is the most appropriate value. For $\lambda = 0$, the graphs looks linear but the three dinosaurs do not conform to the linear pattern suggested

---

[5]Note that when $\lambda = 0$, $Y^\lambda = 1$ for all values of Y. To avoid this problem the transformation $(Y^\lambda - 1)/\lambda$ is used. It can be shown that as λ approaches zero, $(Y^\lambda - 1)/\lambda$ approaches lnY. This transformation is known as the Box-Cox power transformation. For more details, see Carroll and Ruppert (1988).

by the other points. The graph suggests that either the brain weight of the dinosaurs are underestimated and/or their body weight is overestimated.

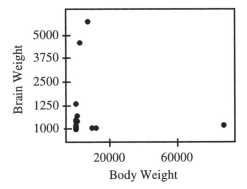

Fig. 6.21   The Brain data: Scatter plots of Brain Weight versus Body Weight.

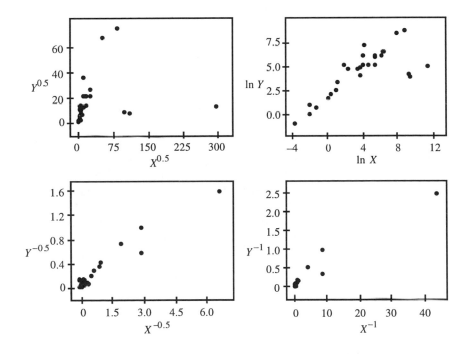

Fig. 6.22   Scatter plots of Y^λ versus X^λ for various values of λ.

Note that in this example we transformed both the response and the predictor variables and that we used the same value of the power for both variables. In other applications, it may be more appropriate to raise each value to a

Table 6.14   The Brain Data: Brain Weight (Grams) and Body Weight (Kilograms).

Name	Brain Weight	Body Weight	Name	Brain Weight	Body Weight
Mountain beaver	8.1	1.35	African elephant	5712.0	6654.00
Cow	423.0	465.00	Triceratops	70.0	9400.00
Gray wolf	119.5	36.33	Rhesus monkey	179.0	6.80
Goat	115.0	27.66	Kangaroo	56.0	35.00
Guinea pig	5.5	1.04	Hamster	1.0	0.12
Diplodocus	50.0	11700.00	Mouse	0.4	0.02
Asian elephant	4603.0	2547.00	Rabbit	12.1	2.50
Donkey	419.0	187.10	Sheep	175.0	55.50
Horse	655.0	521.00	Jaguar	157.0	100.00
Potar monkey	115.0	10.00	Chimpanzee	440.0	52.16
Cat	25.6	3.30	Brachiosaurus	154.5	87000.00
Giraffe	680.0	529.00	Rat	1.9.0	0.28
Gorilla	406.0	207.00	Mole	3.0	0.12
Human	1320.0	62.00	Pig	180.0	192.00

different power and/or to transform only one variable. For further details on data transformation the reader is referred to Carroll and Ruppert (1988) and Atkinson (1985).

6.10   SUMMARY

After fitting a linear model one should examine the residuals for any evidence of heteroscedasticity. Heteroscedasticity is revealed if the residuals tend to increase or decrease with the values of the predictor variable, and is conveniently examined from a plot of the residuals. If heteroscedasticity is present, account should be taken of this in fitting the model. If no account is taken of the unequal error variance, the resulting least squares estimates will not have the maximum precision (smallest variances). Heteroscedasticity can be removed by working with transformed variables. Parameter estimates from the transformed model are then substituted for the appropriate parameters in the original model. The residuals from the appropriately transformed model should show no evidence of heteroscedasticity.

EXERCISES

6.1   Magazine Advertising: In a study of revenue from advertising, data were collected for 41 magazines in 1986 (Table 6.15). The variables observed are number of pages of advertising and advertising revenue. The names of the magazines are listed.

(a)   Fit a linear regression equation relating advertising revenue to advertising pages. Verify that the fit is poor.

(b)   Choose an appropriate transformation of the data and fit the model to the transformed data. Evaluate the fit.

Table 6.15 Advertising Pages (P), in Hundreds, and Advertising Revenue (R), in Millions of Dollars) for 41 Magazines in 1986.

Magazine	P	R	Magazine	P	R
Cosmopolitan	25	50.0	Town and Country	1	7.0
Redbook	15	49.7	True Story	77	6.6
Glamour	20	34.0	Brides	13	6.2
Southern Living	17	30.7	Book Digest Magazine	5	5.8
Vogue	23	27.0	W	7	5.1
Sunset	17	26.3	Yankee	13	4.1
House and Garden	14	24.6	Playgirl	4	3.9
New York Magazine	22	16.9	Saturday Review	6	3.9
House Beautiful	12	16.7	New Woman	3	3.5
Mademoiselle	15	14.6	Ms.	6	3.3
Psychology Today	8	13.8	Cuisine	4	3.0
Life Magazine	7	13.2	Mother Earth News	3	2.5
Smithsonian	9	13.1	1001 Decorating Ideas	3	2.3
Rolling Stone	12	10.6	Self	5	2.3
Modern Bride	1	8.8	Decorating & Craft Ideas	4	1.8
Parents	6	8.7	Saturday Evening Post	4	1.5
Architectural Digest	12	8.5	McCall's Needlework and Craft	3	1.3
Harper's Bazaar	9	8.3	Weight Watchers	3	1.3
Apartment Life	7	8.2	High Times	4	1.0
Bon Appetit	9	8.2	Soap Opera Digest	2	0.3
Gourmet	7	7.3			

(c) You should not be surprised by the presence of a large number of outliers because the magazines are highly heterogeneous and it is unrealistic to expect a single relationship to connect all of them. Delete the outliers and obtain an acceptable regression equation that relates advertising revenue to advertising pages.

6.2 Wind Chill Factor: Table 6.16 gives the effective temperatures (W), which are due to the wind chill effect, for various values of the actual temperatures (T) in still air and windspeed (V). The zero-wind condition is taken as the rate of chilling when one is walking through still air (an apparent wind of four miles per hour (mph)). The National Weather Service originally published the data; we have compiled it from a publication of the Museum of Science of Boston. The temperatures are measured in degrees Fahrenheit (°F), and the wind-speed in mph.

(a) The data in Table 6.16 are not given in a format suitable for direct application of regression programs. You may need to construct another table containing three columns, one column for each of the variables W, T, and V. This table can be found in the book's Web Site.[6]

[6]http://www.ilr.cornell.edu/~hadi/RABE

Table 6.16 Wind Chill Factor (°F) for Various Values of Windspeed, V, in Miles/Hour, and Temperature (°F).

V	Actual Air Temperature (T)											
	50	40	30	20	10	0	-10	-20	-30	-40	-50	-60
5	48	36	27	17	5	-5	-15	-25	-35	-46	-56	-66
10	40	29	18	5	-8	-20	-30	-43	-55	-68	-80	-93
15	35	23	10	-5	-18	-29	-42	-55	-70	-83	-97	-112
20	32	18	4	-10	-23	-34	-50	-64	-79	-94	-108	-121
25	30	15	-1	-15	-28	-38	-55	-72	-88	-105	-118	-130
30	28	13	-5	-18	-33	-44	-60	-76	-92	-109	-124	-134
35	27	11	-6	-20	-35	-48	-65	-80	-96	-113	-130	-137
40	26	10	-7	-21	-37	-52	-68	-83	-100	-117	-135	-140
45	25	9	-8	-22	-39	-54	-70	-86	-103	-120	-139	-143
50	25	8	-9	-23	-40	-55	-72	-88	-105	-123	-142	-145

    (b) Fit a linear relationship between W, T, and V. The pattern of residuals should indicate the inadequacy of the linear model.

    (c) After adjusting W for the effect of T (e.g., keeping T fixed), examine the relationship between W and V. Does the relationship between W and V appear linear?

    (d) After adjusting W for the effect of V, examine the relationship between W and T. Does the relationship appear linear?

    (e) Fit the model

$$W = \beta_0 + \beta_1\, T + \beta_2\, V + \beta_3\, \sqrt{V} + \varepsilon. \tag{6.18}$$

Does the fit of this model appear adequate? The W numbers were produced by the National Weather Service according to the formula (except for rounding errors)

$$W = 0.0817(3.71\sqrt{V} + 5.81 - 0.25V)(T - 91.4) + 91.4. \tag{6.19}$$

Does the formula above give an accurate numerical description of W?

    (f) Can you suggest a model better than those in (6.18) and (6.19)?

6.3 Refer to the Presidential Election Data in Table 5.17, where the response variable V is the proportion of votes obtained by a presidential candidate in United States. Since the response is a proportion, it has a value between 0 and 1. The transformation $Y = \log(V/(1 - V))$ takes the variable V with values between 0 and 1 to a variable Y with values between $-\infty$ to $+\infty$. It is therefore more reasonable to expect that Y satisfies the normality assumption than does V.

    (a) Consider fitting the model

$$Y = \beta_0 + \beta_1 \cdot I + \beta_2 \cdot D + \beta_3 \cdot W + \beta_4 \cdot (G \cdot I)$$

Table 6.17 Annual World Crude Oil Production in Millions of Barrels (1880–1988).

Year	OIL	Year	OIL	Year	OIL
1880	30	1940	2,150	1972	18,584
1890	77	1945	2,595	1974	20,389
1900	149	1950	3,803	1976	20,188
1905	215	1955	5,626	1978	21,922
1910	328	1960	7,674	1980	21,722
1915	432	1962	8,882	1982	19,411
1920	689	1964	10.310	1984	19,837
1925	1,069	1966	12,016	1986	20,246
1930	1,412	1968	14,104	1988	21,338
1935	1,655	1970	16,690		

$$+ \beta_5 \cdot P + \beta_6 \cdot N + \varepsilon, \tag{6.20}$$

which is the same model as in (5.11) but replacing V by Y.

(b) For each of the two models, examine the appropriate residual plots discussed in Chapter 4 to determine which model satisfies the standard assumptions more than the other, the original variable V or the transformed variable Y.

(c) What does the equation in (6.20) imply about the form of the model relating the original variables V in terms of the predictor variables? That is, find the form of the function

$$V = f(\beta_0 + \beta_1 \cdot I + \beta_2 \cdot D + \beta_3 \cdot W + \beta_4 \cdot (C \cdot I)$$
$$+ \beta_5 \cdot P + \beta_6 \cdot N + \varepsilon). \tag{6.21}$$

[Hint: This is a nonlinear function referred to as the *logistic function*, which is discussed in Chapter 12.]

6.4 Oil Production Data: The data in Table 6.17 are the annual world crude oil production in millions of barrels for the period 1880–1988. The data are taken from Moore and McCabe (1993), p. 147.

(a) Construct a scatter plot of the oil production variable (OIL) versus Year and observe that the scatter of points on the graph is not linear. In order to fit a linear model to these data, OIL must be transformed.

(b) Construct a scatter plot of log(OIL) versus Year. The scatter of points now follows a straight line from 1880 to 1973. Political turmoil in the oil-producing regions of the Middle East affected patterns of oil production after 1973.

(c) Fit a linear regression of log(OIL) on Year. Assess the goodness of fit of the model.

(d) Construct the index plot of the standardized residuals. This graph shows clearly that one of the standard assumptions is violated. Which one?

Table 6.18   The Average Price Per Megabyte in Dollars From 1988–1998.

Year	Price	Year	Price
1988	11.54	1994	0.705
1989	9.30	1995	0.333
1990	6.86	1996	0.179
1991	5.23	1997	0.101
1992	3.00	1998	0.068
1993	1.46		

Source: Kindly provided by Jim Porter, Disk/Trends in Wired April 1998.

6.5   One of the remarkable technological developments in computer industry has been the ability to store information densely on hard disk. The cost of storage has steadily declined. Table 6.18 shows the average price per megabyte in dollars from 1988–1998.

(a)   Does a linear time trend describe the data? Define a new variable t by coding 1988 as 1, 1989 as 2, etc.

(b)   Fit the model $P_t = P_0 e^{\beta t}$, where P_t is the price in period t. Does this model describe the data?

(c)   Introduce an indicator variable which takes the value 0 for the years 1988–1991, and 1 for the remaining years. Fit a model to connecting $\log(P_t)$ with time t, the indicator variable, and the variable created by taking the product of time and the indicator variable. Interpret the coefficients of the fitted model.

7

Weighted Least Squares

7.1 INTRODUCTION

So far in our discussion of regression analysis it has been assumed that the underlying regression model is of the form

$$y_i = \beta_0 + \beta_1 x_{i1} + \ldots + \beta_p x_{ip} + \varepsilon_i, \tag{7.1}$$

where the ε_i's are random errors that are independent and identically distributed (i.i.d.) with mean zero and variance σ^2. Various residual plots have been used to check these assumptions (Chapter 4). If the residuals are not consistent with the assumptions, the equation form may be inadequate, additional variables may be required, or some of the observations in the data may be outliers.

There has been one exception to this line of analysis. In the example based on the Supervisor Data of Section 6.5, it is argued that the underlying model does not have residuals that are i.i.d. In particular, the residuals do not have constant variance. For these data, a transformation was applied to correct the situation so that better estimates of the original model parameters could be obtained (better than the ordinary least squares (OLS) method).

In this chapter and in Chapter 8 we investigate situations where the underlying process implies that the errors are not i.i.d. The present chapter deals with the *heteroscedasticity* problem, where the residuals do not have the same variance, and Chapter 8 treats the *autocorrelation* problem, where the residuals are not independent.

In Chapter 6 heteroscedasticity was handled by transforming the variables to stabilize the variance. The *weighted least squares* (WLS) method is equivalent to performing OLS on the transformed variables. The WLS method is presented here both as a way of dealing with heteroscedastic errors and as an estimation method in its own right. For example, WLS perfoms better than OLS in fitting *dose-response curves* (Section 7.5) and *logistic models* (Section 7.5 and Chapter 12).

In this chapter the assumption of equal variance is relaxed. Thus, the ε_i's are assumed to be independently distributed with mean zero and $Var(\varepsilon_i) = \sigma_i^2$. In this case, we use the WLS method to estimate the regression coefficients in (7.1). The WLS estimates of $\beta_0, \beta_1, \ldots, \beta_p$ are obtained by minimizing

$$\sum_{i=1}^{n} w_i (y_i - \beta_0 - \beta_1 x_{i1} - \cdots - \beta_p x_{ip})^2,$$

where w_i are weights inversely proportional to the variances of the residuals (i.e., $w_i = 1/\sigma_i^2$). Note that any observation with a small weight will be severely discounted by WLS in determining the values of $\beta_0, \beta_1, \ldots, \beta_p$. In the extreme case where $w_i = 0$, the effect of WLS is to exclude the ith observation from the estimation process.

Our approach to WLS uses a combination of prior knowledge about the process generating the data and evidence found in the residuals from an OLS fit to detect the heteroscedastic problem. If the weights are unknown, the usual solution prescribed is a two-stage procedure. In Stage 1, the OLS results are used to estimate the weights. In the second stage, WLS is applied using the weights estimated in Stage 1. This is illustrated by examples in the rest of this chapter.

7.2   HETEROSCEDASTIC MODELS

Three different situations in which heteroscedasticity can arise will be distinguished. For the first two situations, estimation can be accomplished in one stage once the source of heteroscedasticity has been identified. The third type is more complex and requires the two-stage estimation procedure mentioned earlier. An example of the first situation is found in Chapter 6 and will be reviewed here. The second situation is described, but no data are analyzed. The third is illustrated with two examples.

7.2.1   Supervisors Data

In Section 6.5, data on the number of workers (X) in an industrial establishment and the number of supervisors (Y) were presented for 27 establishments. The regression model

$$y_i = \beta_0 + \beta_1 x_i + \varepsilon_i \tag{7.2}$$

was proposed. It was argued that the variance of ε_i depends on the size of the establishment as measured by x_i; that is, $\sigma_i^2 = k^2 x_i^2$, where k is a positive constant (see Section 6.5 for details). Empirical evidence for this type of heteroscedasticity is obtained by plotting the standardized residuals versus X. A plot with the characteristics of Figure 7.1 typifies the situation. The residuals tend to have a funnel-shaped distribution, either fanning out or closing in with the values of X. If corrective action is not taken and OLS is applied to the raw data, the resulting estimated coefficients will lack precision in a theoretical sense. In addition, for the type of heteroscedasticity present in these data, the estimated standard errors of the regression coefficients are often understated, giving a false sense of precision. The problem is resolved by using a version of weighted least squares, as described in Chapter 6.

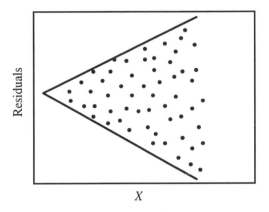

Fig. 7.1   An example of heteroscedastic residuals.

This approach to heteroscedasticity may also be considered in multiple regression models. In (7.1) the variance of the residuals may be affected by only one of the predictor variables. (The case where the variance is a function of more than one predictor variable is discussed later.) Empirical evidence is available from the plots of the standardized residuals versus the suspected variables. For example, if the model is given as (7.1) and it is discovered that the plot of the standardized residuals versus X_2 produces a pattern similar to that shown in Figure 7.1, then one could assume that $Var(\varepsilon_i)$ is proportional to x_{i2}^2, that is, $Var(\varepsilon_i) = k^2 x_{i2}^2$, where $k > 0$. The estimates of the parameters are determined by minimizing

$$\sum_{i=1}^{n} \frac{1}{x_{i2}^2}(y_i - \beta_0 - \beta_1 x_{i1} - \cdots - \beta_p x_{ip})^2.$$

If the software being used has a special weighted least squares procedure, we make the weighting variable equal to $1/x_{i2}^2$. On the other hand, if the software is only capable of performing OLS, we transform the data as described in

Table 7.1   Variables in Cost of Education Survey.

Name	Description
Y	Total annual expense (above tuition)
X_1	Size of city or town where school is located
X_2	Distance to nearest urban center
X_3	Type of school (public or private)
X_4	Size of student body
X_5	Proportion of entering freshman who graduate
X_6	Distance from home

Chapter 6. In other words, we divide both sides of (7.1) by x_{i2} to obtain

$$\frac{y_i}{x_{i2}} = \beta_0 \frac{1}{x_{i2}} + \beta_1 \frac{x_{i1}}{x_{i2}} + \ldots + \beta_p \frac{x_{ip}}{x_{i2}} + \frac{\varepsilon_i}{x_{i2}} .$$

The OLS estimate of the coefficient of the variable $1/X_2$ is the WLS estimate of β_0. The coefficient of the variable X_j/X_2 is an estimate of β_j for all $j \neq 2$. The constant term in this fitting is an estimate of β_2. Refer to Chapter 6 for a detailed discussion of this method applied to simple regression.

7.2.2   College Expense Data

A second type of heteroscedasticity occurs in large-scale surveys where the observations are averages of individual sampling units taken over well-defined groups or clusters. Typically, the average and number of sampling units are reported for each cluster. In some cases, measures of variability such as a standard deviation or range are also reported.

For example, consider a survey of undergraduate college students that is intended to estimate total annual college-related expenses and relate those expenses to characteristics of the institution attended. A list of variables chosen to explain expenses is shown in Table 7.1. Regression analysis with the model

$$Y = \beta_0 + \beta_1 X_1 + \beta_2 X_2 + \cdots + \beta_6 X_6 + \varepsilon \tag{7.3}$$

may be used to study the relationship. In this example, a cluster is equated with a school and an individual sampling unit is a student. Data are collected by selecting a set of schools at random and interviewing a prescribed number of randomly selected students at each school. The response variable, Y, in (7.3) is the average expenditure at the ith school. The predictor variables are characteristics of the school. The numerical values of these variables would be determined from the official statistics published for the school.

The precision of average expenditure is directly proportional to the square root of the sample size on which the average is based. That is, the standard deviation of $\bar{y}_i$ is $\sigma/\sqrt{n_i}$, where n_i represents the number of students interviewed at the ith institution and σ is the standard deviation for annual expense for the population of students. Then the standard deviation of ε_i in the model (7.1) is $\sigma_i = \sigma/\sqrt{n_i}$. Estimation of the regression coefficients is carried out using WLS with weights $w_i = 1/\sigma_i^2$. Since $\sigma_i^2 = \sigma^2/n_i$, the regression coefficients are obtained by minimizing the weighted sum of squared residuals,

$$S = \sum_{i=1}^{n} n_i \left(y_i - \beta_0 - \sum_{j=1}^{6} \beta_j x_{ij} \right)^2. \tag{7.4}$$

Note that the procedure implicitly recognizes that observations from institutions where a large number of students were interviewed as more reliable and should have more weight in determining the regression coefficients than observations from institutions where only a few students were interviewed. The differential precision associated with different observation may be taken as a justification for the weighting scheme.

The estimated coefficients and summary statistics may be computed using a special WLS computer program or by transforming the data and using OLS on the transformed data. Multiplying both sides of (7.1) by $\sqrt{n_i}$, we obtain the new model

$$y_i \sqrt{n_i} = \beta_0 \sqrt{n_i} + \beta_1 x_{i1} \sqrt{n_i} + \cdots + \beta_6 x_{i6} \sqrt{n_i} + \varepsilon_i \sqrt{n_i}. \tag{7.5}$$

The error terms in (7.5), $\varepsilon_i \sqrt{n_i}$ now satisfy the necessary assumption of constant variance. Regression of $y_i \sqrt{n_i}$ against the seven new variables consisting of $\sqrt{n_i}$, and the six transformed predictor variables, $x_{ji} \sqrt{n_i}$ using OLS will produce the desired estimates of the regression coefficients and their standard errors. Note that the regression model in (7.5) has seven predictor variables, a new variable $\sqrt{n_i}$, and the six original predictor variables multiplied by $\sqrt{n_i}$. Note also that there is no constant term in (7.5) because the intercept of the original model, β_0, is now the coefficient of $\sqrt{n_i}$. Thus the regression with the transformed variables must be carried out with the constant term constrained to be zero, that is, we fit a no-intercept model. More details on this point are given in the numerical example in Section 7.4.

7.3 TWO-STAGE ESTIMATION

In the two preceding problems heteroscedasticity was expected at the outset. In the first problem the nature of the process under investigation suggests residual variances that increase with the size of the predictor variable. In the second case, the method of data collection indicates heteroscedasticity. In both cases, homogeneity of variance is accomplished by a transformation. The transformation is constructed directly from information in the raw data. In the problem described in this section, there is also some prior indication that the variances are not equal. But here the exact structure of heteroscedasticity is determined empirically. As a result, estimation of the regression parameters requires two stages.

Detection of heteroscedasticity in multiple regression is not a simple matter. If present it is often discovered as a result of some good intuition on the part of the analyst on how observations may be grouped or clustered. For multiple regression models, the plots of the standardized residuals versus the fitted values and versus each predictor variable can serve as a first step. If the magnitude of the residuals appears to vary systematically with $\hat{y}_i$ or with x_{ij}, heteroscedasticity is suggested. The plot, however, does not necessarily indicate why the variances differ (see the following example).

One direct method for investigating the presence of nonconstant variance is available when there are replicated measurements on the response variable corresponding to a set of fixed values of the predictor variables. For example, in the case of one predictor variable, we may have measurements $y_{11}, y_{21}, \ldots, y_{n_1 1}$ at x_1; $y_{12}, y_{22}, \ldots y_{n_2 2}$ at x_2; and so on, up to $y_{1k}, y_{2k}, \ldots,$ $y_{n_k k}$ at x_k. Taking $k = 5$ for illustrative purposes, a plot of the data appears as Figure 7.2. With this wealth of data, it is not necessary to make restrictive assumptions regarding the nature of heteroscedasticity. It is clear from the graph that the nonconstancy of variance does not follow a simple systematic pattern such as $Var(\varepsilon_i) = k^2 x_i^2$. The variability first decreases as x increases up to x_3, then jumps again at x_4. The regression model could be stated as

$$y_{ij} = \beta_0 + \beta_1 x_j + \varepsilon_{ij}, \quad i = 1, 2, \ldots, n_j; \; j = 1, 2, 3, 4, \qquad (7.6)$$

where $Var(\varepsilon_{ij}) = \sigma_j^2$.

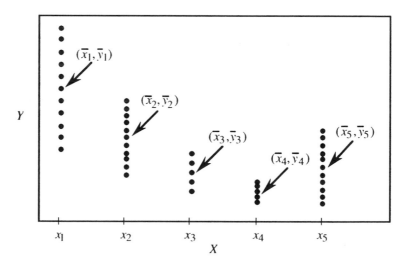

Fig. 7.2 Nonconstant variance with replicated observations.

The observed residual for the ith observation in the jth cluster or group is $e_{ij} = y_{ij} - \hat{y}_{ij}$. Adding and subtracting the mean of the response variable in for the jth cluster, $\bar{y}_j$, we obtain

$$e_{ij} = (y_{ij} - \bar{y}_j) + (\bar{y}_j - \hat{y}_{ij}), \tag{7.7}$$

which shows that the residual is made up of two parts, the difference between y_{ij} and $\bar{y}_j$ and the difference between $\bar{y}_j$ and the point on the regression line, $\hat{y}_{ij}$. The first part is referred to as *pure error*. The second part measures lack of fit. An assessment of heteroscedasticity is based on the pure error.[1] The weights for WLS may be estimated as $w_{ij} = 1/s_j^2$, where

$$s_j^2 = \sum_{i=1}^{n_j}(y_{ij} - \bar{y}_j)^2/(n_j - 1),$$

is the variance of the response variable for the jth group.

When the data are collected in a controlled laboratory setting, the researcher can choose to replicate the observations at any values of the predictor variables. But the presence of replications on the response variable for a given value of X is rather uncommon when data are collected in a nonexperimental setting. When there is only one predictor variable, it is possible that some replications will occur. If there are many predictor variables, it is virtually impossible to imagine coming upon two observations with identical values on all predictor values. However, it may be possible to form pseudoreplications by clustering responses where the predictor values are approximately identical. The reader is referred to Daniel and Wood (1980), where these methods are discussed in considerable detail. A more plausible way to investigate heteroscedasticity in multiple regression is by clustering observations according to prior, natural, and meaningful associations. As an example, we analyze data on state education expenditures. These data were used in Chapter 5.

7.4 EDUCATION EXPENDITURE DATA

The Education Expenditure data were used in Section 5.7 and it was suggested there that these data be looked at across time (the data are available for 1965, 1970, and 1975) to check on the stability of the coefficients. Here we use these data to demonstrate methods of dealing with heteroscedasticity in multiple regression and to analyze the effects of regional characteristics on the regression relationships. For the present analysis we shall work only with the 1975 data. The objective is to get the best representation of the relationship between expenditure on education and the other variables using data for all 50 states. The data are grouped in a natural way, by geographic region. Our assumption is that, although the relationship is structurally the same in

---

[1]The notion of pure error can also be used to obtain a test for lack of fit (see, e.g., Draper and Smith (1998)).

Table 7.2   State Expenditures on Education, Variable List.

Variable	Description
Y	Per capita expenditure on education projected for 1975
X_1	Per capita income in 1973
X_2	Number of residents per thousand under 18 years of age in 1974
X_3	Number of residents per thousand living in urban areas in 1970

each region, the coefficients and residual variances may differ from region to region. The different variances constitute a case of heteroscedasticity that can be treated directly in the analysis. The variable names and definitions appear in Table 7.2 and the data are presented in Table 7.3 and can be found in the book's Web site.[2] The model is

$$Y = \beta_0 + \beta_1 X_1 + \beta_2 X_2 + \beta_3 X_3 + \varepsilon. \tag{7.8}$$

States may be grouped into geographic regions based on the presumption that there exists a sense of regional homogeneity. The four broad geographic regions: (1) Northeast, (2) North Central, (3) South, and (4) West, are used to define the groups. It should be noted that data could be analyzed using indicator variables to look for special effects associated with the regions or to formulate tests for the equality of regressions across regions. However, our objective here is to develop one relationship that can serve as the best representation for all regions and all states. This goal is accomplished by taking regional differences into account through an extension of the method of weighted least squares.

It is assumed that there is a unique residual variance associated with each of the four regions. The variances are denoted as $(c_1\sigma)^2, (c_2\sigma)^2, (c_3\sigma)^2$, and $(c_4\sigma)^2$, where σ is the common part and the c_j's are unique to the regions. According to the principle of weighted least squares, the regression coefficients should be determined by minimizing

$$S_w = S_1 + S_2 + S_3 + S_4,$$

where

$$S_j = \sum_{i=1}^{n_j} \frac{1}{c_j^2} (y_i - \beta_0 - \beta_1 x_{i1} - \beta_2 x_{i2} - \beta_3 x_{i3})^2; \ j = 1, 2, 3, 4. \tag{7.9}$$

Each of S_1 through S_4 corresponds to a region, and the sum is taken over only those states that are in the region. The factors $1/c_j^2$ are the weights that determine how much influence each observation has in estimating the

[2]http://www.ilr.cornell.edu/~hadi/RABE

Table 7.3   Education Expenditure Data.

Row	State	Y	X_1	X_2	X_3	Region
1	ME	235	3944	325	508	1
2	NH	231	4578	323	564	1
3	VT	270	4011	328	322	1
4	MA	261	5233	305	846	1
5	RI	300	4780	303	871	1
6	CT	317	5889	307	774	1
7	NY	387	5663	301	856	1
8	NJ	285	5759	310	889	1
9	PA	300	4894	300	715	1
10	OH	221	5012	324	753	2
11	IN	264	4908	329	649	2
12	IL	308	5753	320	830	2
13	MI	379	5439	337	738	2
14	WI	342	4634	328	659	2
15	MN	378	4921	330	664	2
16	IA	232	4869	318	572	2
17	MO	231	4672	309	701	2
18	ND	246	4782	333	443	2
19	SD	230	4296	330	446	2
20	NB	268	4827	318	615	2
21	KS	337	5057	304	661	2
22	DE	344	5540	328	722	3
23	MD	330	5331	323	766	3
24	VA	261	4715	317	631	3
25	WV	214	3828	310	390	3
26	NC	245	4120	321	450	3
27	SC	233	3817	342	476	3
28	GA	250	4243	339	603	3
29	FL	243	4647	287	805	3
30	KY	216	3967	325	523	3
31	TN	212	3946	315	588	3
32	AL	208	3724	332	584	3
33	MS	215	3448	358	445	3
34	AR	221	3680	320	500	3
35	LA	244	3825	355	661	3
36	OK	234	4189	306	680	3
37	TX	269	4336	335	797	3
38	MT	302	4418	335	534	4
39	ID	268	4323	344	541	4
40	WY	323	4813	331	605	4
41	CO	304	5046	324	785	4
42	NM	317	3764	366	698	4
43	AZ	332	4504	340	796	4
44	UT	315	4005	378	804	4
45	NV	291	5560	330	809	4
46	WA	312	4989	313	726	4
47	OR	316	4697	305	671	4
48	CA	332	5438	307	909	4
49	AK	546	5613	386	484	4
50	HI	311	5309	333	831	4

regression coefficients. The weighting scheme is intuitively justified by arguing that observations that are most erratic (large error variance) should have little influence in determining the coefficients.

The WLS estimates can also be justified by a second argument. The object is to transform the data so that the parameters of the model are unaffected, but the residual variance in the transformed model is constant. The prescribed transformation is to divide each observation by the appropriate c_j, resulting in a regression of Y/c_j on $1/c_j$, X_1/c_j, X_2/c_j, and X_3/c_j.[3] Then the error term, in concept, is also divided by c_j, the resulting residuals have a common variance, σ^2, and the estimated coefficients have all the standard least squares properties.

The values of the c_j's are unknown and must be estimated in the same sense that σ^2 and the β's must be estimated. We propose a two-stage estimation procedure. In the first stage perform a regression using the raw data as prescribed in the model of Equation (7.8). Use the empirical residuals grouped by region to compute an estimate of regional residual variance. For example, in the Northeast, compute $\hat{\sigma}_1^2 = \sum e_i^2/9$, where the sum is taken over the nine residuals corresponding to the nine states in the Northeast. Compute $\hat{\sigma}_2^2, \hat{\sigma}_3^2$, and $\hat{\sigma}_4^2$ in a similar fashion. In the second stage, an estimate of c_j^2 in (7.9) is replaced by $\hat{\sigma}_j^2$.

The regression results for Stage 1 (OLS) using data from all 50 states are given in Table 7.4. Two residual plots are prepared to check on specification. The standardized residuals are plotted versus the fitted values (Figure 7.3) and versus a categorical variable designating region (Figure 7.4). The purpose of Figure 7.3 is to look for patterns in the size and variation of the residuals as a function of the fitted values. The observed scatter of points has a funnel shape, indicating heteroscedasticity. The spread of the residuals in Figure 7.4 is different for the different regions, which also indicates that the variances are not equal. The scatter plots of standardized residual versus each of the predictor variables (Figures 7.5 to 7.7) indicate that the residual variance increases with the values of X_1.

Looking at the standardized residuals and the influence measures in this example is very revealing. The reader can verify that observation 49 (Alaska) is an outlier with a standardized residual value of 3.28. The standardized residual for this observation can actually be seen to be separated from the

---

[3]If we denote a variable with a double subscript, i and j, with j representing region and i representing observation within region, then each variable for an observation in region j is divided by c_j. Note that β_0 is the coefficient attached to the transformed variable $1/c_j$. The transformed model is

$$\frac{y_{ij}}{c_j} = \beta_0 \frac{1}{c_j} + \beta_1 \frac{x_{1ij}}{c_j} + \beta_2 \frac{x_{2ij}}{c_j} + \beta_3 \frac{x_{3ij}}{c_j} + \varepsilon'_{ij}$$

and the variance of ε'_{ij} is σ^2. Notice that the same regression coefficients appear in the transformed model as in the original model. The transformed model is also a no-intercept model.

Table 7.4 Regression Results: State Expenditures on Education ($n = 50$).

Variable	Coefficient	s.e.	t-test	p-value
Constant	−556.568	123.200	−4.52	< 0.0001
X_1	0.072	0.012	6.24	< 0.0001
X_2	1.552	0.315	4.93	< 0.0001
X_3	−0.004	0.051	−0.08	0.9342
$n = 50$	$R^2 = 0.591$	$R_a^2 = 0.565$	$\hat{\sigma} = 40.47$	$d.f. = 46$

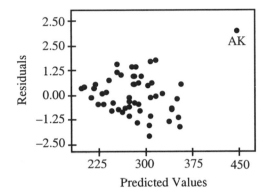

Fig 7.3 Plot of standardized residuals versus fitted values.

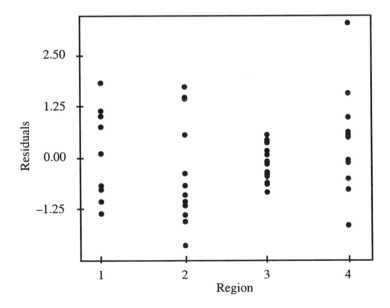

Fig. 7.4 Plot of standardized residuals versus regions.

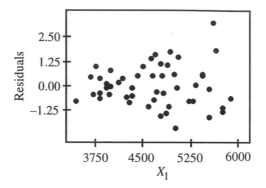

Fig. 7.5   Plot of standardized residuals versus each of the predictor variable X_1.

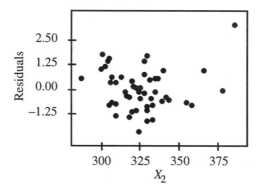

Fig. 7.6   Plot of standardized residuals versus each of the predictor variable X_2.

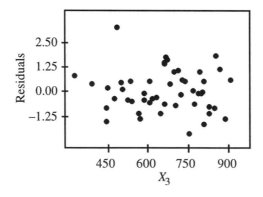

Fig. 7.7   Plot of standardized residuals versus each of the predictor variable X_3.

Table 7.5 Regression Results: State Expenditures on Education ($n = 49$), Alaska Omitted.

Variable	Coefficient	s.e.	t-test	p-value
Constant	-277.577	132.400	-2.10	0.0417
X_1	0.048	0.012	3.98	0.0003
X_2	0.887	0.331	2.68	0.0103
X_3	0.067	0.049	1.35	0.1826
$n = 49$	$R^2 = 0.497$	$R_a^2 = 0.463$	$\hat{\sigma} = 35.81$	$d.f. = 45$

rest of the residuals in Figure 7.3. Observation 44 (Utah) and 49 (Alaska) are high leverage points with leverage values of 0.29 and 0.44, respectively. On examining the influence measures we find only one influential point 49, with a Cook's distance value of 2.13 and a DFIT value of 3.30. Utah is a high leverage point without being influential. Alaska, on the other hand, has high leverage and is also influential. Compared to other states, Alaska represents a very special situation: a state with a very small population and a boom in revenue from oil. The year is 1975! Alaska's education budget is therefore not strictly comparable with those of the other states. Consequently, this observation (Alaska) is excluded from the remainder of the analysis. It represents a special situation that has considerable influence on the regression results, thereby distorting the overall picture.

The data for Alaska may have an undue influence on determining the regression coefficients. To check this possibility, the regression was recomputed with Alaska excluded. The estimated values of the coefficients changed significantly. See Table 7.5. This observation is excluded for the remainder of the analysis because it represents a special situation that has too much influence on the regression results. Plots similar to those of Figures 7.3 and 7.4 are presented as Figures 7.8 and 7.9. With Alaska removed, Figures 7.8 and 7.9 still show indication of heteroscedasticity.

To proceed with the analysis we must obtain the weights. They are computed from the OLS residuals by the method described above and appear in Table 7.6. The WLS regression results appear in Table 7.7 along with the OLS results for comparison. The standardized residuals from the transformed model are plotted in Figures 7.10 and 7.11. There is no pattern in the plot of the standardized residuals versus the fitted values (Figure 7.10). Also, from Figure 7.11, it appears that the spread of residuals by geographic region has evened out compared to Figures 7.4 and 7.9. The WLS solution is preferred to the OLS solution. Referring to Table 7.7, we see that the WLS solution does not fit the historical data as well as the OLS solution when considering

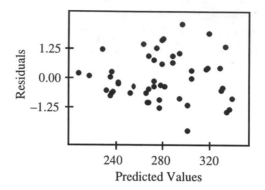

Fig. 7.8   Plot of the standardized residuals versus fitted values (excluding Alaska).

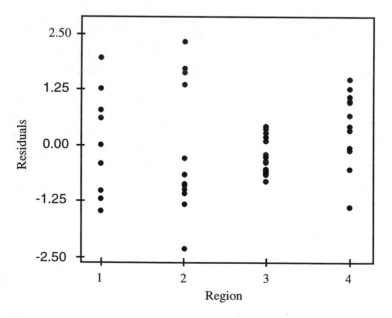

Fig. 7.9   Plot of the standardized residuals versus region (excluding Alaska).

Table 7.6   Weights c_j for Weighted Least Squares.

Region j	n_j	$\hat{\sigma}_j^2$	c_j
Northeast	9	1451.11	1.110
North Central	12	2436.98	1.439
South	16	249.43	0.460
West	12	950.42	0.898

Table 7.7   OLS and WLS Coefficients for Education Data ($n = 49$), Alaska Omitted.

Variable	OLS			WLS		
	Coefficient	s.e.	t	Coefficient	s.e.	t
Constant	−277.577	132.40	−2.10	−315.517	78.18	−4.04
X_1	0.048	0.01	3.98	0.062	0.01	7.92
X_2	0.887	0.33	2.68	0.874	0.20	4.37
X_3	0.067	0.05	1.35	0.029	0.03	0.86
	$R^2 = 0.497$		$\hat{\sigma} = 35.81$	$R^2 = 0.477$		$\hat{\sigma} = 36.50$

$\hat{\sigma}$ or R^2 as indicators of goodness of fit.[4] This result is expected since one of the important properties of OLS is that it provides a solution with minimum $\hat{\sigma}$ or, equivalently, maximum R^2. Our choice of the WLS solution is based on the pattern of the residuals. The difference in the scatter of the standardized residuals when plotted against Region (compare Figures 7.9 and 7.11) shows that WLS has succeeded in taking account of heteroscedasticity.

It is not possible to make a precise test of significance because exact distribution theory for the two-stage procedure used to obtain the WLS solution has not been worked out. If the weights were known in advance rather than as estimates from data, then the statistical tests based on the WLS procedure would be exact. Of course, it is difficult to imagine a situation similar to the one being discussed where the weights would be known in advance. Nevertheless, based on the empirical analysis above, there is a clear suggestion that weighting is required. In addition, since less than 50% of the variation in Y has been explained ($R^2 = 0.477$), the search for other factors must continue. It is suggested that the reader carry out an analysis of these data by introduc-

---

[4]Note that for comparative purposes, $\hat{\sigma}$ for the WLS solution is computed as the square root of

$$\hat{\sigma}^2 = \frac{1}{45} \sum_{i=1}^{n} (y_i - \hat{y}_i)^2,$$

and $\hat{y}_i = -315.517 + 0.062\, x_{i1} + 0.874\, x_{i2} + 0.0294\, x_{i3}$, are the fitted values computed in terms of the WLS estimated coefficients and the weights, c_j; weights play no further role in the computation of $\hat{\sigma}$.

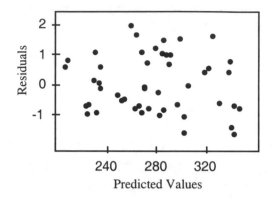

Fig. 7.10   Standardized residuals versus fitted values for WLS solution.

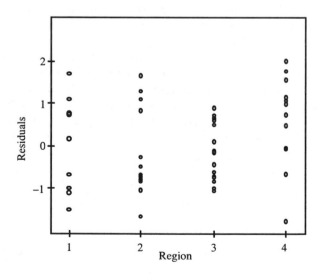

Fig. 7.11   Standardized residuals by geographic region for WLS solution.

ing indicator variables for the four geographical regions. In any model with four categories, as has been pointed out in Chapter 5, only three indicator variables are needed. Heteroscedasticity can often be eliminated by the introduction of indicator variables corresponding to different subgroups in the data.

7.5   FITTING A DOSE-RESPONSE RELATIONSHIP CURVE

An important area for the application of weighted least squares analysis is the fitting of a linear regression line when the response variable Y is a proportion (values between zero and one). Consider the following situation: An experimenter can administer a stimulus at different levels. Subjects are assigned at random to different levels of the stimulus and for each subject a binary response is noted. From this set of observations, a relationship between the stimulus and the proportion responding to the stimulus is constructed. A very common example is in the field of pharmacology, in bioassay, where the levels of stimulus may represent different doses of a drug or poison, and the binary response is death or survival. Another example is the study of consumer behavior where the stimulus is the discount offered and the binary response is the purchase or nonpurchase of some merchandise.

Suppose that a pesticide is tried at k different levels. At the jth level of dosage x_j, let r_j be the number of insects dying out of a total n_j exposed ($j = 1, 2, \ldots, k$). We want to estimate the relationship between dose and the proportion dying. The sample proportion $p_j = r_j/n_j$ is a binomial random variable, with mean value π_j and variance $\pi_j(1 - \pi_j)/n_j$, where π_j is the population probability of death for a subject receiving dose x_j. The relationship between π and X is based on the notion that

$$\pi = f(X), \tag{7.10}$$

where the function $f(\cdot)$ is increasing (or at least not decreasing) with X and is bounded between 0 and 1. The function should satisfy these properties because (1) π being a probability is bounded between 0 and 1, and (2) if the pesticide is toxic, higher doses should decrease the chances of survival (or increase the chances for death) for a subject. These considerations effectively rule out the linear model

$$\pi_j = \alpha + \beta x_j + \varepsilon_j, \tag{7.11}$$

because π_j would be unbounded.

Stimulus-response relationships are generally nonlinear. A nonlinear function which has been found to represent accurately the relationship between dose x_j and the proportion dying is

$$\pi_j = \frac{e^{\beta_0 + \beta_1 x_j}}{1 + e^{\beta_0 + \beta_1 x_j}}. \tag{7.12}$$

The relationship (7.12) is called the *logistic response function* and has the shape shown in Figure 7.12. It is seen that the logistic function is bounded between 0 and 1, and is monotonic. Physical considerations based on concepts of threshold values provide a heuristic justification for the use of (7.12) to represent a stimulus-response relationship (Cox, 1989).

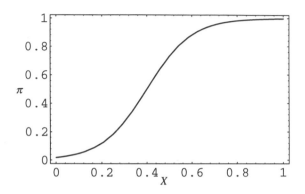

Fig. 7.12   Logistic response function.

The setup described above differs considerably from those of our other examples. In the present situation the experimenter has the control of dosages or stimuli and can use replication to estimate the variability of response at each dose level. This is a designed, experimental study, unlike the others, which were observational or nonexperimental.

The objectives for this type of analysis are not only to determine the nature of dose-response relationship but also to estimate the dosages which induce specified levels of response. Of particular interest is the dosage that produces a response in 50% of the population (median dose).

The logistic model (sometimes called *logit model*) has been used extensively in biological and epidemiological work. For analyzing proportions from binary response data, it is a very appealing model and easy to fit.

An alternative model in which the response function is represented by the cumulative distribution function of the normal probability distribution is also used. The cumulative curve of the normal distribution has a shape similar to that of the logistic function. This model is called the *probit model*, and for details we refer the reader to Finney (1964).

Besides medicine and pharmacology, the logistic model has been used in risk analysis, learning theory, in the study of consumer behavior (choice models) and market promotion studies.

Since the response function in (7.12) is nonlinear, we can work with transformed variables. The transformation is chosen to make the response function linear. However, the transformed variables will have nonconstant variance. Then, we must use the weighted least squares methods for fitting the transformed data.

A whole chapter (Chapter 12) is devoted to the discussion of logistic regression models, for we believe that they have important and varied practical applications. General questions regarding the suitability and fitting of logistic models are considered there.

EXERCISES

7.1 Repeat the analysis in Section 7.4 using the Education Expenditure Data in Table 5.12.

7.2 Repeat the analysis in Section 7.4 using the Education Expenditure Data in Table 5.13.

7.3 Compute the leverage values, the standardized residuals, Cook's distance, and DFIT for the regression model relating Y to the three predictor variables X_1, X_2, and X_3 in Table 7.3. Draw an appropriate graph for each of these measures. From the graph verify that Alaska and Utah are high leverage points, but only Alaska is an influential point.

7.4 Using the Education Expenditure Data in Table 7.3, fit a linear regression model relating Y to the three predictor variables X_1, X_2, and X_3 plus indicator variables for the region. Compare the results of the fitted model with the WLS results obtained in Section 7.4. Test for the equality of regressions across regions.

7.5 Repeat the previous exercise for the data in Table 5.12.

8

The Problem of Correlated Errors

8.1 INTRODUCTION: AUTOCORRELATION

One of the standard assumptions in the regression model is that the error terms ε_i and ε_j, associated with the ith and jth observations, are uncorrelated. Correlation in the error terms suggests that there is additional information in the data that has not been exploited in the current model. When the observations have a *natural* sequential order, the correlation is referred to as *autocorrelation*.

Autocorrelation may occur for several reasons. Adjacent residuals tend to be similar in both temporal and spatial dimensions. Successive residuals in economic time series tend to be positively correlated. Large positive errors are followed by other positive errors, and large negative errors are followed by other negative errors. Observations sampled from adjacent experimental plots or areas tend to have residuals that are correlated since they are affected by similar external conditions.

The symptoms of autocorrelation may also appear as the result of a variable having been omitted from the right-hand side of the regression equation. If successive values of the omitted variable are correlated, the errors from the estimated model will appear to be correlated. When the variable is added to the equation, the apparent problem of autocorrelation disappears. The presence of autocorrelation has several effects on the analysis. These are summarized as follows:

1. Least squares estimates of the regression coefficients are unbiased but are not efficient in the sense that they no longer have minimum variance.

2. The estimate of σ^2 and the standard errors of the regression coefficients may be seriously understated; that is, from the data the estimated standard errors would be much smaller than they actually are, giving a spurious impression of accuracy.

3. The confidence intervals and the various tests of significance commonly employed would no longer be strictly valid.

The presence of autocorrelation can be a problem of serious concern for the preceding reasons and should not be ignored.

We distinguish between two types of autocorrelation and describe methods for dealing with each. The first type is only autocorrelation in appearance. It is due to the omission of a variable that should be in the model. Once this variable is uncovered, the autocorrelation problem is resolved. The second type of autocorrelation may be referred to as pure autocorrelation. The methods of correcting for pure autocorrelation involve a transformation of the data. Formal derivations of the methods can be found in Johnston (1984) and Kmenta (1986).

8.2   CONSUMER EXPENDITURE AND MONEY STOCK

Table 8.1 gives quarterly data from 1952 to 1956 on consumer expenditure (Y) and the stock of money (X), both measured in billions of current dollars for the United States. The data can be found in the book's Web site.[1]

A simplified version of the quantity theory of money suggests a model given by

$$y_t = \beta_0 + \beta_1 x_t + \varepsilon_t, \tag{8.1}$$

where β_0 and β_1 are constants, ε_t the error term. Economists are interested in estimating β_1 and its standard error; β_1 is called the *multiplier* and has crucial importance as an instrument in fiscal and monetary policy. Since the observations are ordered in time, it is reasonable to expect that autocorrelation may be present. A summary of the regression results is given in Table 8.2.

The regression coefficients are significant; the standard error of the slope coefficient is 0.115. For a unit change in the money supply the 95% confidence interval for the change in the aggregate consumer expenditure would be $2.30 \pm 2.10 \times 0.115 = (2.06, 2.54)$. The value of R^2 indicates that roughly 96% of the variation in the consumer expenditure can be accounted for by the variation in money stock. The analysis would be complete if the basic regression assumptions were valid. To check on the model assumption, we examine the residuals. If there are indications that autocorrelation is present, the model should be reestimated after eliminating the autocorrelation.

[1] http://www.ilr.cornell.edu/~hadi/RABE

Table 8.1   Consumer Expenditure and Money Stock.

Year	Quarter	Consumer Expenditure	Money Stock	Year	Quarter	Consumer Expenditure	Money Stock
1952	1	214.6	159.3	1954	3	238.7	173.9
	2	217.7	161.2		4	243.2	176.1
	3	219.6	162.8	1955	1	249.4	178.0
	4	227.2	164.6		2	254.3	179.1
1953	1	230.9	165.9		3	260.9	180.2
	2	233.3	167.9		4	263.3	181.2
	3	234.1	168.3	1956	1	265.6	181.6
	4	232.3	169.7		2	268.2	182.5
1954	1	233.7	170.5		3	270.4	183.3
	2	236.5	171.6		4	275.6	184.3

Source: Friedman and Meiselman (1963), p. 266.

Table 8.2   Results When Consumer Expenditure Is Regressed on Money Stock, X.

Variable	Coefficient	s.e.	t-test	p-value
Constant	-154.72	19.850	-7.79	< 0.0001
X	2.30	0.115	20.08	< 0.0001
$n = 20$	$R^2 = 0.957$	$R_a^2 = 0.955$	$\hat{\sigma} = 3.983$	$d.f. = 18$

For time series data a useful plot for analysis is the index plot (plot of the standardized residuals versus time). The graph is given in Figure 8.1. The pattern of residuals is revealing and is characteristic of situations where the errors are correlated. Residuals of the same sign occur in clusters or bunches. The characteristic pattern would be that several successive residuals are positive, the next several are negative, and so on. From Figure 8.1 we see that the first seven residuals are positive, the next seven negative, and the last six positive. This pattern suggests that the error terms in the model are correlated and some additional analysis is required.

This visual impression can be formally confirmed by counting the number of runs in a plot of the signs of the residuals, the residuals taken in the order of the observations. These types of plots are called *sequence plots*. In our present example the sequence plot of the signs of the residuals is

$$+ + + + + + + - - - - - - - + + + + + +$$

and it indicates three runs. With n_1 residuals positive and n_2 residuals negative, under the hypothesis of randomness the expected number of runs μ and its variance σ^2 would be

$$\mu = \frac{2n_1 n_2}{n_1 + n_2} + 1,$$

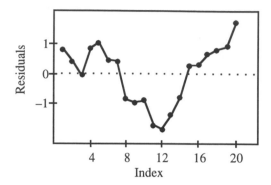

Fig. 8.1 Index plot of the standardized residuals.

$$\sigma^2 = \frac{2n_1 n_2 (2n_1 n_2 - n_1 - n_2)}{(n_1 + n_2)^2 (n_1 + n_2 - 1)}.$$

In our case $n_1 = 13, n_2 = 7$, giving the expected number of runs to be 8.1 and a standard deviation of 1.97. The observed number of runs is three. The deviation of 5.1 from the expected number of runs is more than twice the standard deviation, indicating a significant departure from randomness. This formal *runs test* procedure merely confirms the conclusion arrived at visually that there is a pattern in the residuals.

Many computer packages now have the runs test as an available option. This approximate runs test for confirmation can therefore be easily executed. The runs test as we have described it should not, however, be used for small values of n_1 and n_2 (less than 10). For small values of n_1 and n_2 one needs exact tables of probability to judge significance. For more details on the runs test, the reader should refer to a book on nonparametric statistics such as Lehmann (1975), Gibbons (1993), and Hollander and Wollfe (1999). Besides the graphical analysis, which can be confirmed by the runs test, autocorrelated errors can also be detected by the Durbin-Watson statistic.

8.3 DURBIN-WATSON STATISTIC

The Durbin-Watson statistic is the basis of a popular test of autocorrelation in regression analysis. The test is based on the assumption that successive errors are correlated, namely,

$$\varepsilon_t = \rho \varepsilon_{t-1} + \omega_t, \quad |\rho| < 1, \tag{8.2}$$

where ρ is the correlation coefficient between ε_t and ε_{t-1}, and ω_t is normally independently distributed with zero mean and constant variance. In this case, the errors are said to have *first-order autoregressive structure* or *first-order autocorrelation*. In most situations the error ε_t may have a much more complex correlation structure. The first-order dependency structure, given in (8.2), is taken as a simple approximation to the actual error structure.

The Durbin-Watson statistic is defined as

$$d = \frac{\sum_{t=2}^{n}(e_t - e_{t-1})^2}{\sum_{t=1}^{n} e_t^2},$$

where e_i is the ith ordinary least squares (OLS) residual. The statistics d is used for testing the null hypothesis $H_0 : \rho = 0$ against an alternative $H_1 : \rho > 0$. Note that when $\rho = 0$ in Equation (8.2), the ε's are uncorrelated.

Since ρ is unknown, we estimate the parameter ρ by $\hat{\rho}$, where

$$\hat{\rho} = \frac{\sum_{t=2}^{n} e_t e_{t-1}}{\sum_{t=1}^{n} e_t^2}. \tag{8.3}$$

An approximate relationship between d and $\hat{\rho}$ is

$$d \doteq 2(1 - \hat{\rho}),$$

($\doteq$ means approximately equal to) showing that d has a range of 0 to 4. Since $\hat{\rho}$ is an estimate of ρ, it is clear that d is close to 2 when $\rho = 0$ and near to zero when $\rho = 1$. The closer the sample value of d to 2, the firmer the evidence that there is no autocorrelation present in the error. Evidence of autocorrelation is indicated by the deviation of d from 2. The formal test for positive autocorrelation operates as follows: Calculate the sample statistic d. Then, if

1. $d < d_L$, reject H_0.

2. $d > d_U$, do not reject H_0.

3. $d_L < d < d_U$, the test is inconclusive.

The values of (d_L, d_U) for different percentage points have been tabulated by Durbin and Watson (1951). A table is provided in the Appendix at the end of the book (Tables A.6 and A.7).

Tests for negative autocorrelation are seldom performed. If, however, a test is desired, then instead of working with d, one works with $(4 - d)$ and follows the same procedure as for the testing of positive autocorrelation.

In our Money Stock and Consumer Expenditure data, the value of d is 0.328. From Table A.6, with $n = 20$, $p = 1$ (the number of predictors), and a significance level of 0.05, we have $d_L = 1.20$ and $d_U = 1.41$. Since $d < d_L$, we conclude that the value of d is significant at the 5% level and H_0 is rejected, showing that autocorrelation is present. This essentially reconfirms our earlier conclusion, which was arrived at by looking at the index plot of the residuals.

If d had been larger than $d_U = 1.41$, autocorrelation would not be a problem and no further analysis is needed. When $d_L < d < d_U$, additional analysis of the equation is optional. We suggest that in cases where the Durbin-Watson statistic lies in the inconclusive region, reestimate the equation using the methods described below to see if any major changes occur.

As pointed out earlier, the presence of correlated errors distorts estimates of standard errors, confidence intervals, and statistical tests, and therefore we should reestimate the equation. When autocorrelated errors are indicated, two approaches may be followed. These are (1) work with transformed variables, or (2) introduce additional variables that have time-ordered effects. We illustrate the first approach with the Money Stock data. The second approach is illustrated in Section 8.6.

8.4   REMOVAL OF AUTOCORRELATION BY TRANSFORMATION

When the residual plots and Durbin-Watson statistic indicate the presence of correlated errors, the estimated regression equation should be refitted taking the autocorrelation into account. One method for adjusting the model is the use of a transformation that involves the unknown autocorrelation parameter, ρ. The introduction of ρ causes the model to be nonlinear. The direct application of least squares is not possible. However, there are a number of procedures that may be used to circumvent the nonlinearity (Johnston, 1984). We use the method due to Cochrane and Orcutt (1949).

From model (8.1), ε_t and ε_{t-1} can be expressed as

$$\begin{aligned} \varepsilon_t &= y_t - \beta_0 - \beta_1 x_t, \\ \varepsilon_{t-1} &= y_{t-1} - \beta_0 - \beta_1 x_{t-1}. \end{aligned}$$

Substituting these in (8.2), we obtain

$$y_t - \beta_0 - \beta_1 x_t = \rho(y_{t-1} - \beta_0 - \beta_1 x_{t-1}) + \omega_t.$$

Rearranging terms in the above equation, we get

$$\begin{aligned} y_t - \rho y_{t-1} &= \beta_0(1-\rho) &+ \beta_1(x_t - \rho x_{t-1}) &+ \omega_t, \\ y_t^* &= \beta_0^* &+ \beta_1^* \, x_t^* &+ \omega_t, \end{aligned} \qquad (8.4)$$

where

$$\begin{aligned} y_t^* &= y_t - \rho y_{t-1}, \\ x_t^* &= x_t - \rho x_{t-1}, \\ \beta_0^* &= \beta_0(1-\rho), \\ \beta_1^* &= \beta_1. \end{aligned}$$

Since the ω's are uncorrelated, Equation (8.4) represents a linear model with uncorrelated errors. This suggests that we run an ordinary least squares regression using y_t^* as a response variable and x_t^* as a predictor. The estimates of the parameters in the original equations are

$$\hat{\beta}_0 = \frac{\hat{\beta}_0^*}{1 - \hat{\rho}} \quad \text{and} \quad \hat{\beta}_1 = \hat{\beta}_1^*. \tag{8.5}$$

Therefore, when the errors in model (8.1) have an autoregressive structure as given in (8.2), we can transform both sides of the equation and obtain transformed variables which satisfy the assumption of uncorrelated errors.

The value of ρ is unknown and has to be estimated from the data. Cochrane and Orcutt (1949) have proposed an iterative procedure. The procedure operates as follows:

1. Compute the OLS estimates of β_0 and β_1 by fitting model (8.1) to the data.

2. Calculate the residuals and from the residuals estimate ρ using (8.3).

3. Fit the equation given in (8.4) using the variables $y_t - \hat{\rho}y_{t-1}$ and $x_t - \hat{\rho}x_{t-1}$ as response and predictor variables, respectively, and obtain $\hat{\beta}_0$ and $\hat{\beta}_1$ using (8.5).

4. Examine the residuals of the newly fitted equation. If the new residuals continue to show autocorrelation, repeat the entire procedure using the estimates $\hat{\beta}_0$ and $\hat{\beta}_1$ as estimates of β_0 and β_1 instead of the original least squares estimates. On the other hand, if the new residuals show no autocorrelation, the procedure is terminated and the fitted equation for the original data is:
$$\hat{y}_t = \hat{\beta}_0 + \hat{\beta}_1 x_t.$$

As a practical rule we suggest that if the first application of Cochrane-Orcutt procedure does not yield non-autocorrelated residuals, one should look for alternative methods of removing autocorrelation. We apply the Cochrane-Orcutt procedure to the data given in Table 8.1.

The d value for the original data is 0.328, which is highly significant. The value of $\hat{\rho}$ is 0.751. On fitting the regression equation to the variables $(y_t - 0.751y_{t-1})$ and $(x_t - 0.751x_{t-1})$, we have a d value of 1.43. The value of d_U for $n = 19$ and $p = 1$ is 1.40 at the 5% level. Consequently, $H_0 : \rho = 0$ is not rejected.[2] The fitted equation is

$$\hat{y}_t^* = -53.64 + 2.64x_t^*,$$

which, using (8.5), the fitted equation

$$\hat{y}_t = -215.4 + 2.64x_t,$$

---

[2]The significance level of the test is not exact because $\hat{\rho}$ was used in the estimation process. The d value of 1.43 may be viewed as an index of autocorrelation that indicates an improvement from the previous value of 0.328.

in terms of the original variables. The estimated standard error for the slope is 0.31, as opposed to the least squares estimate of the original equation, which was $y_t = -154.7 + 2.3x_t$ with a standard error for the slope of 0.115. The newly estimated standard error is larger by a factor of almost 3. The residual plots for the fitted equation of the transformed variables are shown in Figure 8.2. The residual plots show less clustering of the adjacent residuals by sign, and the Cochrane-Orcutt procedure has worked to our advantage.

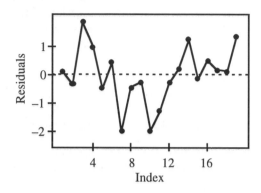

Fig. 8.2 Index plot of standardized residuals after one iteration of the Cochrane-Orcutt method.

8.5   ITERATIVE ESTIMATION WITH AUTOCORRELATED ERRORS

One advantage of the Cochrane-Orcutt procedure is that estimates of the parameters are obtained using standard least squares computations. Although two stages are required, the procedure is relatively simple. A more direct approach is to try to estimate values of ρ, β_0, and β_1 simultaneously. The model is formulated as before requiring the construction of transformed variables $y_t - \rho y_{t-1}$ and $x_t - \rho x_{t-1}$. Parameter estimates are obtained by minimizing the sum of squared errors, which is given as

$$S(\beta_0, \beta_1, \rho) = \sum_{t=2}^{n} [y_t - \rho y_{t-1} - \beta_0(1 - \rho) - \beta_1(x_t - \rho x_{t-1})]^2.$$

If the value of ρ were known, β_0 and β_1 would be easily obtained by regressing $y_t - \rho y_{t-1}$ on $x_t - \rho x_{t-1}$. Final estimates are obtained by searching through many values of ρ until a combination of ρ, β_0 and β_1 is found that minimizes $S(\rho, \beta_0, \beta_1)$. The search could be accomplished using a standard regression computer program, but the process can be much more efficient with an automated search procedure. This method is due to Hildreth and Lu (1960). For a discussion of the estimation procedure and properties of the estimates obtained, see Kmenta (1986).

Table 8.3 Comparison of Regression Estimates.

Method	$\hat{\rho}$	$\hat{\beta}_0$	$\hat{\beta}_1$	$s.e.(\hat{\beta}_1)$
OLS	–	−154.700	2.300	0.115
Cochrane-Orcutt	0.874	−324.440	2.758	0.444
Iterative	0.824	−235.509	2.753	0.436

Once the minimizing values, say $\tilde{\rho}, \tilde{\beta}_0$, and $\tilde{\beta}_1$, have been obtained, the standard error for the estimate of β_1 can be approximated using a version of Equation (2.24) of Chapter 2. The formula is used as though $y_t - \rho y_{t-1}$ were regressed on $x_t - \rho x_{t-1}$ with ρ known; that is, the estimated standard error of $\tilde{\beta}_1$ is

$$s.e.(\tilde{\beta}_1) = \frac{\hat{\sigma}}{\sqrt{\sum[x_t - \tilde{\rho}x_{t-1} - \bar{x}(1 - \tilde{\rho})]^2}} \, ,$$

where $\hat{\sigma}$ is the square root of $S(\tilde{\rho}, \tilde{\beta}_0, \tilde{\beta}_1)/(n - 2)$. When adequate computing facilities are available such that the iterative computations are easy to accomplish, then the latter method is recommended. However, it is not expected that the estimates and standard errors for the iterative method and the two-stage Cochrane-Orcutt method would be appreciably different. The estimates from the three methods, OLS, Cochrane-Orcutt, and iterative for the data of Table 8.1, are given in Table 8.3 for comparison.

8.6 AUTOCORRELATION AND MISSING VARIABLES

The characteristics of the regression residuals that suggest autocorrelation may also be indicative of other aspects of faulty model specification. In the preceding example, the index plot of residuals and the statistical test based on the Durbin-Watson statistic were used to conclude that the residuals are autocorrelated. Autocorrelation is only one of a number of possible explanations for the clustered type of residual plot or low Durbin-Watson value.

In general, a plot of residuals versus any one of the list of potential predictor variables may uncover additional information that can be used to further explain variation in the response variable. When an index plot of residuals shows a pattern of the type described in the preceding example, it is reasonable to suspect that it may be due to the omission of variables that change over time. Certainly, when the residuals appear in clusters alternating above and below the mean value line of zero, when the estimated autocorrelation coefficient is large and the Durbin-Watson statistic is significant, it would appear that the presence of autocorrelation is overwhelmingly supported. We shall see that this conclusion may be incorrect. The observed symptoms would be better interpreted initially as a general indication of some form of model misspecification.

All possible correction procedures should be considered. In fact, it is always better to explore fully the possibility of some additional predictor variables before yielding to an autoregressive model for the error structure. It is more satisfying and probably more useful to be able to understand the source of apparent autocorrelation in terms of an additional variable. The marginal effect of that variable can then be estimated and used in an informative way. The transformations that correct for pure autocorrelation may be viewed as an action of last resort.

8.7   ANALYSIS OF HOUSING STARTS

As an example of a situation where autocorrelation appears artificially because of the omission of another predictor variable, consider the following project undertaken by a midwestern construction industry association. The association wants to have a better understanding of the relationship between housing starts and population growth. They are interested in being able to forecast construction activity. Their approach is to develop annual data on regional housing starts and try to relate these data to potential home buyers in the region. Realizing that it is almost impossible to measure the number of potential house buyers accurately, the researchers settled for the size of the 22- to 44-year-old population group in the region as a variable that reflects the size of potential home buyers. With some diligent work they were able to bring together 25 years of historical data for the region (see Table 8.4). The data in Table 8.4 can be obtained from the book's Web site. Their goal was to get a simple regression relationship between housing starts and population,

$$H_t = \beta_0 + \beta_1 P_t + \varepsilon_t. \tag{8.6}$$

Then using methods that they developed for projecting population changes, they would be able to estimate corresponding changes in the requirements for new houses. The construction association was aware that the relationship between population and housing starts could be very complex. It is even reasonable to suggest that housing affects population growth (by migration) instead of the other way around. Although the proposed model is undoubtedly naive, it serves a useful purpose as a starting point for their analysis.

Analysis

The regression results from fitting model (8.6) to the 25 years of data are given in Table 8.5. The proportion of variation in H accounted for by the variability in P is $R^2 = 0.925$. We also see that an increase in population of 1 million leads to an increase in housing starts of about 71,000. The Durbin-Watson statistic and the index plot of the residuals (Figure 8.3) suggest strong auto-

Table 8.4 Data for Housing Starts (H), Population Size (P) in millions, and Availability for Mortgage Money Index (D).

Row	H	P	D
1	0.09090	2.200	0.03635
2	0.08942	2.222	0.03345
3	0.09755	2.244	0.03870
4	0.09550	2.267	0.03745
5	0.09678	2.280	0.04063
6	0.10327	2.289	0.04237
7	0.10513	2.289	0.04715
8	0.10840	2.290	0.04883
9	0.10822	2.299	0.04836
10	0.10741	2.300	0.05160
11	0.10751	2.300	0.04879
12	0.11429	2.340	0.05523
13	0.11048	2.386	0.04770
14	0.11604	2.433	0.05282
15	0.11688	2.482	0.05473
16	0.12044	2.532	0.05531
17	0.12125	2.580	0.05898
18	0.12080	2.605	0.06267
19	0.12368	2.631	0.05462
20	0.12679	2.658	0.05672
21	0.12996	2.684	0.06674
22	0.13445	2.711	0.06451
23	0.13325	2.738	0.06313
24	0.13863	2.766	0.06573
25	0.13964	2.793	0.07229

Table 8.5 Regression on Housing Starts (H) Versus Population (P).

Variable	Coefficient	s.e.	t-test	p-value
Constant	−0.0609	0.0104	−5.85	< 0.0001
P	0.0714	0.0042	16.90	< 0.0001
$n = 25$	$R^2 = 0.925$	$d = 0.621$	$\hat{\sigma} = 0.0041$	$d.f. = 23$

correlation. However, it is fairly simple to conjecture about other variables that may further explain housing starts and could be responsible for the appearance of autocorrelation. These variables include the unemployment rate, social trends in marriage and family formation, government programs in housing, and the availability of construction and mortgage funds. The first choice was an index that measures the availability of mortgage money for the region. Adding that variable to the equation the model becomes

$$H_t = \beta_0 + \beta_1 P_t + \beta_2 D_t + \varepsilon_t.$$

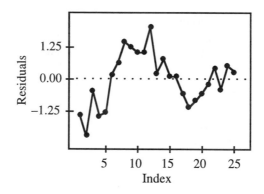

Fig. 8.3 Index plot of standardized residuals from the regression of H_t on P_t for the Housing Starts data.

The introduction of the additional variable has the effect of removing autocorrelation. From Table 8.6 we see that the Durbin-Watson statistic has the new value 1.852, well into the acceptable region. The index plot of the residuals (Figure 8.4) is also improved. The regression coefficients and their corresponding t-values show that there is a significant population effect but that it was overstated by a factor of more than 2 in the first equation. In a certain sense, the effect of changes in the availability of mortgage money for a fixed level of population is more important than a similar change in population.

If each variable in the regression equation is replaced by the standardized version of the variable (the variables transformed so as to have mean 0, and

Table 8.6 Results of the Regression of Housing Starts (H) on Population (P) and Index (D).

Variable	Coefficient	s.e.	t-test	p-value
Constant	−0.0104	0.0103	−1.01	0.3220
P	0.0347	0.0064	5.39	< 0.0001
D	0.7605	0.1216	6.25	< 0.0001
$n = 25$	$R^2 = 0.973$	$d = 1.85$	$\hat{\sigma} = 0.0025$	$d.f. = 22$

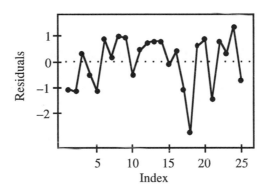

Fig. 8.4 Index plot of the standardized residuals from the regression of H_t on P_t and D_t for the Housing Starts data.

unit variance), the resulting regression equation is

$$\tilde{H}_t = 0.4668\tilde{P}_t + 0.5413\tilde{D}_t \,,$$

where $\tilde{H}$ denotes the standardized value of $H, \tilde{H} = (H - \bar{H})/s_H$. A unit increase in the standardized value of $\tilde{P}_t$ is worth an additional 0.4668 to the standardized value of H_t; that is, if the population increases by standard deviation then H_t increases by 0.4668 standard deviation. Similarly, if D_t increases by 1 standard deviation H_t increases by 0.5413 standard deviation. Therefore, in terms of the standardized variables, the mortgage index is more important (has a larger effect) than population size.

The example on housing starts illustrates two important points. First, a large value of R^2 does not imply that the data have been fitted and explained well. Any pair of variables that show trends over time are usually highly correlated. A large value of R^2 does not necessarily confirm that the relationship between the two variables has been adequately characterized. Second, the Durbin-Watson statistic as well as the residual plots may indicate the presence of autocorrelation among the errors when, in fact, the errors are independent but the omission of a variable or variables has given rise to the observed situation. Even though the Durbin-Watson statistic was designed to detect first-order autocorrelation it can have a significant value when some other model assumptions are violated such as misspecification of the variables to be included in the model. In general, a significant value of the Durbin-Watson statistic should be interpreted as an indication that a problem exists, and both the possibility of a missing variable or the presence of autocorrelation should be considered.

8.8   LIMITATIONS OF DURBIN-WATSON STATISTIC

In the previous examples on Expenditure versus Money Stock and Housing Starts versus Population Size the residuals from the initial regression equations indicated model misspecifications associated with time dependence. In both cases the Durbin-Watson statistic was small enough to conclude that positive autocorrelation was present. The index plot of residuals further confirmed the presence of a time-dependent error term. In each of the two problems the presence of autocorrelation was dealt with differently. In one case (Housing Starts) an additional variable was uncovered that had been responsible for the appearance of autocorrelation, and in the other case (Money Stock) the Cochrane-Orcutt method was used to deal with what was perceived as pure autocorrelation. It should be noted that the time dependence observed in the residuals in both cases is a first-order type of dependence. Both the Durbin-Watson statistic and the pattern of residuals indicate dependence between residuals in adjacent time periods. If the pattern of time dependence is other than first order, the plot of residuals will still be infor-

Table 8.7   Ski Sales Versus PDI.

Variable	Coefficient	s.e.	*t*-test	*p*-value
Constant	12.3921	2.539	4.9	< 0.0001
P	0.1979	0.016	12.4	< 0.0001
$n = 40$	$R^2 = 0.801$	$d = 1.968$	$\hat{\sigma} = 3.019$	$d.f. = 38$

mative. However, the Durbin-Watson statistic is not designed to measure higher-order time dependence and may not yield much valuable information.

As an example we consider the efforts of a company that produces and markets ski equipment in the United States to obtain a simple aggregate relationship of quarterly sales to some leading economic indicator. The indicator chosen is personal disposable income, PDI, in billions of current dollars. The initial model is

$$S_t = \beta_0 + \beta_1 PDI_t + \varepsilon_t,$$

where S_t is ski sales in period t in millions of dollars and PDI_t is the personal disposable income for the same period. Data for 10 years (40 quarters) are available (Table 5.11). The data can be obtained from the book's Web site. The regression output is in Table 8.7 and the index plot of residuals is given in Figure 8.6.

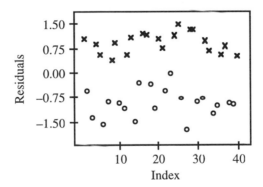

Fig. 8.5 Index plot of the standardized residuals. (Quarters 1 and 4 are indicated by a cross and Quarters 2 and 3 are indicated by a circle.)

At first glance the results in Table 8.7 are encouraging. The proportion of variation in sales accounted for by PDI is 0.80. The marginal contribution of an additional dollar unit of PDI to sales is between \$165,420 and \$230,380 ($\hat{\beta}_1 = 0.1979$) with a confidence coefficient of 95%. In addition, the Durbin-Watson statistic is 1.968, indicating no first-order autocorrelation.

It should be expected that PDI would explain a large proportion of the variation in sales since both variables are increasing over time. Therefore, although the R^2 value of 0.80 is good, it should not be taken as a final evaluation of the model. Also, the Durbin-Watson value is in the acceptable range, but it is clear from Figure 8.5 that there is some sort of time dependence of the residuals. We notice that residuals from the first and fourth quarters are positive, while residuals from the second and third quarters are negative for all the years. Since skiing activities are affected by weather conditions, we suspect that a seasonal effect has been overlooked. The pattern of residuals suggests that there are two seasons that have some bearing on ski sales: the second and third quarters, which correspond to the warm weather season, and the fourth and first quarters, which correspond to the winter season, when skiing is in full progress. This seasonal effect can be simply characterized by defining an indicator (dummy) variable that takes the value 1 for each winter quarter and is set equal to zero for each summer quarter (see Chapter 5). The expanded data set is listed in Table 8.8 and can be obtained from the book's Web site.

8.9   INDICATOR VARIABLES TO REMOVE SEASONALITY

Using the additional seasonal variable, the model is expanded to be

$$S_t = \beta_0 + \beta_1 PDI_t + \beta_2 Z_t + \varepsilon_t, \tag{8.7}$$

where Z_t is the zero-one variable described above and β_2 is a parameter that measures the seasonal effect. Note that the model in (8.7) can be represented by the two models (one for the cold weather quarters where $Z_t = 1$) and the other for the warm quarters where $Z_t = 0$):

$$
\begin{aligned}
\text{Winter season}: \quad S_t &= (\beta_0 + \beta_2) &+ \beta_1\, PDI_t &+ \varepsilon_t, \\
\text{Summer season}: \quad S_t &= \beta_0 &+ \beta_1\, PDI_t &+ \varepsilon_t.
\end{aligned}
$$

Thus, the model represents the assumption that sales can be approximated by a linear function of PDI, in one line for the winter season and one for the summer season. The lines are parallel; that is, the marginal effect of changes in PDI is the same in both seasons. The level of sales, as reflected by the intercept, is different in each season (Figure 8.6).

The regression results are summarized in Table 8.9 and the index plot of the standardized residuals is shown in Figure 8.7. We see that all indications of the seasonal pattern have been removed. Furthermore, the precision of the estimated marginal effect of PDI increased. The confidence interval is now $186,520 to $210,880. Also, the seasonal effect has been quantified and we can say that for a fixed level of PDI the winter season brings between $4,734,109 and $6,194,491 over the summer season (with 95% confidence).

The ski data illustrate two important points concerning autocorrelation. First, the Durbin-Watson statistic is only sensitive to correlated errors when the correlation occurs between adjacent observations (first-order autocorrelation). In the ski data the first-order correlation is -0.001. The second-,

Table 8.8   Disposable Income and Ski Sales, and Seasonal Variables for Years 1964–1973.

Quarter	Sales	PDI	Season
Q1/64	37.0	109	1
Q2/64	33.5	115	0
Q3/64	30.8	113	0
Q4/64	37.9	116	1
Q1/65	37.4	118	1
Q2/65	31.6	120	0
Q3/65	34.0	122	0
Q4/65	38.1	124	1
Q1/66	40.0	126	1
Q2/66	35.0	128	0
Q3/66	34.9	130	0
Q4/66	40.2	132	1
Q1/67	41.9	133	1
Q2/67	34.7	135	0
Q3/67	38.8	138	0
Q4/67	43.7	140	1
Q1/68	44.2	143	1
Q2/68	40.4	147	0
Q3/68	38.4	148	0
Q4/68	45.4	151	1
Q1/69	44.9	153	1
Q2/69	41.6	156	0
Q3/69	44.0	160	0
Q4/69	48.1	163	1
Q1/70	49.7	166	1
Q2/70	43.9	171	0
Q3/70	41.6	174	0
Q4/70	51.0	175	1
Q1/71	52.0	180	1
Q2/71	46.2	184	0
Q3/71	47.1	187	0
Q4/71	52.7	189	1
Q1/72	52.2	191	1
Q2/72	47.0	193	0
Q3/72	47.8	194	0
Q4/72	52.8	196	1
Q1/73	54.1	199	1
Q2/73	49.5	201	0
Q3/73	49.5	202	0
Q4/73	54.3	204	1

Table 8.9   Ski Sales Versus PDI and Seasonal Variables.

Variable	Coefficient	s.e.	t-test	p-value
Constant	9.5402	0.9748	9.79	0.3220
PDI	0.1987	0.0060	32.90	< 0.0001
Z	5.4643	0.3597	15.20	< 0.0001
$n = 40$	$R^2 = 0.972$	$d = 1.772$	$\hat{\sigma} = 1.137$	$d.f. = 37$

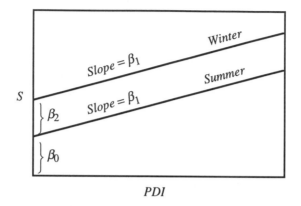

Fig. 8.6   Model for Ski Sales and PDI adjusted for season.

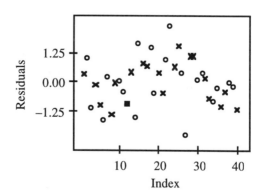

Fig. 8.7 Index plot of the standardized residuals with seasonal variables (quarters indicated).   (Quarters 1 and 4 are indicated by a cross and Quarters 2 and 3 are indicated by a circle.)

fourth-, sixth-, and eighth-order correlations are -0.81, 0.76, -0.71, and 0.73, respectively. The Durbin-Watson test does not show significance in this case. There are other tests that may be used for the detection of higher-order autocorrelations (see Box and Pierce (1970)). But in all cases, the graph of residuals will show the presence of time dependence in the error term when it exists.

Second, when autocorrelation is indicated the model should be refitted. Often the autocorrelation appears because a time-dependent variable is missing from the model. The inclusion of the omitted variable often removes the observed autocorrelation. Sometimes, however, no such variable is present. Then one has to make a differencing type of transformation on the original variables to remove the autocorrelation.

If the observations are not ordered in time, the Durbin-Watson statistic is not strictly relevant. The statistic may still, however, be a useful diagnostic tool. If the data are ordered by an extraneous criterion, for example, an alphabetic listing, the value of the Durbin-Watson statistic should be near 2.0. Small values are suspicious, and the data should be scrutinized very carefully.

Many data sets are ordered on a criterion that may be relevant to the study. A list of cities or companies may be ordered by size. A low value of the Durbin-Watson statistic would indicate the presence of a significant size effect. A measure of size should therefore be included as a predictor variable. Differencing or Cochrane-Orcutt type of differencing would not be appropriate under these conditions.

8.10   REGRESSING TWO TIME SERIES

The data sets analyzed in this chapter all have the common characteristic that they are time series data (i.e., the observations arise in successive periods of time). This is quite unlike the data sets studied in previous chapters (a notable exception being the bacteria data in Chapter 6), where all the observations are generated at the same point in time. The observations in these examples were contemporaneous and gave rise to *cross-sectional data*. When the observations are generated simultaneously (and relate to a single time period), we have cross-sectional data. The contrast between time series and cross-sectional data can be seen by comparing the ski sales data discussed in this chapter (data arising sequentially in time), and the supervisor performance data in Section 3.3, where all the data were collected in an attitude survey and relate to one historical point in time.

Regression analysis of one time series on another is performed extensively in economics, business, public health, and other social sciences. There are some special features in time series data that are not present in cross-sectional data. We draw attention to these features and suggest possible techniques for handling them.

The concept of autocorrelation is not relevant in cross-sectional data. The ordering of the observations is often arbitrary. Consequently, the correlation of adjacent residuals is an artifact of the organization of the data. For time series data, however, autocorrelation is often a significant factor. The presence of autocorrelation shows that there are hidden structures in the data (often time related) which have not been detected. In addition, most time series data exhibit seasonality, and an investigator should look for seasonal patterns. A regular time pattern in the residuals (as in the ski data) will often indicate the presence of seasonality. For quarterly or monthly data, introduction of indicator variables, as has been pointed out, is a satisfactory solution. For quarterly data, four indicator variables would be needed but only three used in the analysis (see the discussion in Chapter 4). For monthly data, we will need 12 indicator variables but use only 11, to avoid problems of collinearity (this is discussed in Chapter 5). Not all of the indicator variables will be significant and some of them may well be deleted in the final stages of the analysis.

In attempting to find a relationship between y_t and $x_{1t}, x_{2t}, \ldots, x_{pt}$ one may expand the set of predictor variables by including lagged values of the predictor variables. A model such as

$$y_t = \beta_0 + \beta_1 x_{1t} + \beta_2 x_{1t-1} + \beta_3 x_{2t} + \varepsilon_t$$

is meaningful in an analysis of time series data but not with cross-sectional data. The model given above implies that the value of Y in a given period is affected not only by the values of X_1 and X_2 of that period but also by the value of X_1 in the preceding period (i.e., there is a lingering effect of X_1 on Y for one period). Variables lagged by more than one period are also possibilities and could be included in the set of predictor variables.

Time series data are also likely to contain trends. Data in which time trends are likely to occur are often analyzed by including variables that are direct functions of time (t). Variables such as t and t^2 are included in the list of predictor variables. They are used to account for possible linear or quadratic trend. Simple first differencing $(y_t - y_{t-1})$, or more complex lagging of the type $(y_t - a y_{t-1})$ as in the Cochrane-Orcutt procedure, are also possibilities. For a fuller discussion, the reader should consult a book on time series analysis such as Shumway (1988), Hamilton (1994).

To summarize, when performing regression analysis with time series data the analyst should be watchful for autocorrelation and seasonal effects, which are often present in the data. The possibility of using lagged predictor variables should also be explored.

EXERCISES

8.1 Fit model (8.6) to the data in Table 8.4.

    (a) Compute the Durbin-Watson statistic d. What conclusion regarding the presence of autocorrelation would you draw from d?

(b) Compare the number of runs to their expected value and standard deviation when fitting model (8.6) to the data in Table 8.4. What conclusion regarding the presence of autocorrelation would you draw from this comparison?

8.2 Oil Production Data: Refer to the oil production data in Table 6.17. The index plot of the residuals obtained after fitting a linear regression of log(OIL) on Year show a clear cyclical pattern.

(a) Compute the Durbin-Watson statistic d. What conclusion regarding the presence of autocorrelation would you draw from d?

(b) Compare the number of runs to their expected value and standard deviation. What conclusion regarding the presence of autocorrelation would you draw from this comparison?

8.3 Refer to the Presidential Election Data in Table 5.17. Since the data come over time (for 1916–1996 election years), one might suspect the presence of the autocorrelation problem when fitting the model in (5.11) to the data.

(a) Do you agree? Explain.

(b) Would adding a time trend (e.g., year) as an additional predictor variable improve or exacerbate the autocorrelation? Explain.

8.4 Dow Jones Industrial Average (DJIA): Tables 8.10 and 8.11 contain the values of the daily DJIA for all the trading days in 1996. The data can be found in the book's Web site.[3] DJIA is a very popular financial index and is meant to reflect the level of stock prices in the New York Stock Exchange. The Index is composed of 30 stocks. The variable Day denotes the trading day of the year. There were 262 trading days in 1996, and as such the variable Day goes from 1 to 262.

(a) Fit a linear regression model connecting DJIA with Day using all 262 trading days in 1996. Is the linear trend model adequate? Examine the residuals for time dependencies.

(b) Regress $DJIA_{(t)}$ against $DJIA_{(t-1)}$, that is, regress DJIA against its own value lagged by one period. Is this an adequate model? Are there any evidences of autocorrelation in the residuals?

(c) The variability (volatility) of the daily DJIA is large, and to accommodate this phenomenon the analysis is carried out on the logarithm of DJIA. Repeat the above exercises using log(DJIA) instead of DJIA. Are your conclusions similar? Do you notice any differences?

8.5 Refer again to the DJIA data in Exercise 8.4.

(a) Use the form of the model you found adequate in Exercise 8.4 and refit the model but using only the trading days in the first six

[3]http://www.ilr.cornell.edu/~hadi/RABE

Table 8.10 DJIA Data for the First Six Months of 1996.

Day	Date	DJIA	Day	Date	DJIA	Day	Date	DJIA
1	1/1/96	5117.12	45	3/1/96	5536.56	89	5/2/96	5498.27
2	1/2/96	5177.45	46	3/4/96	5600.15	90	5/3/96	5478.03
3	1/3/96	5194.07	47	3/5/96	5642.42	91	5/6/96	5464.31
4	1/4/96	5173.84	48	3/6/96	5629.77	92	5/7/96	5420.95
5	1/5/96	5181.43	49	3/7/96	5641.69	93	5/8/96	5474.06
6	1/8/96	5197.68	50	3/8/96	5470.45	94	5/9/96	5475.14
7	1/9/96	5130.13	51	3/11/96	5581.00	95	5/10/96	5518.14
8	1/10/96	5032.94	52	3/12/96	5583.89	96	5/13/96	5582.60
9	1/11/96	5065.10	53	3/13/96	5568.72	97	5/14/96	5624.71
10	1/12/96	5061.12	54	3/14/96	5586.06	98	5/15/96	5625.44
11	1/15/96	5043.78	55	3/15/96	5584.97	99	5/16/96	5635.05
12	1/16/96	5088.22	56	3/18/96	5683.60	100	5/17/96	5687.50
13	1/17/96	5066.90	57	3/19/96	5669.51	101	5/20/96	5748.82
14	1/18/96	5124.35	58	3/20/96	5655.42	102	5/21/96	5736.26
15	1/19/96	5184.68	59	3/21/96	5626.88	103	5/22/96	5778.00
16	1/22/96	5219.36	60	3/22/96	5636.64	104	5/23/96	5762.12
17	1/23/96	5192.27	61	3/25/96	5643.86	105	5/24/96	5762.86
18	1/24/96	5242.84	62	3/26/96	5670.60	106	5/27/96	5762.86
19	1/25/96	5216.83	63	3/27/96	5626.88	107	5/28/96	5709.67
20	1/26/96	5271.75	64	3/28/96	5630.85	108	5/29/96	5673.83
21	1/29/96	5304.98	65	3/29/96	5587.14	109	5/30/96	5693.41
22	1/30/96	5381.21	66	4/1/96	5637.72	110	5/31/96	5643.18
23	1/31/96	5395.30	67	4/2/96	5671.68	111	6/3/96	5624.71
24	2/1/96	5405.06	68	4/3/96	5689.74	112	6/4/96	5665.71
25	2/2/96	5373.99	69	4/4/96	5682.88	113	6/5/96	5697.48
26	2/5/96	5407.59	70	4/5/96	5682.88	114	6/6/96	5667.19
27	2/6/96	5459.61	71	4/8/96	5594.37	115	6/7/96	5697.11
28	2/7/96	5492.12	72	4/9/96	5560.41	116	6/10/96	5687.87
29	2/8/96	5539.45	73	4/10/96	5485.98	117	6/11/96	5668.66
30	2/9/96	5541.62	74	4/11/96	5487.07	118	6/12/96	5668.29
31	2/12/96	5600.15	75	4/12/96	5532.59	119	6/13/96	5657.95
32	2/13/96	5601.23	76	4/15/96	5592.92	120	6/14/96	5649.45
33	2/14/96	5579.55	77	4/16/96	5620.02	121	6/17/96	5652.78
34	2/15/96	5551.37	78	4/17/96	5549.93	122	6/18/96	5628.03
35	2/16/96	5503.32	79	4/18/96	5551.74	123	6/19/96	5648.35
36	2/19/96	5503.32	80	4/19/96	5535.48	124	6/20/96	5659.43
37	2/20/96	5458.53	81	4/22/96	5564.74	125	6/21/96	5705.23
38	2/21/96	5515.97	82	4/23/96	5588.59	126	6/24/96	5717.79
39	2/22/96	5608.46	83	4/24/96	5553.90	127	6/25/96	5719.27
40	2/23/96	5630.49	84	4/25/96	5566.91	128	6/26/96	5682.70
41	2/26/96	5565.10	85	4/26/96	5567.99	129	6/27/96	5677.53
42	2/27/96	5549.21	86	4/29/96	5573.41	130	6/28/96	5654.63
43	2/28/96	5506.21	87	4/30/96	5569.08			
44	2/29/96	5485.62	88	5/1/96	5575.22			

Table 8.11   DJIA Data for the Second Six Months of 1996.

Day	Date	DJIA	Day	Date	DJIA	Day	Date	DJIA
131	7/1/96	5729.98	175	8/30/96	5616.21	219	10/31/96	6029.38
132	7/2/96	5720.38	176	9/2/96	5616.21	220	11/1/96	6021.93
133	7/3/96	5703.02	177	9/3/96	5648.39	221	11/4/96	6041.68
134	7/4/96	5703.02	178	9/4/96	5656.90	222	11/5/96	6081.18
135	7/5/96	5588.14	179	9/5/96	5606.96	223	11/6/96	6177.71
136	7/8/96	5550.83	180	9/6/96	5659.86	224	11/7/96	6206.04
137	7/9/96	5581.86	181	9/9/96	5733.84	225	11/8/96	6219.82
138	7/10/96	5603.65	182	9/10/96	5727.18	226	11/11/96	6255.60
139	7/11/96	5520.50	183	9/11/96	5754.92	227	11/12/96	6266.04
140	7/12/96	5510.56	184	9/12/96	5771.94	228	11/13/96	6274.24
141	7/15/96	5349.51	185	9/13/96	5838.52	229	11/14/96	6313.00
142	7/16/96	5358.76	186	9/16/96	5889.20	230	11/15/96	6348.03
143	7/17/96	5376.88	187	9/17/96	5888.83	231	11/18/96	6346.91
144	7/18/96	5464.18	188	9/18/96	5877.36	232	11/19/96	6397.60
145	7/19/96	5426.82	189	9/19/96	5867.74	233	11/20/96	6430.02
146	7/22/96	5390.94	190	9/20/96	5888.46	234	11/21/96	6418.47
147	7/23/96	5346.55	191	9/23/96	5894.74	235	11/22/96	6471.76
148	7/24/96	5354.69	192	9/24/96	5874.03	236	11/25/96	6547.79
149	7/25/96	5422.01	193	9/25/96	5877.36	237	11/26/96	6528.41
150	7/26/96	5473.06	194	9/26/96	5868.85	238	11/27/96	6499.34
151	7/29/96	5434.59	195	9/27/96	5872.92	239	11/28/96	6499.34
152	7/30/96	5481.93	196	9/30/96	5882.17	240	11/29/96	6521.70
153	7/31/96	5528.91	197	10/1/96	5904.90	241	12/2/96	6521.70
154	8/1/96	5594.75	198	10/2/96	5933.97	242	12/3/96	6442.69
155	8/2/96	5679.83	199	10/3/96	5932.85	243	12/4/96	6422.94
156	8/5/96	5674.28	200	10/4/96	5992.86	244	12/5/96	6437.10
157	8/6/96	5696.11	201	10/7/96	5979.81	245	12/6/96	6381.94
158	8/7/96	5718.67	202	10/8/96	5966.77	246	12/9/96	6463.94
159	8/8/96	5713.49	203	10/9/96	5930.62	247	12/10/96	6473.25
160	8/9/96	5681.31	204	10/10/96	5921.67	248	12/11/96	6402.52
161	8/12/96	5704.98	205	10/11/96	5969.38	249	12/12/96	6303.71
162	8/13/96	5647.28	206	10/14/96	6010.00	250	12/13/96	6304.87
163	8/14/96	5666.88	207	10/15/96	6004.78	251	12/16/96	6268.35
164	8/15/96	5665.78	208	10/16/96	6020.81	252	12/17/96	6308.33
165	8/16/96	5689.45	209	10/17/96	6059.20	253	12/18/96	6346.77
166	8/19/96	5699.44	210	10/18/96	6094.23	254	12/19/96	6473.64
167	8/20/96	5721.26	211	10/21/96	6090.87	255	12/20/96	6484.40
168	8/21/96	5689.82	212	10/22/96	6061.80	256	12/23/96	6489.02
169	8/22/96	5733.47	213	10/23/96	6036.46	257	12/24/96	6522.85
170	8/23/96	5722.74	214	10/24/96	5992.48	258	12/26/06	6522.85
171	8/26/96	5693.89	215	10/25/96	6007.02	259	12/26/96	6546.68
172	8/27/96	5711.27	216	10/28/96	5972.73	260	12/27/96	6560.91
173	8/28/96	5712.38	217	10/29/96	6007.02	261	12/30/96	6549.37
174	8/29/96	5647.65	218	10/30/96	5993.23	262	12/31/96	6448.27

months of 1996 (the 130 days in Table 8.10). Compute the residual mean square.

(b) Use the above model to predict the daily DJIA for the first fifteen trading days in July 1996 (Table 8.11). Compare your pedictions with the actual values of the DJIA in Table 8.11 by computing the *prediction errors*, which is the difference between the actual values of the DJIA for the first 15 days of July, 1996 and their corresponding values predicted by the model.

(c) Compute the average of the squared prediction errors and compare with the residual mean square.

(d) Repeat the above exercise but using the model to predict the daily DJIA for the second half of the year (132 days).

(e) Explain the results you obtained above in the light of the scatter plot the DJIA versus Day.

8.6 Continuing with modeling the DJIA data in Exercises 8.4 and 8.5. A simplified version of the so-called *random walk model* of stock prices states that the best prediction of the stock index at Day t is the value of the index at Day $t - 1$. In regression model terms it would mean that for the models fitted in Exercises 8.4 and 8.5 the constant term is 0, and the regression coefficient is 1.

(a) Carry out the appropriate statistical tests of significance. (Test the values of the coefficients individually and then simultaneously.) Which test is the appropriate one: the individual or the simultaneous?

(b) The random walk theory implies that the first differences of the index (the difference between successive values) should be independently normally distributed with zero mean and constant variance. Examine the first differences of DJIA and log(DJIA) to see if the this hypothesis holds.

(c) DJIA is widely available. Collect the latest values available to see if the findings for 1996 hold for the latest period.

9

Analysis of Collinear Data

9.1 INTRODUCTION

Interpretation of the multiple regression equation depends implicitly on the assumption that the predictor variables are not strongly interrelated. It is usual to interpret a regression coefficient as measuring the change in the response variable when the corresponding predictor variable is increased by one unit and all other predictor variables are held constant. This interpretation may not be valid if there are strong linear relationships among the predictor variables. It is always conceptually possible to increase the value of one variable in an estimated regression equation while holding the others constant. However, there may be no information about the result of such a manipulation in the estimation data. Moreover, it may be impossible to change one variable while holding all others constant in the process being studied. When these conditions exist, simple interpretation of the regression coefficient as a marginal effect is lost.

When there is a complete absence of linear relationship among the predictor variables, they are said to be *orthogonal*. In most regression applications the predictor variables are not orthogonal. Usually, the lack of orthogonality is not serious enough to affect the analysis. However, in some situations the predictor variables are so strongly interrelated that the regression results are ambiguous. Typically, it is impossible to estimate the unique effects of individual variables in the regression equation. The estimated values of the coefficients are very sensitive to slight changes in the data and to the addition or deletion of variables in the equation. The regression coefficients have large sampling errors, which affect both inference and forecasting that is based on the regression model.

The condition of severe nonorthogonality is also referred to as the problem of collinear data, or *multicollinearity*. The problem can be extremely difficult to detect. It is not a specification error that may be uncovered by exploring regression residual. In fact, multicollinearity is not a modeling error. It is a condition of deficient data. In any event, it is important to know when multicollinearity is present and to be aware of its possible consequences. It is recommended that one should be very cautious about any and all substantive conclusions based on a regression analysis in the presence of multicollinearity.

This chapter focuses on three questions:

- How does multicollinearity affect statistical inference and forecasting?

- How can multicollinearity be detected?

- What can be done to resolve the difficulties associated with multicollinearity?

When analyzing data, these questions cannot be answered separately. If multicollinearity is a potential problem, the three issues must be treated simultaneously by necessity.

The discussion begins with two examples. They have been chosen to demonstrate the effects of multicollinearity on inference and forecasting, respectively. A treatment of methods for detecting multicollinearity follows and the chapter concludes with a presentation of methods for resolving problems of multicollinearity. The obvious prescription to collect better data is considered, but the discussion is mostly directed at improving interpretation of the existing data. Alternatives to the ordinary least squares estimation method that perform efficiently in the presence of multicollinearity are considered in Chapter 10.

9.2   EFFECTS ON INFERENCE

This first example demonstrates the ambiguity that may result when attempting to identify important predictor variables from among a linearly dependent collection of predictor variables. The context of the example is borrowed from research on equal opportunity in public education as reported by Coleman et al. (1966), Mosteller and Moynihan (1972), and others.

In conjunction with the Civil Rights Act of 1964, the Congress of the United States ordered a survey "concerning the lack of availability of equal educational opportunities for individuals by reason of race, color, religion or national origin in public educational institutions...." Data were collected from a cross-section of school districts throughout the country. In addition to reporting summary statistics on variables such as level of student achievement and school facilities, regression analysis was used to try to establish factors that are the most important determinants of achievement. The data for this example consist of measurements taken in 1965 for 70 schools selected at random. The data consist of variables that measure student achievement, school facilities, and faculty credentials. The objective is to evaluate the effect of school inputs on achievement.

Assume that an acceptable index has been developed to measure those aspects of the school environment that would be expected to affect achievement. The index includes evaluations of the physical plant, teaching materials, special programs, training and motivation of the faculty, and so on. Achievement can be measured by using an index constructed from standardized test scores. There are also other variables that may affect the relationship between school inputs and achievement. Students' performances may be affected by their home environments and the influence of their peer group in the school. These variables must be accounted for in the analysis before the effect of school inputs can be evaluated. We assume that indexes have been constructed for these variables that are satisfactory for our purposes. The data are given in Tables 9.1 and 9.2, and can also be found in the book's Web site.[1]

Adjustment for the two basic variables (achievement and school) can be accomplished by using the regression model

$$\text{ACHV} = \beta_0 + \beta_1 \cdot \text{FAM} + \beta_2 \cdot \text{PEER} + \beta_3 \cdot \text{SCHOOL} + \varepsilon. \qquad (9.1)$$

The contribution of the school variable can be tested using the t-value for β_3. Recall that the t-value for β_3 tests whether SCHOOL is necessary in the equation when FAM and PEER are already included. Effectively, the model above is being compared to

$$\text{ACHV} - \beta_1 \cdot \text{FAM} - \beta_2 \cdot \text{PEER} = \beta_0 + \beta_3 \cdot \text{SCHOOL} + \varepsilon, \qquad (9.2)$$

that is, the contribution of the school variable is being evaluated after adjustment for FAM and PEER. Another view of the adjustment notion is obtained by manipulating Equation (9.1) to form

$$\text{ACHV} - \beta_1 \cdot \text{FAM} - \beta_2 \cdot \text{PEER} = \beta_0 + \beta_3 \cdot \text{SCHOOL} + \varepsilon.$$

The left-hand side is an adjusted achievement index where adjustment is accomplished by subtracting the linear contributions of FAM and PEER. The equation is in the form of a regression of the adjusted achievement score on the SCHOOL variable. This representation is used only for the sake of interpretation. The estimated β's are obtained from the original model given in Equation (9.1). The regression results are summarized in Table 9.3 and a plot of the residuals against the predicted values of ACHV appears as Figure 9.1.

---

[1] http://www.ilr.cornell.edu/~hadi/RABE

Table 9.1   First 50 Observations of the Equal Educational Opportunity (EEO) Data; Standardized Indexes.

Row	ACHV	FAM	PEER	SCHOOL
1	−0.43148	0.60814	0.03509	0.16607
2	0.79969	0.79369	0.47924	0.53356
3	−0.92467	−0.82630	−0.61951	−0.78635
4	−2.19081	−1.25310	−1.21675	−1.04076
5	−2.84818	0.17399	−0.18517	0.14229
6	−0.66233	0.20246	0.12764	0.27311
7	2.63674	0.24184	−0.09022	0.04967
8	2.35847	0.59421	0.21750	0.51876
9	−0.91305	−0.61561	−0.48971	−0.63219
10	0.59445	0.99391	0.62228	0.93368
11	1.21073	1.21721	1.00627	1.17381
12	1.87164	0.41436	0.71103	0.58978
13	−0.10178	0.83782	0.74281	0.72154
14	−2.87949	−0.75512	−0.64411	−0.56986
15	3.92590	−0.37407	−0.13787	−0.21770
16	4.35084	1.40353	1.14085	1.37147
17	1.57922	1.64194	1.29229	1.40269
18	3.95689	−0.31304	−0.07980	−0.21455
19	1.09275	1.28525	1.22441	1.20428
20	−0.62389	−1.51938	−1.27565	−1.36598
21	−0.63654	−0.38224	−0.05353	−0.35560
22	−2.02659	−0.19186	−0.42605	−0.53718
23	−1.46692	1.27649	0.81427	0.91967
24	3.15078	0.52310	0.30720	0.47231
25	−2.18938	−1.59810	−1.01572	−1.48315
26	1.91715	0.77914	0.87771	0.76496
27	−2.71428	−1.04745	−0.77536	−0.91397
28	−6.59852	−1.63217	−1.47709	−1.71347
29	0.65101	0.44328	0.60956	0.32833
30	−0.13772	−0.24972	0.07876	−0.17216
31	−2.43959	−0.33480	−0.39314	−0.37198
32	−3.27802	−0.20680	−0.13936	0.05626
33	−2.48058	−1.99375	−1.69587	−1.87838
34	1.88639	0.66475	0.79670	0.69865
35	5.06459	−0.27977	0.10817	−0.26450
36	1.96335	−0.43990	−0.66022	−0.58490
37	0.26274	−0.05334	−0.02396	−0.16795
38	−2.94593	−2.06699	−1.31832	−1.72082
39	−1.38628	−1.02560	−1.15858	−1.19420
40	−0.20797	0.45847	0.21555	0.31347
41	−1.07820	0.93979	0.63454	0.69907
42	−1.66386	−0.93238	−0.95216	−1.02725
43	0.58117	−0.35988	−0.30693	−0.46232
44	1.37447	−0.00518	0.35985	0.02485
45	−2.82687	−0.18892	−0.07959	0.01704
46	3.86363	0.87271	0.47644	0.57036
47	−2.64141	−2.06993	−1.82915	−2.16738
48	0.05387	0.32143	−0.25961	0.21632
49	0.50763	−1.42382	−0.77620	−1.07473
50	0.64347	−0.07852	−0.21347	−0.11750

Table 9.2 Last 20 Observations of Equal Educational Opportunity (EEO) Data; Standardized Indexes.

Row	ACHV	FAM	PEER	SCHOOL
51	2.49414	−0.14925	−0.03192	−0.36598
52	0.61955	0.52666	0.79149	0.71369
53	0.61745	−1.49102	−1.02073	−1.38103
54	−1.00743	−0.94757	−1.28991	−1.24799
55	−0.37469	0.24550	0.83794	0.59596
56	−2.52824	−0.41630	−0.60312	−0.34951
57	0.02372	1.38143	1.54542	1.59429
58	2.51077	1.03806	0.91637	0.97602
59	−4.22716	−0.88639	−0.47652	−0.77693
60	1.96847	1.08655	0.65700	0.89401
61	1.25668	−1.95142	−1.94199	−1.89645
62	−0.16848	2.83384	2.47398	2.79222
63	−0.34158	1.86753	1.55229	1.80057
64	−2.23973	−1.11172	−0.69732	−0.80197
65	3.62654	1.41958	1.11481	1.24558
66	0.97034	0.53940	0.16182	0.33477
67	3.16093	0.22491	0.74800	0.66182
68	−1.90801	1.48244	1.47079	1.54283
69	0.64598	2.05425	1.80369	1.90066
70	−1.75915	1.24058	0.64484	0.87372

Table 9.3 EEO Data: Regression Results.

Variable	Coefficient	s.e.	t-test	p-value
Constant	−0.070	0.251	−0.28	0.7810
FAM	1.101	1.411	0.78	0.4378
PEER	2.322	1.481	1.57	0.1218
SCHOOL	−2.281	2.220	−1.03	0.3080
$n = 70$	$R^2 = 0.206$	$R_a^2 = 0.170$	$\hat{\sigma} = 2.07$	$d.f. = 66$

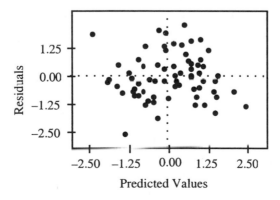

Fig. 9.1 Standardized residuals against fitted values of ACHV.

Checking first the residual plot we see that there are no glaring indications of misspecification. The point located in the lower left of the graph has a residual value that is about 2.5 standard deviations from the mean of zero and should possibly be looked at more closely. However, when it is deleted from the sample, the regression results show almost no change. Therefore, the observation has been retained in the analysis.

From Table 9.3 we see that about 20% of the variation in achievement score is accounted for by the three predictors jointly ($R^2 = 0.206$). The F-value is 5.72 based on 3 and 66 degrees of freedom and is significant at better than the 0.01 level. Therefore, even though the total explained variation is estimated at only 20%, it is accepted that FAM, PEER, and SCHOOL are valid predictor variables. However, the individual t-values are all small. In total, the summary statistics say that the three predictors taken together are important but from the t-values, it follows that any one predictor may be deleted from the model provided the other two are retained.

These results are typical of a situation where extreme multicollinearity is present. The predictor variables are so highly correlated that each one may serve as a proxy for the others in the regression equation without affecting the total explanatory power. The low t-values confirm that any one of the predictor variables may be dropped from the equation. Hence the regression analysis has failed to provide any information for evaluating the importance of school inputs on achievement. The culprit is clearly multicollinearity. The pairwise correlation coefficients of the three predictor variables and the corresponding scatter plots (Figure 9.2), all show strong linear relationships among all pairs of predictor variables. All pairwise correlation coefficients are high. In all scatter plots, all the observations lie close to the straight line through the average values of the corresponding variables.

Multicollinearity in this instance could have been expected. It is the nature of these three variables that each is determined by and helps to determine the others. It is not unreasonable to conclude that there are not three variables but in fact only one. Unfortunately, that conclusion does not help to answer the original question about the effects of school facilities on achievement. There remain two possibilities. First, multicollinearity may be present because the sample data are deficient, but can be improved with additional observations. Second, multicollinearity may be present because the interrelationships among the variables are an inherent characteristic of the process under investigation. Both situations are discussed in the following paragraphs.

In the first case the sample should have been selected to insure that the correlations between the predictor variables were not large. For example, in the scatter plot of FAM versus SCHOOL (the graph in the top right corner in Figure 9.2), there are no schools in the sample with values in the upper left or lower right regions of the graph. Hence there is no information in the sample on achievement when the value of FAM is high and SCHOOL is low, or FAM is low and SCHOOL is high. But it is only with data collected under these two conditions that the individual effects of FAM and SCHOOL on ACHV

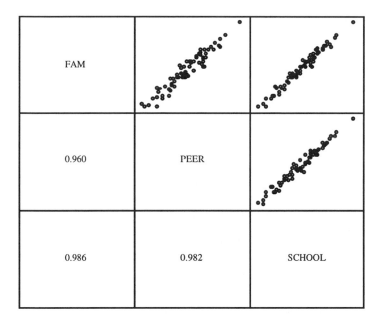

Fig. 9.2 Pairwise scatter plots of the three predictor variables FAM, PEER, and SCHOOL; and the corresponding pairwise correlation coefficients.

Table 9.4   Data Combinations for Three Predictor Variables.

Combination	Variable		
	FAM	PEER	SCHOOL
1	+	+	+
2	+	+	−
3	+	−	+
4	−	+	+
5	+	−	−
6	−	+	−
7	−	−	+
8	−	−	−

can be determined. For example, assume that there were some observations in the upper left quadrant of the graph. Then it would at least be possible to compare average ACHV for low and high values of SCHOOL when FAM is held constant.

Since there are three predictor variables in the model, then there are eight distinct combinations of data that should be included in the sample. Using + to represent a value above the average and − to represent a value below the average, the eight possibilities are represented in Table 9.4.

The large correlations that were found in the analysis suggest that only combinations 1 and 8 are represented in the data. If the sample turned out this way by chance, the prescription for resolving the multicollinearity problem is to collect additional data on some of the other combinations. For example, data based on combinations 1 and 2 alone could be used to evaluate the effect of SCHOOL on ACHV holding FAM and PEER at a constant level, both above average. If these were the only combinations represented in the data, the analysis would consist of the simple regression of ACHV against SCHOOL. The results would give only a partial answer, namely, an evaluation of the school-achievement relationship when FAM and PEER are both above average.

The prescription for additional data as a way to resolve multicollinearity is not a panacea. It is often not possible to collect more data because of constraints on budgets, time, and staff. It is always better to be aware of impending data deficiencies beforehand. Whenever possible, the data should be collected according to design. Unfortunately, prior design is not always feasible. In surveys, or observational studies such as the one being discussed, the values of the predictor variables are usually not known until the sampling unit is selected for the sample and some costly and time-consuming measurements are developed. Following this procedure, it is fairly difficult to ensure that a balanced sample will be obtained.

The second reason that multicollinearity may appear is because the relationships among the variables are an inherent characteristic of the process being sampled. If FAM, PEER, and SCHOOL exist in the population only as data combinations 1 and 8 of Table 9.4, it is not possible to estimate the individual effects of these variables on achievement. The only recourse for continued analysis of these effects would be to search for underlying causes that may explain the interrelationships of the predictor variables. Through this process, one may discover other variables that are more basic determinants affecting equal opportunity in education and achievement.

9.3   EFFECTS ON FORECASTING

We shall examine the effects of multicollinearity in forecasting when the forecasts are based on a multiple regression equation. A historical data set with observations indexed by time is used to estimate the regression coefficients. Forecasts of the response variable are produced by using future values of the predictor variables in the estimated regression equation. The future values of the predictor variables must be known or forecasted from other data and models. We shall not treat the uncertainty in the forecasted predictor variables. In our discussion it is assumed that the future values of the predictor variables are given.

We have chosen an example based on aggregate data concerning import activity in the French economy. The data have been analyzed by Malinvaud

Table 9.5 Data on French Economy.

YEAR	IMPORT	DOPROD	STOCK	CONSUM
49	15.9	149.3	4.2	108.1
50	16.4	161.2	4.1	114.8
51	19.0	171.5	3.1	123.2
52	19.1	175.5	3.1	126.9
53	18.8	180.8	1.1	132.1
54	20.4	190.7	2.2	137.7
55	22.7	202.1	2.1	146.0
56	26.5	212.4	5.6	154.1
57	28.1	226.1	5.0	162.3
58	27.6	231.9	5.1	164.3
59	26.3	239.0	0.7	167.6
60	31.1	258.0	5.6	176.8
61	33.3	269.8	3.9	186.6
62	37.0	288.4	3.1	199.7
63	43.3	304.5	4.6	213.9
64	49.0	323.4	7.0	223.8
65	50.3	336.8	1.2	232.0
66	56.6	353.9	4.5	242.9

Source: Malinvaud (1968).

(1968). Our discussion follows his presentation. The variables are imports (IMPORT), domestic production (DOPROD), stock formation (STOCK), and domestic consumption (CONSUM), all measured in billions of French francs for the years 1949 through 1966. The data are given in Table 9.5 and can be obtained from the book's Web site. The model being considered is

$$\text{IMPORT} = \beta_0 + \beta_1 \cdot \text{DOPROD} + \beta_2 \cdot \text{STOCK} + \beta_3 \cdot \text{CONSUM} + \varepsilon. \quad (9.3)$$

The regression results appear as Table 9.6. The index plot of residuals (Figure 9.3) shows a distinctive pattern, suggesting that the model is not well specified. Even though multicollinearity appears to be present ($R^2 = 0.973$ and all t-values small), it should not be pursued further in this model. Multicollinearity should only be attacked after the model specification is satisfactory. The difficulty with the model is that the European Common Market began operations in 1960, causing changes in import-export relationships. Since our objective in this chapter is to study the effects of multicollinearity, we shall not complicate the model by attempting to capture the behavior after 1959. We shall assume that it is now 1960 and look only at the 11 years 1949–1959. The regression results for those data are summarized in Table 9.7. The residual plot is now satisfactory (Figure 9.4).

The value of $R^2 = 0.99$ is high. However, the coefficient of DOPROD is negative and not statistically significant, which is contrary to prior expectation. We believe that if STOCK and CONSUM were held constant, an increase in DOPROD would cause an increase in IMPORT, probably for raw materials or manufacturing equipment. Multicollinearity is a possibility here and in fact is the case. The simple correlation between CONSUM and DOPROD is 0.997. Upon further investigation it turns out that CONSUM has been

Table 9.6 Import data (1949–1966): Regression Results.

ANOVA Table				
Source	Sum of Squares	d.f.	Mean Square	F-test
Regression	2576.92	3	858.974	168
Residuals	71.39	14	5.099	
Coefficients Table				
Variable	Coefficient	s.e.	t-test	p-value
Constant	−19.725	4.125	−4.78	0.0003
DOPROD	0.032	0.187	0.17	0.8656
STOCK	0.414	0.322	1.29	0.2195
CONSUM	0.243	0.285	0.85	0.4093
$n = 18$	$R^2 = 0.973$	$R_a^2 = 0.967$	$\hat{\sigma} = 2.258$	$d.f. = 14$

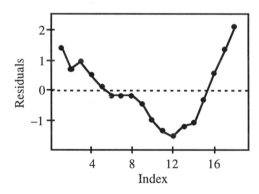

Fig. 9.3 Import data (1949–1966): Index plot of the standardized residuals.

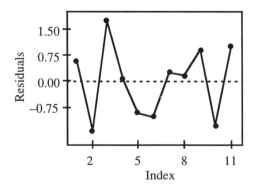

Fig. 9.4 Import data (1949–1959): Index plot of the standardized residuals.

Table 9.7 Import data (1949–1959): Regression Results.

ANOVA Table				
Source	Sum of Squares	d.f.	Mean Square	F-test
Regression	204.776	3	68.2587	286
Residuals	1.673	7	0.2390	

Coefficients Table				
Variable	Coefficient	s.e.	t-test	p-value
Constant	−10.128	1.212	−8.36	< 0.0001
DOPROD	−0.051	0.070	−0.73	0.4883
STOCK	0.587	0.095	6.20	0.0004
CONSUM	0.287	0.102	2.81	0.0263
$n = 11$	$R^2 = 0.992$	$R_a^2 = 0.988$	$\hat{\sigma} = 0.4889$	$d.f. = 7$

about two-thirds of DOPROD throughout the 11-year period. The estimated relationship between the two quantities is

$$\text{CONSUM} = 6.259 \mid 0.686 \cdot \text{DOPROD}.$$

Even in the presence of such severe multicollinearity the regression equation may produce some good forecasts. From Table 9.7, the forecasting equation is
$$\text{IMPORT} = -10.13 - 0.051 \cdot \text{DOPROD} + 0.587 \cdot \text{STOCK} + 0.287 \cdot \text{CONSUM}.$$

Recall that the fit to the historical data is very good and the residual variation appears to be purely random. To forecast we must be confident that the character and strength of the overall relationship will hold into future periods. This matter of confidence is a problem in all forecasting models whether or not multicollinearity is present. For the purpose of this example we assume that the overall relationship does hold into future periods.[2] Implicit in this assumption is the relationship between DOPROD and CONSUM. The forecast will be accurate as long as the future values of DOPROD, STOCK, and CONSUM have the relationship that CONSUM is approximately equal to two-thirds of DOPROD.

For example, let us forecast the change in IMPORT next year corresponding to an increase in DOPROD of 10 units while holding STOCK and CONSUM

---

[2]For the purpose of convenient exposition we ignore the difficulties that arise because of our previous finding that the formation of the European Common Market has altered the relationship since 1960. But we are impelled to advise the reader that changes in structure make forecasting a very delicate endeavor even when the historical fit is excellent.

at their current levels. The resulting forecast is

$$\text{IMPORT}_{1960} = \text{IMPORT}_{1959} - 0.051(10),$$

which means that IMPORT will decrease by -0.51 units. However, if the relationship between DOPROD and CONSUM is kept intact, CONSUM will increase by $10(2/3) = 6.67$ units and the forecasted result is

$$\text{IMPORT}_{1960} = \text{IMPORT}_{1959} - 0.51 + 0.287 \times 6.67 = \text{IMPORT}_{1959} + 1.4.$$

IMPORT actually increases by 1.4 units, a more satisfying result and probably a better forecast. The case where DOPROD increases alone corresponds to a change in the basic structure of the data that were used to estimate the model parameters and cannot be expected to produce meaningful forecasts.

In summary, the two examples demonstrate that multicollinear data can seriously limit the use of regression analysis for inference and forecasting. Extreme care is required when attempting to interpret regression results when multicollinearity is suspected. In Section 9.4 we discuss methods for detecting extreme collinearity among predictor variables.

9.4 DETECTION OF MULTICOLLINEARITY

In the preceding examples some of the ideas for detecting multicollinearity were already introduced. In this section we review those ideas and introduce additional criteria that indicate collinearity. Multicollinearity is associated with unstable estimated regression coefficients. This situation results from the presence of strong linear relationships among the predictor variables. It is not a problem of misspecification. Therefore, the empirical investigation of problems that result from a collinear data set should begin only after the model has been satisfactorily specified. However, there may be some indications of multicollinearity that are encountered during the process of adding, deleting, and transforming variables or data points in search of the good model. Indication of multicollinearity that appear as instability in the estimated coefficients are as follows:

- Large changes in the estimated coefficients when a variable is added or deleted.

- Large changes in the coefficients when a data point is altered or dropped.

Once the residual plots indicate that the model has been satisfactorily specified, multicollinearity may be present if:

- The algebraic signs of the estimated coefficients do not conform to prior expectations; or

Table 9.8 Import Data (1949–1959): Regression Coefficients for All Possible Regressions.

Regression	Constant	DOPROD	STOCK	CONSUM
		Variable		
1	−6.558	0.146	−	−
2	19.611	−	0.691	−
3	−8.013	−	−	0.214
4	−8.440	0.145	0.622	−
5	−8.884	−0.109	−	0.372
6	−9.743	−	0.596	0.212
7	−10.128	−0.051	0.587	0.287

- Coefficients of variables that are expected to be important have large standard errors (small t-values).

For the IMPORT data discussed previously, the coefficient of DOPROD was negative and not significant. Both results are contrary to prior expectations. The effects of dropping or adding a variable can be seen in Table 9.8. There we see that the presence or absence of certain variables has a large effect on the other coefficients. For the EEO data (Tables 9.1 and 9.2) the algebraic signs are all correct, but their standard errors are so large that none of the coefficients are statistically significant. It was expected that they would all be important.

The presence of multicollinearity is also indicated by the size of the correlation coefficients that exist among the predictor variables. A large correlation between a pair of predictor variables indicates a strong linear relationship between those two variables. The correlations for the EEO data (Figure 9.2) are large for all pairs of predictor variables. For the IMPORT data, the correlation coefficient between DOPROD and CONSUM is 0.997.

The source of multicollinearity may be more subtle than a simple relationship between two variables. A linear relation can involve many of the predictor variables. It may not be possible to detect such a relationship with a simple correlation coefficient. As an example, we shall look at an analysis of the effects of advertising expenditures (A_t), promotion expenditures (P_t), and sales expense (E_t) on the aggregate sales of a firm in period t. The data represent a period of 23 years during which the firm was operating under fairly stable conditions. The data are given in Table 9.9 and can be obtained from the book's Web site.

The proposed regression model is

$$S_t = \beta_0 + \beta_1 A_t + \beta_2 P_t + \beta_3 E_t + \beta_4 A_{t-1} + \beta_5 P_{t-1} + \varepsilon_t, \qquad (9.4)$$

where A_{t-1} and P_{t-1} are the lagged one-year variables. The regression results are given in Table 9.10. The plot of residuals versus fitted values and the index

Table 9.9 Annual Data on Advertising, Promotions, Sales Expenses, and Sales (Millions of Dollars).

Row	S_t	A_t	P_t	E_t	A_{t-1}	P_{t-1}
1	20.11371	1.98786	1.0	0.30	2.01722	0.0
2	15.10439	1.94418	0.0	0.30	1.98786	1.0
3	18.68375	2.19954	0.8	0.35	1.94418	0.0
4	16.05173	2.00107	0.0	0.35	2.19954	0.8
5	21.30101	1.69292	1.3	0.30	2.00107	0.0
6	17.85004	1.74334	0.3	0.32	1.69292	1.3
7	18.87558	2.06907	1.0	0.31	1.74334	0.3
8	21.26599	1.01709	1.0	0.41	2.06907	1.0
9	20.48473	2.01906	0.9	0.45	1.01709	1.0
10	20.54032	1.06139	1.0	0.45	2.01906	0.9
11	26.18441	1.45999	1.5	0.50	1.06139	1.0
12	21.71606	1.87511	0.0	0.60	1.45999	1.5
13	28.69595	2.27109	0.8	0.65	1.87511	0.0
14	25.83720	1.11191	1.0	0.65	2.27109	0.8
15	29.31987	1.77407	1.2	0.65	1.11191	1.0
16	24.19041	0.95878	1.0	0.65	1.77407	1.2
17	26.58966	1.98930	1.0	0.62	0.95878	1.0
18	22.24466	1.97111	0.0	0.60	1.98930	1.0
19	24.79944	2.26603	0.7	0.60	1.97111	0.0
20	21.19105	1.98346	0.1	0.61	2.26603	0.7
21	26.03441	2.10054	1.0	0.60	1.98346	0.1
22	27.39304	1.06815	1.0	0.58	2.10054	1.0

plot of residuals (Figures 9.5 and 9.6), as well as other plots of the residuals versus the predictor variables (not shown), do not suggest any problems of misspecification. Furthermore, the correlation coefficients between the predictor variables are small (Table 9.11). However, if we do a little experimentation to check the stability of the coefficients by dropping the contemporaneous advertising variable A from the model, many things change. The coefficient of P_t drops from 8.37 to 3.70; the coefficients of lagged advertising A_{t-1} and lagged promotions P_{t-1} change signs. But the coefficient of sales expense is stable and R^2 does not change much.

The evidence suggests that there is some type of relationship involving the contemporaneous and lagged values of the advertising and promotions variables. The regression of A_t on P_t, A_{t-1}, and P_{t-1} returns an R^2 of 0.973. The equation takes the form

$$\hat{A}_t = 4.63 - 0.87P_t - 0.86A_{t-1} - 0.95P_{t-1}.$$

Upon further investigation into the operations of the firm, it was discovered that close control was exercised over the expense budget during those 23 years of stability. In particular, there was an approximate rule imposed on the budget that the sum of A_t, A_{t-1}, P_t, and P_{t-1} was to be held to approximately five units over every two-year period. The relationship

$$A_t + P_t + A_{t-1} + P_{t-1} \doteq 5$$

Table 9.10   Regression Results for the Advertising Data.

ANOVA Table				
Source	Sum of Squares	d.f.	Mean Square	F-test
Regression	307.572	5	61.514	35.3
Residuals	27.879	16	1.742	

Coefficients Table				
Variable	Coefficient	s.e.	t-test	p-value
Constant	−14.194	18.715	−0.76	0.4592
A	5.361	4.028	1.33	0.2019
P	8.372	3.586	2.33	0.0329
E	22.521	2.142	10.51	< 0.0001
A_{t-1}	3.855	3.578	1.08	0.2973
P_{t-1}	4.125	3.895	1.06	0.3053
$n = 22$	$R^2 = 0.917$	$R_a^2 = 0.891$	$\hat{\sigma} = 1.320$	$d.f. = 16$

Table 9.11   Pairwise Correlation Coefficients for the Advertising Data.

	A_t	P_t	E_t	A_{t-1}	P_{t-1}
A_t	1.000				
P_t	−0.357	1.000			
E_t	−0.129	0.063	1.000		
A_{t-1}	−0.140	−0.316	−0.166	1.000	
P_{t-1}	−0.496	−0.296	0.208	−0.358	1.000

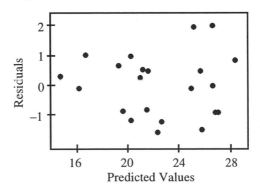

Fig. 9.5   Standardized residuals versus fitted values of Sales.

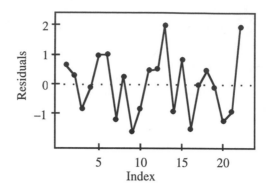

Fig. 9.6 Index plot of the standardized residuals.

is the cause of the multicollinearity.

A thorough investigation of multicollinearity will involve examining the value of R^2 that results from regressing each of the predictor variables against all the others. The relationship between the predictor variables can be judged by examining a quantity called the *variance inflation factor* (VIF). Let R_j^2 be the square of the multiple correlation coefficient that results when the predictor variable X_j is regressed against all the other predictor variables. Then the variance inflation for X_j is

$$\text{VIF}_j = \frac{1}{1 - R_j^2}, \quad j = 1, \ldots, p, \tag{9.5}$$

where p is the number of predictor variables. It is clear that if X_j has a strong linear relationship with the other predictor variables, R_j^2 would be close to 1, and VIF_j would be large. Values of variance inflation factors greater than 10 is often taken as a signal that the data have collinearity problems.

In absence of any linear relationship between the predictor variables (i.e., if the predictor variables are orthogonal), R_j^2 would be zero and VIF_j would be one. The deviation of VIF_j value from 1 indicates departure from orthogonality and tendency toward collinearity. The value of VIF_j also measures the amount by which the variance of the jth regression coefficient is increased due to the linear association of X_j with other predictor variables relative to the variance that would result if X_j were not related to them linearly. This explains the naming of this particular diagnostic.

As R_j^2 tends toward 1, indicating the presence of a linear relationship in the predictor variables, the VIF for $\hat{\beta}_j$ tends to infinity. It is suggested that a VIF in excess of 10 is an indication that multicollinearity may be causing problems in estimation.

The precision of an ordinary least squares (OLS) estimated regression coefficient is measured by its variance, which is proportional to σ^2, the variance of the error term in the regression model. The constant of proportionality is the VIF. Thus, the VIFs may be used to obtain an expression for the expected squared distance of the OLS estimators from their true values. Denoting the square of the distance by D^2, it can be shown that, on average,

$$D^2 = \sigma^2 \sum_{j=1}^{p} \text{VIF}_j.$$

This distance is another measure of precision of the least squares estimators. The smaller the distance, the more accurate are the estimates. If the predictor variables were orthogonal, the VIFs would all be 1 and D^2 would be $p\sigma^2$. It follows that the ratio

$$\frac{\sigma^2 \sum_{i=1}^{p} \text{VIF}_i}{p\sigma^2} = \frac{\sum_{i=1}^{p} \text{VIF}_i}{p} = \overline{\text{VIF}},$$

which shows that the average of the VIFs measures the squared error in the OLS estimators relative to the size of that error if the data were orthogonal. Hence, $\overline{\text{VIF}}$ may also be used as an index of multicollinearity.

Most computer packages now furnish values of VIF_j routinely. Some have built-in messages when high values of VIF_j are observed. In any regression analysis the values of VIF_j should always be examined to avoid the pitfalls resulting from fitting a regression model to collinear data by least squares.

In each of the three examples (EEO, Import, and Advertising) we have seen evidence of collinearity. The VIF_j's and their average values for these data sets are given in Table 9.12. For the EEO data the values of VIF_j range from 30.2 to 83.2, showing that all three variables are strongly intercorrelated and that dropping one of the variables will not eliminate collinearity. The average value of VIF of 50.3 indicates that the squared error in the OLS estimators is 50 times as large as it would be if the predictor variables were orthogonal.

For the Import data, the squared error in the OLS estimators is 313 times as large as it would be if the predictor variables were orthogonal. However, the VIF_j's indicate that domestic production and consumption are strongly correlated but are not correlated with the STOCK variable. A regression equation containing either CONSUM or DOPROD along with STOCK will eliminate collinearity.

For the Advertising data, VIF_E (for the variable E) is 1.1, indicating that this variable is not correlated with the remaining predictor variables. The VIF_j's for the other four variables are large, ranging from 26.6 to 44.1. This indicates that there is a strong linear relationship among the four variables, a fact that we have already noted. Here the prescription might be to regress sales S_t against E_t and three of the remaining four variables $(A_t, P_t, A_{t-1}, S_{t-1})$ and examine the resulting VIF_j's to see if collinearity has been eliminated.

Table 9.12   Variance Inflation Factors for Three Data Sets.

EEO		Import		Advertising	
Variable	VIF	Variable	VIF	Variable	VIF
FAM	37.6	DOPROD	469.7	A_t	37.4
PEER	30.2	STOCK	1.0	P_t	33.5
SCHOOL	83.2	CONSUM	469.4	E_t	1.1
				A_{t-1}	26.6
				P_{t-1}	44.1
Average	50.3	Average	313.4	Average	28.5

9.5   CENTERING AND SCALING

The indicators of multicollinearity that have been described so far can all be obtained using standard regression computations. There is another, more unified way to analyze multicollinearity which requires some calculations that are not usually included in standard regression packages. The analysis follows from the fact that every linear regression model can be restated in terms of a set of orthogonal predictor variables. These new variables are obtained as linear combinations of the original predictor variables. They are referred to as the *principal components* of the set of predictor variables (Seber, 1984; Johnson and Wichern, 1992).

To develop the method of principal components, we may first need to *center* and/or *scale* the variables. We have been mainly dealing with regression models of the form

$$Y = \beta_0 + \beta_1 X_1 + \ldots + \beta_p X_p + \varepsilon, \tag{9.6}$$

which are models with a constant term β_0. But we have also seen situations where fitting the *no-intercept* model

$$Y = \beta_1 X_1 + \ldots + \beta_p X_p + \varepsilon \tag{9.7}$$

is necessary (see, e.g., Chapters 3 and 7). When dealing with constant term models, it is convenient to center and scale the variables, but when dealing with a no-intercept model, we need only to scale the variables.

9.5.1   Centering and Scaling in Intercept Models

If we are fitting an intercept model as in (9.6), we need to center and scale the variables. A *centered* variable is obtained by subtracting from each observation the mean of all observations. For example, the centered response variable is $(Y - \bar{y})$ and the centered jth predictor variable is $(X_j - \bar{x}_j)$. The mean of a centered variable is zero. The centered variables can also be scaled. Two types of scaling are usually needed: *unit length scaling* and *standardizing*. Unit length scaling of the response variable Y and the jth predictor variable

X_j is obtained as follows:

$$
\begin{aligned}
\tilde{Z}_y &= \frac{Y - \bar{y}}{L_y}, \\
\tilde{Z}_j &= \frac{X_j - \bar{x}_j}{L_j}, \quad j = 1, \ldots, p,
\end{aligned}
\tag{9.8}
$$

where $\bar{y}$ is the mean of Y, $\bar{x}_j$ is the mean of X_j, and

$$
L_y = \sqrt{\sum_{i=1}^{n}(y_i - \bar{y})^2}, \text{ and } L_j = \sqrt{\sum_{i=1}^{n}(x_{ij} - \bar{x}_j)^2}, \; j = 1, \ldots, p.
\tag{9.9}
$$

The quantities L_y is referred to as the *length* of the centered variable $Y - \bar{y}$ because it measures the size or the magnitudes of the observations in $Y - \bar{y}$. Similarly, L_j measure the length of the variable $X_j - \bar{x}_j$. The variables $\tilde{Z}_y$ and $\tilde{Z}_j$ in (9.8) have zero means and unit lengths, hence this type of scaling is called unit length scaling. In addition, unit length scaling has the following property:

$$
\mathrm{Cor}(X_j, X_k) = \sum_{i=1}^{n} z_{ij} z_{ik}.
\tag{9.10}
$$

That is, the correlation coefficient between the original variables, X_j and X_k, can be computed easily as the sum of the products of the scaled versions Z_j and Z_k.

The second type of scaling is called standardizing, which is defined by

$$
\begin{aligned}
\tilde{Y} &= \frac{Y - \bar{y}}{s_y}, \\
\tilde{X}_j &= \frac{X_j - \bar{x}_j}{s_j}, \quad j = 1, \ldots, p,
\end{aligned}
\tag{9.11}
$$

where

$$
s_y = \sqrt{\frac{\sum_{i=1}^{n}(y_i - \bar{y})^2}{n-1}}, \text{ and } s_j = \sqrt{\frac{\sum_{i=1}^{n}(x_{ij} - \bar{x}_j)^2}{n-1}}, \; j = 1, \ldots, p,
\tag{9.12}
$$

are standard deviations of the response and jth predictor variable, respectively. The standardized variables $\tilde{Y}$ and $\tilde{X}_j$ in (9.11) have means zero and unit standard deviations.

Since correlations are unaffected by shifting or scaling the data, it is both sufficient and convenient to deal with either the unit length scaled or the standardized versions of the variables. The variances and covariances of a set of p variables, X_1, ..., X_p, can be neatly displayed as a squared array of numbers called a *matrix*. This matrix is known as the *variance-covariance matrix*. The elements on the diagonal that runs from the upper-left corner to the lower-right corner of the matrix are known as the *diagonal elements*. The elements on the diagonal of a variance-covariance matrix are the variances and the elements off the diagonal are the covariances.[3] The variance-covariance matrix of the three predictor variables in the Import data for the years 1949–1959 is:

$$
\begin{array}{c}
\\
\text{DOPROD} \\
\text{STOCK} \\
\text{CONSUM}
\end{array}
\begin{array}{ccc}
\text{DOPROD} & \text{STOCK} & \text{CONSUM} \\
\left(\begin{array}{ccc}
899.971 & 1.279 & 617.326 \\
1.279 & 2.720 & 1.214 \\
617.326 & 1.214 & 425.779
\end{array} \right)
\end{array}.
$$

Thus, for example, $Var(\text{DOPROD}) = 899.971$, which is in the first diagonal element, and $Cov(\text{DOPROD}, \text{CONSUM}) = 617.326$, which is the value is the intersection of the first row and third column (or the third row and first column).

Similarly, the pairwise correlation coefficients can be displayed in matrix known as the *correlation matrix*. The correlation matrix of the three predictor variables in the Import data is:

$$
\begin{array}{c}
\\
\text{DOPROD} \\
\text{STOCK} \\
\text{CONSUM}
\end{array}
\begin{array}{ccc}
\text{DOPROD} & \text{STOCK} & \text{CONSUM} \\
\left(\begin{array}{ccc}
1.000 & 0.026 & 0.997 \\
0.026 & 1.000 & 0.036 \\
0.997 & 0.036 & 1.000
\end{array} \right)
\end{array}. \tag{9.13}
$$

This is the same as the variance-covariance matrix of the standardized predictor variables. Thus, for example, $Cor(\text{DOPROD}, \text{CONSUM}) = 0.997$, which indicates that the two variables are highly correlated. Note that all the diagonal elements of the correlation matrix are equal to one.

Recall that a set of variables is said to be orthogonal if there exists no linear relationships among them. If the standardized predictor variables are orthogonal, their matrix of variances and covariances consists of one for the diagonal elements and zero for the off-diagonal elements.

9.5.2   Scaling in No-Intercept Models

If we are fitting a no-intercept model as in (9.7), we do not center the data because centering has the effect of including a constant term in the model. This can be seen from:

$$
Y - \bar{y} = \beta_1(X_1 - \bar{x}_1) + \ldots + \beta_p(X_p - \bar{x}_p) + \varepsilon. \tag{9.14}
$$

rearranging terms, we obtain

$$
\begin{aligned}
Y &= \bar{y} - (\beta_1\bar{x}_1 + \ldots + \beta_p\bar{x}_p) + \beta_1 X_1 + \ldots + \beta_p X_p + \varepsilon \\
&= \beta_0 + \beta_1 X_1 + \ldots + \beta_p X_p + \varepsilon, \tag{9.15}
\end{aligned}
$$

---

[3] Readers not familiar with matrix algebra may benefit from reading the book, *Matrix Algebra As a Tool*, by Hadi (1996).

where $\beta_0 = \bar{y} - (\beta_1 \bar{x}_1 + \ldots + \beta_p \bar{x}_p)$. Although a constant term does not appear in an explicit form in (9.14), it is clearly seen in (9.15). Thus, when we deal with no-intercept models, we need only to scale the data. The scaled variables are defined by:

$$
\begin{aligned}
\tilde{Z}_y &= \frac{Y}{L_y}, \\
\tilde{Z}_j &= \frac{X_j}{L_j}, \ j = 1, \ldots, p,
\end{aligned}
\tag{9.16}
$$

where

$$
L_y = \sqrt{\sum_{i=1}^{n} y_i^2}, \text{ and } L_j = \sqrt{\sum_{i=1}^{n} x_{ij}^2}, \ j = 1, \ldots, p.
\tag{9.17}
$$

The scaled variables in (9.16) have unit lengths but do not necessarily have means zero. Nor do they satisfy (9.10) unless the original variables have zero means.

We should mention here that centering (when appropriate) and/or scaling can be done without loss of generality because the regression coefficients of the original variables can be recovered from the regression coefficients of the transformed variables. For example, if we fit a regression model to centered data, the obtained regression coefficients $\hat{\beta}_1, \ldots, \hat{\beta}_p$ are the same as the estimates obtained from fitting the model to the original data. The estimate of the constant term when using the centered data will always be zero. The estimate of the constant term for an intercept model can be obtained from:

$$
\hat{\beta}_0 = \bar{y} - (\hat{\beta}_1 \bar{x}_1 + \ldots + \hat{\beta}_p \bar{x}_p).
$$

Scaling, however, will change the values of the estimated regression coefficients. For example, the relationship between the estimates, $\hat{\beta}_1, \ldots, \hat{\beta}_p$, obtained from using the original data and the those obtained using the standardized data is given by

$$
\begin{aligned}
\hat{\beta}_j &= (s_y/s_j)\hat{\theta}_j, \qquad j = 1, 2, 3, 4, p, \\
\hat{\beta}_0 &= \bar{y} - \sum_{j=1}^{5} \hat{\beta}_j \bar{x}_j,
\end{aligned}
\tag{9.18}
$$

where $\hat{\beta}_j$ and $\hat{\theta}_j$ are the jth estimated regression coefficients obtained when using the original and standardized data, respectively. Similar formulas can be obtained when using unit length scaling instead of standardizing.

We shall make extensive use of the centered and/or scaled variables in the rest of this chapter and in Chapter 10.

9.6   PRINCIPAL COMPONENTS APPROACH

As we mentioned in the previous section, the principal components approach to the detection of multicollinearity is based on the fact that any set of p

variables can be transformed to a set of p orthogonal variables. The new orthogonal variables are known as the *principal components* (PCs) and are denoted by $C_1, \ldots, C_p$. Each variable C_j is a linear combination of the variables $\tilde{X}_1, \ldots, \tilde{X}_p$ in (9.11). That is,

$$C_j = v_{1j}\tilde{X}_1 + v_{2j}\tilde{X}_2 + \ldots + v_{pj}\tilde{X}_p, \quad j = 1, 2, \ldots, p. \qquad (9.19)$$

The linear combinations are chosen so that the variables $C_1, \ldots, C_p$ are orthogonal.[4] The variance-covariance matrix of the PCs is of the form:

$$
\begin{array}{c}
C_1 \\
C_2 \\
\vdots \\
C_p
\end{array}
\begin{array}{cccc}
C_1 & C_2 & \cdots & C_p \\
\begin{pmatrix}
\lambda_1 & 0 & \cdots & 0 \\
0 & \lambda_2 & \cdots & 0 \\
\vdots & \vdots & \ddots & \vdots \\
0 & 0 & \cdots & \lambda_p
\end{pmatrix}
\end{array}.
$$

All the off-diagonal elements are zero because the PCs are orthogonal. The value on the jth diagonal element, λ_j is the variance of C_j, the jth PC. The PCs are arranged so that $\lambda_1 \geq \lambda_2 \geq \cdots \geq \lambda_p$, that is, the first PC has the largest variance and the last PC has the smallest variance. The λ's are called *eigenvalues* of the correlation matrix of the predictor variables $X_1, \ldots, X_p$. The coefficients involved in the creation of C_j in (9.19) can be neatly arranged in a column like

$$
\begin{pmatrix}
v_{1j} \\
v_{2j} \\
\vdots \\
v_{pj}
\end{pmatrix},
$$

which is known as the *eigenvector* associated with the jth eigenvalue λ_j. If any one of the λ's is exactly equal to zero, there is a perfect linear relationship among the original variables, which is an extreme case of multicollinearity. If one of the λ's is much smaller than the others (and near zero), multicollinearity is present. The number of near zero λ's is equal to the number of different sets of multicollinearity that exist in the data. So, if there is only one near zero λ, there is only one set of multicollinearity; if there are two near zero λ's, there are two sets of different multicollinearity; and so on.

The eigenvalues of the correlation matrix in (9.13) are $\lambda_1 = 1.999$, $\lambda_2 = 0.998$, and $\lambda_3 = 0.003$. The corresponding eigenvectors are

$$
\begin{pmatrix}
0.706 \\
0.044 \\
0.707
\end{pmatrix},
\quad
\begin{pmatrix}
-0.036 \\
0.999 \\
-0.026
\end{pmatrix},
\quad
\begin{pmatrix}
-0.707 \\
-0.007 \\
0.707
\end{pmatrix}.
$$

---

[4] A description of this technique employing matrix algebra is given in the Appendix to this chapter.

Table 9.13   The PCs for the Import Data (1949–1959).

Year	C_1	C_2	C_3
49	−2.1258	0.6394	−0.0204
50	−1.6189	0.5561	−0.0709
51	−1.1153	−0.0726	−0.0216
52	−0.8944	−0.0821	0.0110
53	−0.6449	−1.3064	0.0727
54	−0.1907	−0.6591	0.0266
55	0.3593	−0.7438	0.0427
56	0.9726	1.3537	0.0627
57	1.5600	0.9635	0.0233
58	1.7677	1.0146	−0.0453
59	1.9304	−1.6633	−0.0809

Thus, the PCs for the Import data for the years 1949–1959 are:

$$
\begin{aligned}
C_1 &= 0.706\ \tilde{X}_1 + 0.044\ \tilde{X}_2 + 0.707\ \tilde{X}_3, \\
C_2 &= -0.036\ \tilde{X}_1 + 0.999\ \tilde{X}_2 - 0.026\ \tilde{X}_3, \\
C_3 &= -0.707\ \tilde{X}_1 - 0.007\ \tilde{X}_2 + 0.707\ \tilde{X}_3.
\end{aligned}
\tag{9.20}
$$

These PCs are given in Table 9.13. The variance-covariance matrix of the new variables is

$$
\begin{array}{c}
\begin{array}{ccc} C_1 & C_2 & C_3 \end{array} \\
\begin{array}{c} C_1 \\ C_2 \\ C_3 \end{array}
\left(
\begin{array}{ccc}
1.999 & 0 & 0 \\
0 & 0.998 & 0 \\
0 & 0 & 0.003
\end{array}
\right).
\end{array}
$$

The PCs lack simple interpretation since each is, in a sense, a mixture of the original variables. However, these new variables provide a unified approach for obtaining information about multicollinearity and serve as the basis of one of the alternative estimation techniques described in Chapter 10.

For the Import data, the small value of $\lambda_3 = 0.003$ points to multicollinearity. The other data sets considered in this chapter also have informative eigenvalues. For the EEO data, $\lambda_1 = 2.952$, $\lambda_2 = 0.040$, and $\lambda_3 = 0.008$. For the advertising data, $\lambda_1 = 1.701$, $\lambda_2 = 1.288$, $\lambda_3 = 1.145$, $\lambda_4 = 0.859$, and $\lambda_5 = 0.007$. In each case the presence of a small eigenvalue is indicative of multicollinearity.

A measure of the overall multicollinearity of the variables can be obtained by computing the *condition number* of the correlation matrix. The condition number is defined by

$$
\kappa = \sqrt{\frac{\text{maximum eigenvalue of the correlation matrix}}{\text{minimum eigenvalue of the correlation matrix}}} = \sqrt{\frac{\lambda_1}{\lambda_p}}.
$$

The condition number will always be greater than 1. A large condition number indicates evidence of strong collinearity. The harmful effects of collinearity in the data become strong when the values of the condition number exceeds 15 (which means that λ_1 is more than 225 times λ_p). The condition numbers for the three data sets EEO, Import, and advertising data are 19.20, 25.81, and 15.59, respectively. The cutoff value of 15 is not based on any theoretical considerations, but arises from empirical observation. Corrective action should always be taken when the condition number of the correlation matrix exceeds 30.

Another empirical criterion for the presence of multicollinearity is given by the sum of the reciprocals of the eigenvalues, that is,

$$\sum_{j=1}^{p} \frac{1}{\lambda_j} . \tag{9.21}$$

If this sum is greater than, say, five times the number of predictor variables, multicollinearity is present.

One additional piece of information is available through this type of analysis. Since λ_j is the variance of the jth PC, if λ_j is approximately zero, the corresponding PC, C_j, is approximately equal to a constant. It follows that the equation defining the PC gives some idea about the type of relationship among the predictor variables that is causing multicollinearity. For example, in the Import data, $\lambda_3 = 0.003 \doteq 0$. Therefore, C_3 is approximately constant. The constant is the mean value of C_3 which is zero. The PCs all have means of zero since they are linear functions of the standardized variables and each standardized variable has a zero mean. Therefore

$$C_3 = -0.707\, \tilde{X}_1 - 0.007\, \tilde{X}_2 + 0.707\, \tilde{X}_3 \doteq 0.$$

Rearranging the terms yields

$$\tilde{X}_1 \doteq \tilde{X}_3, \tag{9.22}$$

where the coefficient of $\tilde{X}_2$ (-0.007) has been approximated as zero. Equation (9.22) represents the approximate relationship that exists between the standardized versions of CONSUM and DOPROD. This result is consistent with our previous finding based on the high simple correlation coefficient ($r = 0.997$) between the predictor variables CONSUM and DOPROD. (The reader can confirm this high value of r by examining the scatter plot of CONSUM versus DOPROD.) Since λ_3 is the only small eigenvalue, the analysis of the PCs tells us that the dependence structure among the predictor variables as reflected in the data is no more complex than the simple relationship between CONSUM and DOPROD as given in Equation (9.22).

For the advertising data, the smallest eigenvalue is $\lambda_5 = 0.007$. The corresponding PC is

$$C_5 = 0.514\, \tilde{X}_1 + 0.489\, \tilde{X}_2 - 0.010\, \tilde{X}_3 + 0.428\, \tilde{X}_4 + 0.559\, \tilde{X}_5. \tag{9.23}$$

Setting C_5 to zero and solving for $\tilde{X}_1$ leads to the approximate relationship,

$$\tilde{X}_1 \doteq -0.951\, \tilde{X}_2 - 0.833\, \tilde{X}_4 - 1.087\, \tilde{X}_5, \tag{9.24}$$

where we have taken the coefficient of $\tilde{X}_3$ to be approximately zero. This equation reflects our earlier findings about the relationship between A_t, P_t, A_{t-1}, and P_{t-1}. Furthermore, since $\lambda_4 = 0.859$ and the other λ's are all large, we can be confident that the relationship involving A_t, P_t, A_{t-1}, and P_{t-1} in (9.24) is the only source of multicollinearity in the data.

Throughout this section, investigations concerning the presence of multicollinearity have been based on judging the magnitudes of various indicators, either a correlation coefficient or an eigenvalue. Although we speak in terms of large and small, there is no way to determine these threshold values. The size is relative and is used to give an indication either that everything seems to be in order or that something is amiss. The only reasonable criterion for judging size is to decide whether the ambiguity resulting from the perceived multicollinearity is of material importance in the underlying problem.

We should also caution here that the data analyzed may contain one or few observations that can have an undue influence on the various measures of collinearity (e.g., correlation coefficients, eigenvalues, or the condition number). These observations are called *collinearity-influential observations*. For more details the reader is referred to Hadi (1988).

9.7 IMPOSING CONSTRAINTS

We have noted that multicollinearity is a condition associated with deficient data and not due to misspecification of the model. It is assumed that the form of the model has been carefully structured and that the residuals are acceptable before questions of multicollinearity are considered. Since it is usually not practical and often impossible to improve the data, we shall focus our attention on methods of better interpretation of the given data than would be available from a direct application of least squares. In this section, rather than trying to interpret individual regression coefficients, we shall attempt to identify and estimate informative linear functions of the regression coefficients. Alternative estimating methods for the individual coefficients are treated in Chapter 10.

Before turning to the problem of searching the data for informative linear functions of the regression coefficients, one additional point concerning model specification must be discussed. A subtle step in specifying a relationship that can have a bearing on multicollinearity is acknowledging the presence of theoretical relationships among the regression coefficients. For example in the model for the Import data,

$$\text{IMPORT} = \beta_0 + \beta_1 \cdot \text{DOPROD} + \beta_2 \cdot \text{STOCK} + \beta_3 \cdot \text{CONSUM} + \varepsilon, \quad (9.25)$$

one may argue that the marginal effects of DOPROD and CONSUM are equal. That is, on the basis of economic reasoning, and before looking at the data, it is decided the $\beta_1 = \beta_3$ or equivalently, $\beta_1 - \beta_3 = 0$. As described in Section

Table 9.14   Regression Results of Import Data (1949–1959) with the Constraint $\beta_1 = \beta_3$.

Variable	Coefficient	s.e.	t-test	p-value
Constant	−9.007	1.245	−7.23	< 0.0001
STOCK	0.612	0.109	5.60	0.0005
NEWVAR	0.086	0.004	24.30	< 0.0001
$n = 11$	$R^2 = 0.987$	$R_a^2 = 0.984$	$\hat{\sigma} = 0.5693$	$d.f. = 8$

3.9.3, the model in (9.25) becomes

$$\begin{aligned} \text{IMPORT} &= \beta_0 + \beta_1 \cdot \text{DOPROD} + \beta_2 \cdot \text{STOCK} + \beta_1 \cdot \text{CONSUM} + \varepsilon, \\ &= \beta_0 + \beta_2 \cdot \text{STOCK} + \beta_1 \left(\text{DOPROD} + \text{CONSUM} \right) + \varepsilon. \end{aligned}$$

Thus, the common value of β_1 and β_3 is estimated by regressing IMPORT on STOCK and a new variable constructed as NEWVAR = DOPROD + CONSUM. The new variable has significance only as a technical manipulation to extract an estimate of the common value of β_1 and β_3. The results of the regression appear in Table 9.14. The correlation between the two predictor variables, STOCK and NEWVAR, is 0.0299 and the eigenvalues are $\lambda_1 = 1.030$ and $\lambda_2 = 0.970$. There is no longer any indication of multicollinearity. The residual plots against time and the fitted values indicate that there are no other problems of specification (Figures 9.7 and 9.8, respectively). The estimated model is

$$\begin{aligned} \text{IMPORT} = \ &{-9.007} + 0.086 \cdot \text{DOPROD} + 0.612 \cdot \text{STOCK} \\ &+ 0.086 \cdot \text{CONSUM}. \end{aligned}$$

Note that following the methods outlined in Section 3.9.3, it is also possible to test the constraint, $\beta_1 = \beta_3$, as a hypothesis. Even though the argument for $\beta_1 = \beta_3$ may have been imposed on the basis of existing theory, it is still interesting to evaluate the effect of the constraint on the explanatory power of the full model. The values of R^2 for the full and restricted models are 0.992 and 0.987, respectively. The F-ratio for testing $H_0(\beta_1 = \beta_3)$ is 3.36 with 1 and 8 degrees of freedom. Both results suggest that the constraint is consistent with the data.

The constraint that $\beta_1 = \beta_3$ is, of course, only one example of the many types of constraints that may be used when specifying a regression model. The general class of possibilities is found in the set of linear constraints described in Chapter 3. Constraints are usually justified on the basis of underlying theory. They may often resolve what appears to be a problem of multicollinearity. In addition, any particular constraint may be viewed as a testable hypothesis and judged by the methods described in Chapter 3.

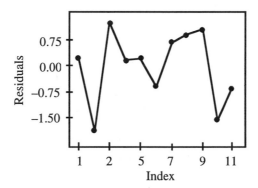

Fig. 9.7 Index plot of the standardized residuals. Import data (1949–1959) with the constraint $\beta_1 = \beta_3$.

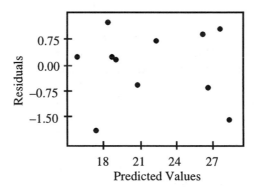

Fig. 9.8 Standardized residuals against fitted values of Import data (1949–1959) with the constraint $\beta_1 = \beta_3$.

9.8   SEARCHING FOR LINEAR FUNCTIONS OF THE β'S

We assume that the model

$$Y = \beta_0 + \beta_1 X_1 + \cdots + \beta_p X_p + \varepsilon$$

has been carefully specified so that the regression coefficients appearing are of primary interest for policy analysis and decision making. We have seen that the presence of multicollinearity may prevent individual β's from being accurately estimated. However, as demonstrated below, it is always possible to estimate some linear functions of the β's accurately (Silvey, 1969). The obvious questions are: Which linear functions can be estimated, and of those that can be estimated, which are of interest in the analysis? In this section we use the data to help identify those linear functions that can be accurately estimated and, at the same time, have some value in the analysis.

First we shall demonstrate in an indirect way that there are always linear functions of the β's that can be accurately estimated.[5] Consider once again the Import data. We have argued that there is a historical relationship between CONSUM and DOPROD that is approximated as CONSUM = (2/3) DOPROD. Replacing CONSUM in the original model,

$$\text{IMPORT} = \beta_0 + (\beta_1 + \frac{2}{3}\beta_3) \cdot \text{DOPROD} + \beta_2 \cdot \text{STOCK} + \varepsilon. \qquad (9.26)$$

Equivalently stated, by dropping CONSUM from the equation we are able to obtain accurate estimates of $\beta_1 + (2/3)\beta_3$ and β_2. Multicollinearity is no longer present. The correlation between DOPROD and STOCK is 0.026. The results are given in Table 9.15. R^2 is almost unchanged and the residual plots (not shown) are satisfactory. In this case we have used information in addition to the data to argue that the coefficient of DOPROD in the regression of IMPORT on DOPROD and STOCK is the linear combination $\beta_1 + (2/3)\beta_3$. Also, we have demonstrated that this linear function can be estimated accurately even though multicollinearity is present in the data. Whether or not it is useful to know the value of $\beta_1 + (2/3)\beta_3$, of course, is another question. At least it is important to know that the estimate of the coefficient of DOPROD in this regression is not measuring the pure marginal effect of DOPROD, but includes part of the effect of CONSUM.

The above example demonstrates in an indirect way that there are always linear functions of the β's that can be accurately estimated. However, there is a constructive approach for identifying the linear combinations of the β's that can be accurately estimated. We shall use the advertising data introduced in Section 9.4 to demonstrate the method. The concepts are less intuitive than those found in the other sections of the chapter. We have attempted to keep

---

[5]Refer to the Appendix to this chapter for further treatment of this problem.

Table 9.15 Regression Results When Fitting Model (9.26) to the Import Data (1949–1959).

Variable	Coefficient	s.e.	t-test	p-value
Constant	−8.440	1.435	−5.88	0.0004
DOPROD	0.145	0.007	20.70	< 0.0001
STOCK	0.622	0.128	4.87	0.0012
$n = 11$	$R^2 = 0.983$	$R_a^2 = 0.978$	$\hat{\sigma} = 0.667$	$d.f. = 8$

things simple. A formal development of this problem is given in the Appendix to this chapter.

We begin with the linear transformation introduced in Section 9.6 that takes the standardized predictor variables into a new orthogonal set of variables. The standardized versions of the five predictor variables are denoted by $\tilde{X}_1, \ldots, \tilde{X}_5$. The standardized response variable, sales, is denoted by $\tilde{Y}$. The transformation that takes $X_1, \ldots, X_5$ into the new set of orthogonal variables $C_1, \ldots, C_5$ is

$$
\begin{aligned}
C_1 &= 0.532\tilde{X}_1 - 0.232\tilde{X}_2 - 0.389\tilde{X}_3 + 0.395\tilde{X}_4 - 0.595\tilde{X}_5, \\
C_2 &= -0.024\tilde{X}_1 + 0.825\tilde{X}_2 - 0.022\tilde{X}_3 - 0.260\tilde{X}_4 - 0.501\tilde{X}_5, \\
C_3 &= -0.668\tilde{X}_1 + 0.158\tilde{X}_2 - 0.217\tilde{X}_3 + 0.692\tilde{X}_4 - 0.057\tilde{X}_5, \\
C_4 &= 0.074\tilde{X}_1 - 0.037\tilde{X}_2 + 0.895\tilde{X}_3 + 0.338\tilde{X}_4 - 0.279\tilde{X}_5, \\
C_5 &= -0.514\tilde{X}_1 - 0.489\tilde{X}_2 + 0.010\tilde{X}_3 - 0.428\tilde{X}_4 - 0.559\tilde{X}_5.
\end{aligned}
\tag{9.27}
$$

The coefficients in the equation defining C_1 are the components of the eigenvector corresponding to the largest eigenvalue of the correlation matrix of the predictor variables. Similarly, the coefficients defining C_2 through C_5 are components of the eigenvectors corresponding to the remaining eigenvalues in order by size. The variables $C_1, \ldots, C_5$ are the PCs associated with the standardized versions of the predictors variables, as described in the preceding Section 9.6.

The regression model stated, as given in (9.4) in terms of the original variables is

$$
S_t = \beta_0 + \beta_1 A_t + \beta_2 P_t + \beta_3 E_t + \beta_4 A_{t-1} + \beta_5 P_{t-1} + \varepsilon_t.
\tag{9.28}
$$

In terms of standardized variables, the equation is written as

$$
\tilde{Y} = \theta_1 \tilde{A}_t + \theta_2 \tilde{P}_t + \theta_3 \tilde{E}_t + \theta_4 \tilde{A}_{t-1} + \theta_5 \tilde{P}_{t-1} + \varepsilon',
\tag{9.29}
$$

where $\tilde{A}_t$ denotes the standardized version of the variable A_t. The regression coefficients in Equation (9.29) are often referred to as the *beta coefficients*. They represent marginal effects of the predictor variables in standard deviation units. For example, θ_1 measures the change in standardized units of sales

(S) corresponding to an increase of one standard deviation unit in advertising (A).

Let $\hat{\beta}_j$ be the least squares estimate of β_j when model (9.28) is fit to the data. Similarly, let $\hat{\theta}_j$ be the least squares estimate of θ_j obtained from fitting model (9.29). Then $\hat{\beta}_j$ and $\hat{\theta}_j$ are related by

$$
\begin{aligned}
\hat{\beta}_j &= (s_y/s_j)\hat{\theta}_j, \qquad j = 1,2,3,4,5, \\
\hat{\beta}_0 &= \bar{y} - \sum_{j=1}^{5} \hat{\beta}_j \bar{x}_j,
\end{aligned}
\tag{9.30}
$$

where $\bar{y}$ is the mean of Y and s_y and s_j are standard deviations of the response and jth predictor variable, respectively.

Equation (9.29) has an equivalent form, given as

$$
\tilde{Y} = \alpha_1 C_1 + \alpha_2 C_2 + \alpha_3 C_3 + \alpha_4 C_4 + \alpha_5 C_5 + \varepsilon'.
\tag{9.31}
$$

The equivalence of Equations (9.29) and (9.31) results from the relationship between the $\tilde{X}$'s and C's in Equations (9.27) and the relationship between the α's and θ's and their estimated values, $\hat{\alpha}$'s and $\hat{\theta}$'s, given as

$$
\begin{aligned}
\hat{\theta}_1 &= 0.532\hat{\alpha}_1 - 0.024\hat{\alpha}_2 - 0.668\hat{\alpha}_3 + 0.074\hat{\alpha}_4 - 0.514\hat{\alpha}_5, \\
\hat{\theta}_2 &= -0.232\hat{\alpha}_1 + 0.825\hat{\alpha}_2 + 0.158\hat{\alpha}_3 - 0.037\hat{\alpha}_4 - 0.489\hat{\alpha}_5, \\
\hat{\theta}_3 &= -0.389\hat{\alpha}_1 - 0.022\hat{\alpha}_2 - 0.217\hat{\alpha}_3 + 0.895\hat{\alpha}_4 + 0.010\hat{\alpha}_5, \\
\hat{\theta}_4 &= 0.395\hat{\alpha}_1 - 0.260\hat{\alpha}_2 + 0.692\hat{\alpha}_3 + 0.338\hat{\alpha}_4 - 0.428\hat{\alpha}_5, \\
\hat{\theta}_5 &= -0.595\hat{\alpha}_1 - 0.501\hat{\alpha}_2 - 0.057\hat{\alpha}_3 - 0.279\hat{\alpha}_4 - 0.559\hat{\alpha}_5.
\end{aligned}
\tag{9.32}
$$

Note that the transformation involves the same weights that are used to define Equation (9.27). The advantage of the transformed model is that the PCs are orthogonal. The precision of the estimated regression coefficients as measured by the variance of the $\hat{\alpha}$'s is easily evaluated. The estimated variance of $\hat{\alpha}_j$ is $\hat{\sigma}^2/\lambda_j$. It is inversely proportional to the ith eigenvalue. All but $\hat{\alpha}_5$ may be accurately estimated since only λ_5 is small. (Recall that $\lambda_1 = 1.701, \lambda_2 = 1.288, \lambda_3 = 1.145, \lambda_4 = 0.859$, and $\lambda_5 = 0.007$.)

Our interest in the $\hat{\alpha}$'s is only as a vehicle for analyzing the $\hat{\theta}$'s. From the representation of Equation (9.32) it is a simple matter to compute and analyze the variances and, in turn, the standard errors of the $\hat{\theta}$'s. The variance of $\hat{\theta}_j$ is

$$
Var(\hat{\theta}_j) = \sum_{i=1}^{p} v_{ij}^2 Var(\hat{\alpha}_i), \quad j = 1,\ldots,p,
\tag{9.33}
$$

where v_{ij} is the coefficient of $\hat{\alpha}_i$ in the j Equation in (9.32). Since the estimated variance of $\hat{\alpha}_i = \hat{\sigma}^2/\lambda_i$, where $\hat{\sigma}^2$ is the residual mean square, (9.33) becomes

$$
Var(\hat{\theta}_j) = \hat{\sigma}^2 \sum_{i=1}^{p} \frac{v_{ij}^2}{\lambda_i}.
\tag{9.34}
$$

For example, the estimated variance of $\hat{\theta}_1$ is

$$\hat{\sigma}^2 \left[\frac{(0.532)^2}{\lambda_1} + \frac{(-0.024)^2}{\lambda_2} + \frac{(-0.668)^2}{\lambda_3} + \frac{(0.074)^2}{\lambda_4} + \frac{(-0.514)^2}{\lambda_5} \right]. \quad (9.35)$$

Recall that $\lambda_1 \geq \lambda_2 \geq \cdots \geq \lambda_5$ and only λ_5 is small, ($\lambda_5 = 0.007$). Therefore, it is only the last term in the expression for the variance that is large and could destroy the precision of $\hat{\theta}_1$. Since expressions for the variances of the other $\hat{\theta}_j$'s are similar to Equation (9.35), a requirement for small variance is equivalent to the requirement that the coefficient of $1/\lambda_5$ be small. Scanning the equations that define the transformation from $\{\hat{\alpha}_i\}$ to $\{\hat{\theta}_j\}$, we see that $\hat{\theta}_3$ is the most precise estimate since the coefficient of $1/\lambda_5$ in the variance expression for $\hat{\theta}_3$ is $(-0.01)^2 = 0.0001$.

Expanding this type of analysis, it may be possible to identify meaningful linear functions of the θ's that can be more accurately estimated than individual θ's. For example, we may be more interested in estimating $\theta_1 - \theta_2$ than θ_1 and θ_2 separately. In the sales model, $\theta_1 - \theta_2$ measures the increment to sales that corresponds to increasing the current year's advertising budget, X_1, by one unit and simultaneously reducing the current year's promotions budget, X_2, by one unit. In other words, $\theta_1 - \theta_2$ represents the effect of a shift in the use of resources in the current year. The estimate of $\theta_1 - \theta_2$ is $\hat{\theta}_1 - \hat{\theta}_2$. The variance of this estimate is obtained simply by subtracting the equation for $\hat{\theta}_2$ and $\hat{\theta}_1$ in (9.32) and using the resulting coefficients of the $\hat{\alpha}$'s as before. That is,

$$\hat{\theta}_1 - \hat{\theta}_2 = 0.764\hat{\alpha}_1 - 0.849\hat{\alpha}_2 - 0.826\hat{\alpha}_3 + 0.111\hat{\alpha}_4 - 0.025\hat{\alpha}_5,$$

from which we obtain the estimated variance of $(\hat{\theta}_1 - \hat{\theta}_2)$ as

$$(0.764)^2 \, Var(\hat{\alpha}_1) + (-0.849)^2 \, Var(\hat{\alpha}_2) + (-0.826)^2 \, Var(\hat{\alpha}_3)$$
$$+ (0.111)^2 \, Var(\hat{\alpha}_4) + (-0.025)^2 \, Var(\hat{\alpha}_1), \quad (9.36)$$

or, equivalently, as

$$\hat{\sigma}^2 \left[\frac{(0.764)^2}{\lambda_1} + \frac{(-0.849)^2}{\lambda_2} + \frac{(-0.826)^2}{\lambda_3} + \frac{(0.111)^2}{\lambda_4} + \frac{(-0.025)^2}{\lambda_5} \right]. \quad (9.37)$$

The small coefficient of $1/\lambda_5$ makes it possible to estimate $\theta_1 - \theta_2$ accurately. Generalizing this procedure we see that any linear function of the θ's that results in a small coefficient for $1/\lambda_5$ in the variance expression can be estimated with precision.

9.9   COMPUTATIONS USING PRINCIPAL COMPONENTS

The computations required for this analysis involve something in addition to a standard least squares computer program. The raw data must be processed

Table 9.16   Regression Results Obtained From Fitting the Model in (9.29).

Variable	Coefficient	s.e.	t-test	p-value
$\tilde{X}_1$	0.583	0.438	1.33	0.2019
$\tilde{X}_2$	0.973	0.417	2.33	0.0329
$\tilde{X}_3$	0.786	0.075	10.50	< 0.0001
$\tilde{X}_4$	0.395	0.367	1.08	0.2973
$\tilde{X}_5$	0.503	0.476	1.06	0.3053
$n = 22$	$R^2 = 0.917$	$R_a^2 = 0.891$	$\hat{\sigma} = 0.3303$	$d.f. = 16$

through a principal components subroutine that operates on the correlation matrix of the predictor variables in order to compute the eigenvalues and the transformation weights found in Equations (9.32). Most regression packages produce the estimated beta coefficients as part of the standard output.

For the advertising data, the estimates $\hat{\theta}_1, \ldots, \hat{\theta}_5$ can be computed in two equivalent ways. They can be obtained directly from a regression of the standardized variables as represented in Equation (9.29). The results of this regression are given in Table 9.16. Alternatively, we can fit the model in (9.31) by the least squares regression of the standardized response variable on the the five PCs and obtain the estimates $\hat{\alpha}_1, \ldots, \hat{\alpha}_5$. The results of this regression are shown in Table 9.17. Then, we use (9.32) to obtain $\hat{\theta}_1, \ldots, \hat{\theta}_5$. For example,

$$\hat{\theta}_1 = (0.532)(-0.346019) + (-0.024)(0.417889) + (-0.668)(-0.151328)$$
$$+ (0.074)(0.659946) + (-0.514)(-1.22026) = 0.5830.$$

Using the coefficients in (9.32), the standard error of $\hat{\theta}_1, \ldots, \hat{\theta}_5$ can be computed. For example the estimated variance of $\hat{\theta}_1$ is

$$(0.532 \times s.e.(\hat{\alpha}_1))^2 + (-0.024 \times s.e.(\hat{\alpha}_2))^2 + (-0.668 \times s.e.(\hat{\alpha}_3))^2$$
$$+ (0.074 \times s.e.(\hat{\alpha}_4))^2 + (-0.514 \times s.e.(\hat{\alpha}_5))^2 = (0.532 \times 0.0529)^2$$
$$+ (-0.024 \times 0.0635)^2 + (-0.668 \times 0.0674)^2 + (0.074 \times 0.0780)^2$$
$$(-0.514 \times 0.8456)^2 = 0.1918,$$

which means that the standard error of $\hat{\theta}_1$ is

$$s.e.(\hat{\theta}_1) = \sqrt{0.1918} = 0.438.$$

It should be noted that the t-values for testing β_j and θ_j equal to zero are identical. The beta coefficient, θ_j is a scaled version of β_j. When constructing t-values as either $\hat{\beta}_j/s.e.(\hat{\beta}_j)$, or $\hat{\theta}_j/s.e.(\hat{\theta}_j)$, the scale factor is canceled.

The estimate of $\theta_1 - \theta_2$ is 0.583 $-0.973 = -0.390$. The variance of $\hat{\theta}_1 - \hat{\theta}_2$ can be computed from Equation (9.36) as 0.008. A 95% confidence interval

Table 9.17 Regression Results Obtained From Fitting the Model in (9.31).

Variable	Coefficient	s.e.	t-test	p-value
C_1	-0.346	0.053	-6.55	< 0.0001
C_2	0.418	0.064	6.58	< 0.0001
C_3	-0.151	0.067	-2.25	0.0391
C_4	0.660	0.078	8.46	< 0.0001
C_5	-1.220	0.846	-1.44	0.1683
$n = 22$	$R^2 = 0.917$	$R_a^2 = 0.891$	$\hat{\sigma} = 0.3303$	$d.f. = 16$

for $\theta_1 - \theta_2$ is $-0.390 \pm 2.12\sqrt{0.008}$ or -0.58 to -0.20. That is, the effect of shifting one unit of expenditure from promotions to advertising in the current year is a loss of between 0.20 and 0.58 standardized sales unit.

There are other linear functions that may also be accurately estimated. Any function that produces a small coefficient for $1/\lambda_5$ in the variance expression is a possibility. For example, Equations (9.31) suggest that all differences involving $\hat{\theta}_1, \hat{\theta}_2, \hat{\theta}_4$, and $\hat{\theta}_5$ can be considered. However, some of the differences are meaningful in the problem, whereas others are not. For example, the difference $(\theta_1 - \theta_2)$ is meaningful, as described previously. It represents a shift in current expenditures from promotions to advertising. The difference $\theta_1 - \theta_4$ is not particularly meaningful. It represents a shift from current advertising expenditure to a previous year's advertising expenditure. A shift of resources backward in time is impossible. Even though $\theta_1 - \theta_4$ could be accurately estimated, it is not of interest in the analysis of sales.

In general, when the weights in Equation (9.32) are displayed and the corresponding values of the eigenvalues are known, it is always possible to scan the weights and identify those linear functions of the original regression coefficients that can be accurately estimated. Of those linear functions that can be accurately estimated, only some will be of interest for the problem being studied.

To summarize, where multicollinearity is indicated and it is not possible to supplement the data, it may still be possible to estimate some regression coefficients and some linear functions accurately. To investigate which coefficients and linear functions can be estimated, we recommend the analysis (transformation to principal components) that has just been described. This method of analysis will not overcome multicollinearity if it is present. There will still be regression coefficients and functions of regression coefficients that cannot be estimated. But the recommended analysis will indicate those functions that are estimable and indicate the structural dependencies that exist among the predictor variables.

Table 9.18   Variables for the Gasoline Consumption Data in Table 9.19.

Variable	Definition
Y	Miles/gallon
X_1	Displacement (cubic inches)
X_2	Horsepower (feet/pound)
X_3	Torque (feet/pound)
X_4	Compression ratio
X_5	Rear axle ratio
X_6	Carburetor (barrels)
X_7	Number of transmission speeds
X_8	Overall length (inches)
X_9	Width (inches)
X_{10}	Weight (pounds)
X_{11}	Type of transmission (1 = automatic; 0 = manual)

9.10   BIBLIOGRAPHIC NOTES

The principal components techniques used in this chapter are derived in most books on multivariate statistical analysis. It should be noted that principal components analysis involves only the predictor variables. The analysis is aimed at characterizing and identifying dependencies (if they exist) among the predictor variables. For a comprehensive discussion of principal components, the reader is referred to Johnson and Wichern (1992) or Seber (1984). Several statistical software packages are now commercially available to carry out the analysis described in this chapter.

EXERCISES

9.1   In the analysis of the Advertising data in Section 9.4 it is suggested that the regression of sales S_t against E_t and three of the remaining four variables $(A_t, P_t, A_{t-1}, S_{t-1})$ may resolve the collinearity problem. Run the four suggested regressions and, for each of them, examine the resulting VIF_j's to see if collinearity has been eliminated.

9.2   Gasoline Consumption: To study the factors that determine the gasoline consumption of cars, data were collected on 30 models of cars. Besides the gasoline consumption (Y), measured in miles per gallon for each car, 11 other measurements representing physical and mechanical characteristics are given. The source of the data in Table 9.19 is *Motor Trend* magazine for the year 1975. Definitions of variables are given in Table 9.18. We wish to determine whether the data set is collinear.

   (a)   Compute the correlation matrix of the predictor variables X_1, ..., X_{11} and the corresponding pairwise scatter plots. Identify any evidence of collinearity.

Table 9.19 Gasoline Consumption and Automotive Variables.

Y	X_1	X_2	X_3	X_4	X_5	X_6	X_7	X_8	X_9	X_{10}	X_{11}
18.9	350.0	165	260	8.00	2.56	4	3	200.3	69.9	3910	1
17.0	350.0	170	275	8.50	2.56	4	3	199.6	72.9	3860	1
20.0	250.0	105	185	8.25	2.73	1	3	196.7	72.2	3510	1
18.3	351.0	143	255	8.00	3.00	2	3	199.9	74.0	3890	1
20.1	225.0	95	170	8.40	2.76	1	3	194.1	71.8	3365	0
11.2	440.0	215	330	8.20	2.88	4	3	184.5	69.0	4215	1
22.1	231.0	110	175	8.00	2.56	2	3	179.3	65.4	3020	1
21.5	262.0	110	200	8.50	2.56	2	3	179.3	65.4	3180	1
34.7	89.7	70	81	8.20	3.90	2	4	155.7	64.0	1905	0
30.4	96.9	75	83	9.00	4.30	2	5	165.2	65.0	2320	0
16.5	350.0	155	250	8.50	3.08	4	3	195.4	74.4	3885	1
36.5	85.3	80	83	8.50	3.89	2	4	160.6	62.2	2009	0
21.5	171.0	109	146	8.20	3.22	2	4	170.4	66.9	2655	0
19.7	258.0	110	195	8.00	3.08	1	3	171.5	77.0	3375	1
20.3	140.0	83	109	8.40	3.40	2	4	168.8	69.4	2700	0
17.8	302.0	129	220	8.00	3.00	2	3	199.9	74.0	3890	1
14.4	500.0	190	360	8.50	2.73	4	3	224.1	79.8	5290	1
14.9	440.0	215	330	8.20	2.71	4	3	231.0	79.7	5185	1
17.8	350.0	155	250	8.50	3.08	4	3	196.7	72.2	3910	1
16.4	318.0	145	255	8.50	2.45	2	3	197.6	71.0	3660	1
23.5	231.0	110	175	8.00	2.56	2	3	179.3	65.4	3050	1
21.5	360.0	180	290	8.40	2.45	2	3	214.2	76.3	4250	1
31.9	96.9	75	83	9.00	4.30	2	5	165.2	61.8	2275	0
13.3	460.0	223	366	8.00	3.00	4	3	228.0	79.8	5430	1
23.9	133.6	96	120	8.40	3.91	2	5	171.5	63.4	2535	0
19.7	318.0	140	255	8.50	2.71	2	3	215.3	76.3	4370	1
13.9	351.0	148	243	8.00	3.25	2	3	215.5	78.5	4540	1
13.3	351.0	148	243	8.00	3.26	2	3	216.1	78.5	4715	1
13.8	360.0	195	295	8.25	3.15	4	3	209.3	77.4	4215	1
16.5	350.0	165	255	8.50	2.73	4	3	185.2	69.0	3660	1

(b) Compute the eigenvalues, eigenvectors, and the condition number of the correlation matrix. Is multicollinearity present in the data?

(c) Identify the variables involved in multicollinearity by examining the eigenvectors corresponding to small eigenvalues.

(d) Regress Y on the 11 predictor variables and compute the VIF for each of the predictors. Which predictors are affected by the presence of collinearity?

9.3 Refer to the Presidential Election Data in Table 5.17 and consider fitting a model relating V to all the variables (including a time trend representing year of election) plus as many interaction terms involving two or three variables as you possibly can.

(a) What is the maximum number of terms (coefficients) in a linear regression model that you can fit to these data? [*Hint*: Consider the number of observations in the data.]

(b) Examine the predictor variables in the above model for the presence of multicollinearity. (Compute the correlation matrix, the condition number, and the VIFs.)

(c) Identify the subsets of variables involved in collinearity. Attempt to solve the multicollinearity problem by deleting some of the variables involved in multicollinearity.

(d) Fit a model relating V to the set of predictors you found to be free from multicollinearity.

Appendix: Principal Components

In this appendix we present the principal components approach to the detection of multicollinearity using matrix notation.

A. The Model

The regression model can be expressed as

$$\mathbf{Y} = \mathbf{Z}\boldsymbol{\theta} + \boldsymbol{\varepsilon}, \tag{A.1}$$

where $\mathbf{Y}$ is an $n \times 1$ vector of observations on the response variable, $\mathbf{Z} = (\mathbf{Z}_1, \ldots, \mathbf{Z}_p)$ is an $n \times p$ matrix of n observations on p predictor variables, $\boldsymbol{\theta}$ is a $p \times 1$ vector of regression coefficients and $\boldsymbol{\varepsilon}$ is an $n \times 1$ vector of random errors. It is assumed that $E(\boldsymbol{\varepsilon}) = \mathbf{0}$, $E(\boldsymbol{\varepsilon}\boldsymbol{\varepsilon}^T) = \sigma^2 \mathbf{I}$, where $\mathbf{I}$ is the identity matrix of order n. It is also assumed, without loss of generality, that $\mathbf{Y}$ and $\mathbf{Z}$ have been centered and scaled so that $\mathbf{Z}^T\mathbf{Z}$ and $\mathbf{Z}^T\mathbf{Y}$ are matrices of correlation coefficients.

There exist square matrices, $\boldsymbol{\Lambda}$ and $\mathbf{V}$ satisfying[6]

$$\mathbf{V}^T(\mathbf{Z}^T\mathbf{Z})\mathbf{V} = \boldsymbol{\Lambda} \quad \text{and} \quad \mathbf{V}^T\mathbf{V} = \mathbf{V}\mathbf{V}^T = \mathbf{I}. \tag{A.2}$$

The matrix $\boldsymbol{\Lambda}$ is diagonal with the ordered eigenvalues of $\mathbf{Z}^T\mathbf{Z}$ on the diagonal. These eigenvalues are denoted by $\lambda_1 \geq \lambda_2 \geq \cdots \geq \lambda_p$. The columns of $\mathbf{V}$ are the normalized eigenvectors corresponding to $\lambda_1, \ldots, \lambda_p$. Since $\mathbf{V}\mathbf{V}^T = \mathbf{I}$, the regression model in (A.1) can be restated in terms of the PCs as

$$\mathbf{Y} = \mathbf{Z}\mathbf{V}\mathbf{V}^T\boldsymbol{\theta} + \boldsymbol{\varepsilon} = \mathbf{C}\boldsymbol{\alpha} + \boldsymbol{\varepsilon}, \tag{A.3}$$

where

$$\mathbf{C} = \mathbf{Z}\mathbf{V}; \quad \text{and} \quad \boldsymbol{\alpha} = \mathbf{V}^T\boldsymbol{\theta}. \tag{A.4}$$

The matrix $\mathbf{C}$ contains p columns $\mathbf{C}_1, \ldots, \mathbf{C}_p$, each of which is a linear functions of the predictor variables $\mathbf{Z}_1, \ldots, \mathbf{Z}_p$. The columns of $\mathbf{C}$ are orthogonal and are referred to as principal components (PCs) of the predictor variables $\mathbf{Z}_1, \ldots, \mathbf{Z}_p$. The columns of $\mathbf{C}$ satisfy $\mathbf{C}_j^T\mathbf{C}_j = \lambda_j$ and $\mathbf{C}_i^T\mathbf{C}_j = 0$ for $i \neq j$.

The PCs and the eigenvalues may be used to detect and analyze collinearity in the predictor variables. The restatement of the regression model given in Equation (A.3) is a reparameterization of Equation (A.1) in terms of orthogonal predictor variables. The λ's may be viewed as sample variances of the PCs. If $\lambda_i = 0$, all observations on the ith PC are also zero. Since the jth PC is a linear function of $\mathbf{Z}_1, \ldots, \mathbf{Z}_p$, when $\lambda_j = 0$ an exact linear dependence exists among the predictor variables. It follows that when λ_j is small (approximately equal to zero) there is an approximate linear relationship among the predictor variables. That is, a small eigenvalue is an indicator of multicollinearity. In addition, from Equation (A.4) we have

$$\mathbf{C}_j = \sum_{i=1}^{p} v_{ij}Z_i,$$

which identifies the exact form of the linear relationship that is causing the multicollinearity.

B. Precision of Linear Functions of $\hat{\boldsymbol{\theta}}$

Denoting $\hat{\boldsymbol{\alpha}}$ and $\hat{\boldsymbol{\theta}}$ as the least squares estimators for $\boldsymbol{\alpha}$ and $\boldsymbol{\theta}$, respectively, it can be shown that $\hat{\boldsymbol{\alpha}} = \mathbf{V}^T\hat{\boldsymbol{\theta}}$, and conversely, $\hat{\boldsymbol{\theta}} = \mathbf{V}\hat{\boldsymbol{\alpha}}$. With $\hat{\boldsymbol{\alpha}} = (\mathbf{C}^T\mathbf{C})^{-1}\mathbf{C}^T\mathbf{Y}$, it follows that the variance-covariance matrix of $\hat{\boldsymbol{\alpha}}$ is $\mathbf{V}(\hat{\boldsymbol{\alpha}}) = \boldsymbol{\Lambda}^{-1}\sigma^2$, and the corresponding matrix for $\hat{\boldsymbol{\theta}}$ is $\mathbf{V}(\hat{\boldsymbol{\theta}}) = \mathbf{V}\boldsymbol{\Lambda}^{-1}\mathbf{V}^T\sigma^2$. Let $\mathbf{L}$ be an arbitrary $p \times 1$ vector of constants. The linear function $\delta = \mathbf{L}^T\boldsymbol{\theta}$ has least squares estimator $\hat{\delta} = \mathbf{L}^T\hat{\boldsymbol{\theta}}$ and variance

$$Var(\hat{\delta}) = \mathbf{L}^T\mathbf{V}\boldsymbol{\Lambda}^{-1}\mathbf{V}^T\mathbf{L}\sigma^2. \tag{A.5}$$

[6]See, for example, Strang (1988) or Hadi (1996).

Let $\mathbf{V}_j$ be the jth column of $\mathbf{V}$. Then $\mathbf{L}$ can be represented as

$$\mathbf{L} = \sum_{j=1}^{p} r_j \mathbf{V}_j$$

for appropriately chosen constants $r_1, \ldots, r_p$. Then (A.5) becomes $Var(\hat{\delta}) = \mathbf{R}^T \boldsymbol{\Lambda}^{-1} \mathbf{R} \sigma^2$ or, equivalently,

$$Var(\hat{\delta}) = \left(\sum_{j=1}^{p} \frac{r_j^2}{\lambda_j} \right) \sigma^2, \tag{A.6}$$

where $\boldsymbol{\Lambda}^{-1}$ is the inverse of $\boldsymbol{\Lambda}$.

To summarize, the variance of $\hat{\delta}$ is a linear combination of the reciprocals of the eigenvalues. It follows that $\hat{\delta}$ will have good precision either if none of the eigenvalues are near zero or if r_j^2 is at most the same magnitude as λ_j when λ_j is small. Furthermore, it is always possible to select a vector, $\mathbf{L}$, and thereby a linear function of $\hat{\boldsymbol{\theta}}$, so that the effect of one or few small eigenvalues is eliminated and $\mathbf{L}^T \hat{\boldsymbol{\theta}}$ has a small variance. Refer to Silvey (1969) for a more complete development of these concepts.

10

Biased Estimation of Regression Coefficients

10.1 INTRODUCTION

It was demonstrated in Chapter 9 that when multicollinearity is present in a set of predictor variables, the ordinary least squares estimates of the individual regression coefficients tend to be unstable and can lead to erroneous inferences. In this chapter, two alternative estimation methods that provide a more informative analysis of the data than the OLS method when multicollinearity is present are considered. The estimators discussed here are biased but tend to have more precision (as measured by mean square error) than the OLS estimators (see Draper and Smith (1998), McCallum (1970), and Hoerl and Kennard (1970)). These alternative methods do not reproduce the estimation data as well as the OLS method; the sum of squared residuals is not as small and, equivalently, the multiple correlation coefficient is not as large. However, the two alternatives have the potential to produce more precision in the estimated coefficients and smaller prediction errors when the predictions are generated using data other than those used for estimation.

Unfortunately, the criteria for deciding when these methods give better results than the OLS method depend on the true but unknown values of the model regression coefficients. That is, there is no completely objective way to decide when OLS should be replaced in favor of one of the alternatives. Nevertheless, when multicollinearity is suspected, the alternative methods of analysis are recommended. The resulting estimated regression coefficients may suggest a new interpretation of the data that, in turn, can lead to a better understanding of the process under study.

The two specific alternatives to OLS that are considered are (1) principal components regression and (2) ridge regression. Principal components analysis was introduced in Chapter 9. It is assumed that the reader is familiar with that material. It will be demonstrated that the principal components estimation method can be interpreted in two ways; one interpretation relates to the nonorthogonality of the predictor variables, the other has to do with constraints on the regression coefficients. Ridge regression also involves constraints on the coefficients. The ridge method is introduced in this chapter and it is applied again in Chapter 11 to the problem of variable selection. Both methods, principal components and ridge regression, are examined using the French import data that were analyzed in Chapter 9.

10.2   PRINCIPAL COMPONENTS REGRESSION

The model under consideration is

$$\text{IMPORT} = \beta_0 + \beta_1 \cdot \text{DOPROD} + \beta_2 \cdot \text{STOCK} + \beta_3 \cdot \text{CONSUM} + \varepsilon. \quad (10.1)$$

The variables are defined in Section 9.3. Let $\bar{y}$ and $\bar{x}_j$ be the means of Y and X_j, respectively. Also, let s_y and s_j be the standard deviations of Y and X_j, respectively. The model of Equation (10.1) stated in terms of standardized variables (see Section 9.5) is

$$\tilde{Y} = \theta_1 \tilde{X}_1 + \theta_2 \tilde{X}_2 + \theta_3 \tilde{X}_3 + \varepsilon', \quad (10.2)$$

where $\tilde{Y} = (y_i - \bar{y})/s_y$ is the standardized version of the response variable (IMPORT) and $\tilde{X}_j = (x_{ij} - \bar{x}_j)/s_j$ is the standardized version of the jth predictor variable. Many regression packages produce values for both the regular and standardized regression coefficients in (10.1) and (10.2), respectively. The estimated coefficients satisfy

$$
\begin{aligned}
\beta_j &= (s_y/s_j)\theta_j, & j = 1, 2, 3, \\
\beta_0 &= \bar{y} - \beta_1 \bar{x}_1 - \beta_2 \bar{x}_2 - \beta_3 \bar{x}_3.
\end{aligned}
\quad (10.3)
$$

The principal components of the standardized predictor variables are (see Equation (9.20))

$$
\begin{aligned}
C_1 &= & 0.706\, \tilde{X}_1 &+& 0.044\, \tilde{X}_2 &+& 0.707\, \tilde{X}_3, \\
C_2 &= & -0.036\, \tilde{X}_1 &+& 0.999\, \tilde{X}_2 &-& 0.026\, \tilde{X}_3, \\
C_3 &= & -0.707\, \tilde{X}_1 &-& 0.007\, \tilde{X}_2 &+& 0.707\, \tilde{X}_3.
\end{aligned}
\quad (10.4)
$$

These principal components were given in Table 9.13. The model in (10.2) may be written in terms of the principal components as

$$\tilde{Y} = \alpha_1 C_1 + \alpha_2 C_2 + \alpha_3 C_3 + \varepsilon'. \quad (10.5)$$

Table 10.1   Regression Results of Fitting Model (10.2) to the Import Data (1949–1959).

Variable	Coefficient	s.e.	t-test	p-value
$\tilde{X}_1$	-0.339	0.464	-0.73	0.4883
$\tilde{X}_2$	0.213	0.034	6.20	0.0004
$\tilde{X}_3$	1.303	0.464	2.81	0.0263
$n = 11$	$R^2 = 0.992$	$R_a^2 = 0.988$	$\hat{\sigma} = 0.034$	$d.f. = 7$

Table 10.2   Regression Results of Fitting Model (10.5) to the Import Data (1949–1959).

Variable	Coefficient	s.e.	t-test	p-value
C_1	0.690	0.024	28.70	< 0.0001
C_2	0.191	0.034	5.62	0.0008
C_3	1.160	0.656	1.77	0.1204
$n = 11$	$R^2 = 0.992$	$R_a^2 = 0.988$	$\hat{\sigma} = 0.034$	$d.f. = 7$

The equivalence of (10.2) and (10.5) follows since there is a unique relationship between the α's and θ's. In particular,

$$
\begin{aligned}
\alpha_1 &= 0.706\theta_1 + 0.044\theta_2 + 0.707\theta_3, \\
\alpha_2 &= -0.036\theta_1 + 0.999\theta_2 - 0.026\theta_3, \\
\alpha_3 &= -0.707\theta_1 - 0.007\theta_2 + 0.707\theta_3.
\end{aligned}
\tag{10.6}
$$

Conversely,

$$
\begin{aligned}
\theta_1 &= 0.706\alpha_1 - 0.036\alpha_2 - 0.707\alpha_3, \\
\theta_2 &= 0.044\alpha_1 + 0.999\alpha_2 - 0.007\alpha_3, \\
\theta_3 &= 0.707\alpha_1 - 0.026\alpha_2 + 0.707\alpha_3.
\end{aligned}
\tag{10.7}
$$

These same relationships hold for the least squares estimates, the $\hat{\alpha}$'s and $\hat{\theta}$'s of the α's and θ's, respectively. Therefore, the $\hat{\alpha}$'s and $\hat{\theta}$'s may be obtained by the regression of $\tilde{Y}$ against the principal components C_1, C_2, and C_3, or against the original standardized variables. The regression results of fitting models (10.2) and (10.5) to the import data are shown in Tables 10.1 and 10.2. From Table 10.1, the estimates of θ_1, θ_2, and θ_3 are -0.339, 0.213, and 1.303, respectively. Similarly, from Table 10.2, the estimates of α_1, α_2, and α_3 are 0.690, 0.191, and 1.160, respectively. The results in one of these tables can be obtained from the other table using (10.6) and (10.7).

Although Equations (10.2) and (10.5) are equivalent, the C's in (10.5) are orthogonal. Observe, however, that the regression relationship given in terms of the principal components (Equation (10.5)) is not easily interpreted. The predictor variables of that model are linear combinations of the original predictor variables. The α's, unlike the θ's, do not have simple interpretations as marginal effects of the original predictor variables. Therefore, we use principal components regression only as a means for analyzing the multicollinearity problem. The final estimation results are always restated in terms of the θ's for interpretation.

10.3   REMOVING DEPENDENCE AMONG THE PREDICTORS

It has been mentioned that the principal components regression has two interpretations. We shall first use the principal components technique to reduce multicollinearity in the estimation data. The reduction is accomplished by using less than the full set of principal components to explain the variation in the response variable. Note that when all three principal components are used, the OLS solution is reproduced exactly by applying Equations (10.7).

The C's have sample variances $\lambda_1 = 1.999, \lambda_2 = 0.998$, and $\lambda_3 = 0.003$, respectively. Recall that the λ's are the eigenvalues of the correlation matrix of DOPROD, STOCK, and CONSUM. Since C_3 has variance equal to 0.003, the linear function defining C_3 is approximately equal to zero and is the source of multicollinearity in the data. We exclude C_3 and consider regressions of $\tilde{Y}$ against C_1 alone as well as against C_1 and C_2. We consider the two possible regression models

$$\tilde{Y} = \alpha_1 C_1 + \varepsilon \tag{10.8}$$

and

$$\tilde{Y} = \alpha_1 C_1 + \alpha_2 C_2 + \varepsilon. \tag{10.9}$$

Both models lead to estimates for all three of the original coefficients, θ_1, θ_2, and θ_3. The estimates are biased since some information (C_3 in Equation (10.9), C_2 and C_3 in Equation (10.8)) has been excluded in both cases.

The estimated values of α_1 or α_1 and α_2 may be obtained by regressing $\tilde{Y}$ in turn against C_1 and then against C_1 and C_2. However, a simpler computational method is available that exploits the orthogonality of C_1 C_2 and C_3.[1] For example, the same estimated value of α_1 will be obtained from regression using (10.5), (10.8), or (10.9). Similarly, the value of α_2 may be obtained from (10.5) or (10.9). It also follows that if we have the OLS estimates of the θ's, estimates of the α's may be obtained from Equations (10.6). Then principal components regression estimates of the θ's corresponding to (10.8) and (10.9) can be computed by referring back to Equations (10.7) and setting the appropriate α's to zero. The following example clarifies the process.

---

[1] In any regression equation where the full set of potential predictor variables under consideration are orthogonal, the estimated values of regression coefficients are not altered when subsets of these variables are either introduced or deleted.

Using $\alpha_1 = 0.690$ and $\alpha_2 = \alpha_3 = 0$ in Equations (10.7) yields estimated θ's corresponding to regression on only the first principal component, that is,

$$\begin{aligned} \hat{\theta}_1 &= 0.706 \times 0.690 = 0.487, \\ \hat{\theta}_2 &= 0.044 \times 0.690 = 0.030, \\ \hat{\theta}_3 &= 0.707 \times 0.690 = 0.487, \end{aligned} \qquad (10.10)$$

which yields

$$\tilde{Y} = 0.487\tilde{X}_1 + 0.030\tilde{X}_2 + 0.487\tilde{X}_3.$$

The estimates using the first two principal components, as in (10.9), are obtained in a similar fashion using $\alpha_1 = 0.690$, $\alpha_2 = 0.191$, and $\alpha_3 = 0$ in (10.7). The estimated of the regression coefficients, β_0, β_1, β_2, and β_3, of the original variables in Equation (10.1), can be obtained by substituting θ_1, θ_2, and θ_3 in (10.3).

The estimates of the standardized and original regression coefficients using the three principal components models are shown in Table 10.3. It is evident that using different numbers of principal components gives substantially different results. It has already been argued that the OLS estimates are unsatisfactory. The negative coefficient of $\tilde{X}_1$ (DOPROD) is unexpected and cannot be sensibly interpreted. Furthermore, there is extensive multicollinearity which enters through the principal component, C_3. This variable has almost zero variance ($\lambda_3 = 0.003$) and is therefore approximately equal to zero. Of the two remaining principal components, it is fairly clear that the first one is associated with the combined effect of DOPROD and CONSUM. The second principal component is uniquely associated with STOCK. This conclusion is apparent in Table 10.3. The coefficients of DOPROD and CONSUM are completely determined from the regression of IMPORT on C_1 alone. These coefficients do not change when C_2 is used. The addition of C_2 causes the coefficient of STOCK to increase from 0.083 to 0.609. Also, R^2 increases from 0.952 to 0.988. Selecting the model based on the first two principal components, the resulting equation stated in original units is

$$\begin{aligned} \text{IMPORT} = \quad & -9.106 + 0.073 \cdot \text{DOPROD} \\ & + 0.609 \cdot \text{STOCK} + 0.106 \cdot \text{CONSUM}. \end{aligned} \qquad (10.11)$$

It provides a different and more plausible representation of the IMPORT relationship than was obtained from the OLS results. In addition, the analysis has led to an explicit quantification (in standardized variables) of the linear dependency in the predictor variables. We have $C_3 = 0$ or equivalently (from Equations (10.4))

$$-0.707\tilde{X}_1 - 0.007\tilde{X}_2 + 0.707\tilde{X}_3 \doteq 0.$$

The standardized values of DOPROD and CONSUM are essentially equal. This information can be useful qualitatively and quantitatively if Equation (10.11) is used for forecasting or for analyzing policy decisions.

Table 10.3 Estimated Regression Coefficients for the Standardized and Original Variables Using Different Numbers of Principal Components for IMPORT Data (1949–1959).

Variable	First PC Equation (10.8)		First and Second PCs Equation (10.9)		All PCs Equation (10.5)	
	Stand.	Original	Stand.	Original	Stand.	Original
Constant	0	−7.735	0	−9.106	0	−10.130
DOPROD	0.487	0.074	0.480	0.073	−0.339	−0.051
STOCK	0.030	0.083	0.221	0.609	0.213	0.587
CONSUM	0.487	0.107	0.483	0.106	1.303	0.287
$\hat{\sigma}$	0.232		0.121		0.108	
R^2	0.952		0.988		0.992	

10.4   CONSTRAINTS ON THE REGRESSION COEFFICIENTS

There is a second interpretation of the results of the principal components regression equation. The interpretation is linked to the notion of imposing constraints on the θ's which was introduced in Chapter 9. The estimates for Equation (10.9) were obtained by setting α_3 equal to zero in Equations (10.7). From (10.6), $\alpha_3 = 0$ implies that

$$- 0.707\theta_1 - 0.007\theta_2 + 0.707\theta_3 = 0 \qquad (10.12)$$

or $\theta_1 \doteq \theta_3$. In original units, Equation (10.12) becomes

$$- 6.60\beta_1 + 4.54\beta_3 = 0 \qquad (10.13)$$

or $\beta_1 = 0.69\beta_3$. Therefore, the estimates obtained by regression on C_1 and C_2 could have been obtained using OLS as in Chapter 9 with a linear constraint on the coefficients given by Equation (10.13).

Recall that in Chapter 9 we conjectured that $\beta_1 = \beta_3$ as a prior constraint on the coefficients. It was argued that the constraint was the result of a qualitative judgment based on knowledge of the process under study. It was imposed without looking at the data. Now, using the data, we have found that principal components regression on C_1 and C_2 gives a result that is equivalent to imposing the constraint of Equation (10.13). The result suggests that the marginal effect of domestic production on imports is about 69% of the marginal effect of domestic consumption on imports.

To summarize, the method of principal components regression provides both alternative estimates of the regression coefficients as well as other useful information about the underlying process that is generating the data. The structure of linear dependence among the predictor variables is made explicit. Principal components with small variances (eigenvalues) exhibit the linear relationships among the original variables that are the source of multicollinearity. Also elimination of multicollinearity by dropping one or more principal components from the regression is equivalent to imposing constraints on the regression coefficients. It provides a constructive way of identifying those constraints that are consistent with the proposed model and the information contained in the data.

10.5   PRINCIPAL COMPONENTS REGRESSION: A CAUTION

We have seen in Chapter 9 that principal components analysis is an effective tool for the detection of multicollinearity. In this chapter we have used the principal components as an alternative to the least squares method to obtain estimates of the regression coefficients in the presence of multicollinearity. The method has worked to our advantage in the Import data, where the first two of the three principal components have succeeded in capturing most of the variability in the response variable (see Table 10.3). This analysis is not guaranteed to work for all data sets. In fact, the principal components regression can fail in accounting for the variability in the response variable. To illustrate this point Hadi and Ling (1998) use a data set known as the Hald's data and a constructed response variable U. The original data can be found in Draper and Smith (1998), p. 348. It can also be found in the book's Web site.[2] The data set has four predictor variables. The response variable U and the four PCs, $C_1, \ldots, C_4$, corresponding to the four predictor variables are given in Table 10.4. The variable U is already in a standardized form. The sample variances of the four PCs are $\lambda_1 = 2.2357$, $\lambda_2 = 1.5761$, $\lambda_3 = 0.1866$, and $\lambda_4 = 0.0016$. The condition number, $\kappa = \sqrt{\lambda_1/\lambda_4} = \sqrt{2.236/0.002} = 37$, is large, indicating the presence of multicollinearity in the original data.

The regression results obtained from fitting the model

$$U = \alpha_1 C_1 + \alpha_2 C_2 + \alpha_3 C_3 + \alpha_4 C_4 + \varepsilon \tag{10.14}$$

to the data are shown in Table 10.5. The coefficient of the last PC, C_4, is highly significant and all other three coefficients are not significant. Now if we drop C_4, the PC with the smallest variance, we obtain the results in Table 10.6. As it is clear from a comparison of Tables 10.5 and 10.6, all four PCs capture almost all the variability in U, while the first three account for none of the variability in U. Therefore, one should be careful before dropping any of the PCs.

Another problem with the principal component regression is that the results can be unduly influenced by the presence of high leverage point and outliers (see Chapter 4 for detailed discussion of outliers and influence). This is because the PCs are computed from the correlation matrix, which itself can be seriously affected by outliers in the data. A scatter plot of the response variable versus each of the PCs and the pairwise scatter plots of the PCs versus each other would point out outliers if they are present in the data. The

---

[2]http://www.ilr.cornell.edu/~hadi/RABE

Table 10.4   A Response Variable U and a Set of Principal Components of Four Predictor Variables.

U	C_1	C_2	C_3	C_4
0.955	1.467	1.903	−0.530	0.039
−0.746	2.136	0.238	−0.290	−0.030
−2.323	−1.130	0.184	−0.010	−0.094
−0.820	0.660	1.577	0.179	−0.033
0.471	−0.359	0.484	−0.740	0.019
−0.299	−0.967	0.170	0.086	−0.012
0.210	−0.931	−2.135	−0.173	0.008
0.558	2.232	−0.692	0.460	0.023
−1.119	0.352	−1.432	−0.032	−0.045
0.496	−1.663	1.828	0.851	0.020
0.781	1.641	−1.295	0.494	0.031
0.918	−1.693	−0.392	−0.020	0.037
0.918	−1.746	−0.438	−0.275	0.037

Table 10.5   Regression Results Using All Four PCs of Hald's Data.

Variable	Coefficient	s.e.	t-test	p-value
C_1	−0.002	0.001	−1.45	0.1842
C_2	−0.002	0.002	−1.77	0.1154
C_3	0.002	0.005	0.49	0.6409
C_4	24.761	0.049	502.00	< 0.0001
$n = 13$	$R^2 = 1.00$	$R_a^2 = 1.00$	$\hat{\sigma} = 0.0069$	$d.f. = 8$

Table 10.6   Regression Results Using the First Three PCs of Hald's Data.

Variable	Coefficient	s.e.	t-test	p-value
C_1	−0.001	0.223	−0.01	0.9957
C_2	−0.000	0.266	−0.00	0.9996
C_3	0.002	0.772	0.00	0.9975
$n = 13$	$R^2 = 0.00$	$R_a^2 = -0.33$	$\hat{\sigma} = 1.155$	$d.f. = 9$

scatter plot of U versus each of the PCs (Figure 10.1) show that there are no outliers in the data and U is related only to C_4, which is consistent with the results in Tables 10.5 and 10.6. The pairwise scatter plots of the PCs versus each other (not shown) also show no outliers in the data. For other possible pitfalls of principal components regression see Hadi and Ling (1998).

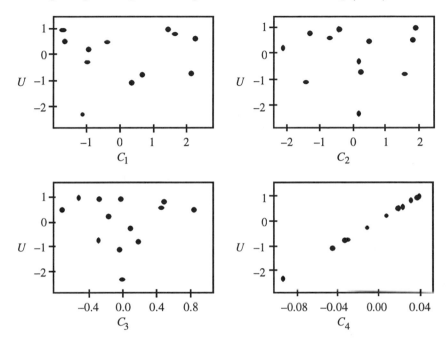

Fig. 10.1 Scatter plots of U versus each of the PCs of the Hald's data.

10.6   RIDGE REGRESSION

Ridge regression[3] provides another alternative estimation method that may be used to advantage when the predictor variables are highly collinear. There are a number of alternative ways to define and compute ridge estimates (see the Appendix to this chapter). We have chosen to present the method associated with the *ridge trace*. It is a graphical approach and may be viewed as an exploratory technique. Ridge analysis using the ridge trace represents a unified approach to problems of detection and estimation when multicollinearity is suspected. The estimators produced are biased but tend to have a smaller mean squared error than OLS estimators (Hoerl and Kennard, 1970).

---

[3] Hoerl (1959) named the method *ridge regression* because of its similarity to ridge analysis used in his earlier work to study second-order response surfaces in many variables.

Ridge estimates of the regression coefficients may be obtained by solving a slightly altered form of the normal equations (introduced in Chapter 3). Assume that the standardized form of the regression model is given as:

$$\tilde{Y} = \theta_1 \tilde{X}_1 + \theta_2 \tilde{X}_2 + \cdots + \theta_p \tilde{X}_p + \varepsilon'. \tag{10.15}$$

The estimating equations for the ridge regression coefficients are

$$
\begin{aligned}
(1+k)\theta_1 &+ r_{12}\,\theta_2 &+ \cdots &+ r_{1p}\,\theta_p &= r_{1y}, \\
r_{21}\,\theta_1 &+ (1+k)\theta_2 &+ \cdots &+ r_{2p}\,\theta_p &= r_{2y}, \\
&\vdots & \vdots \quad \vdots && \vdots \\
r_{p1}\,\theta_1 &+ r_{p2}\,\theta_2 &+ \cdots &+ (1+k)\theta_p &= r_{py},
\end{aligned}
\tag{10.16}
$$

where r_{ij} is the correlation between the ith and jth predictor variables and r_{iy} is the correlation between the ith predictor variable and the response variable $\tilde{Y}$. The solution to (10.16), $\hat{\theta}_1, \ldots, \hat{\theta}_p$, is the set of estimated ridge regression coefficients. The ridge estimates may be viewed as resulting from a set of data that has been slightly altered. See the Appendix to this chapter for a formal treatment.

The essential parameter that distinguishes ridge regression from OLS is k. Note that when $k = 0$, the $\hat{\theta}$'s are the OLS estimates. The parameter k may be referred to as the bias parameter. As k increases from zero, bias of the estimates increases. On the other hand, the *total variance* (the sum of the variances of the estimated regression coefficients), is

$$\text{Total Variance}(k) = \sum_{j=1}^{p} Var(\hat{\theta}_j(k)) = \sigma^2 \sum_{j=1}^{p} \frac{\lambda_j}{(\lambda_j + k)^2}, \tag{10.17}$$

which is a decreasing function of k. The formula in (10.17) shows the effect of the ridge parameter on the total variance of the ridge estimates of the regression coefficients. Substituting $k = 0$ in (10.17), we obtain

$$\text{Total Variance}(0) = \sigma^2 \sum_{j=1}^{p} \frac{1}{\lambda_j}, \tag{10.18}$$

which shows the effect of small eigenvalue on the total variance of the OLS estimates of the regression coefficients.

As k continues to increase without bound, the regression estimates all tend toward zero.[4] The idea of ridge regression is to pick a value of k for which the reduction in total variance is not exceeded by the increase in bias.

It has been shown that there is a positive value of k for which the ridge estimates will be stable with respect to small changes in the estimation data

[4]Because the ridge method tends to shrink the estimates of the regression coefficients toward zero, ridge estimators are sometimes generically referred to as *shrinkage estimators*.

(Hoerl and Kennard, 1970). In practice, a value of k is chosen by computing $\hat{\theta}_1, \ldots, \hat{\theta}_p$ for a range of k values between 0 and 1 and plotting the results against k. The resulting graph is known as the *ridge trace* and is used to select an appropriate value for k. Guidelines for choosing k are given in the following example.

10.7   ESTIMATION BY THE RIDGE METHOD

A method for detecting multicollinearity that comes out of ridge analysis deals with the instability in the estimated coefficients resulting from slight changes in the estimation data. The instability may be observed in the *ridge trace*. The ridge trace is a simultaneous graph of the regression coefficients, $\hat{\theta}_1, \ldots, \hat{\theta}_p$, plotted against k for various values of k such as 0.001, 0.002, and so on. Figure 10.2 is the ridge trace for the IMPORT data. The graph is constructed from Table 10.7, which has the ridge estimated coefficients for 29 values of k ranging from 0 to 1. Typically, the values of k are chosen to be concentrated near the low end of the range. If the estimated coefficients show large fluctuations for small values of k, instability has been demonstrated and multicollinearity is probably at work.

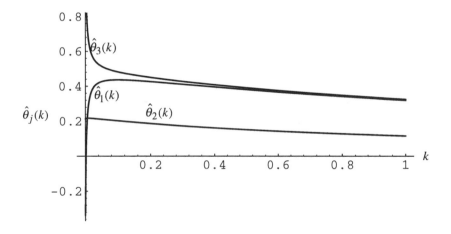

Fig. 10.2   Ridge trace: IMPORT data (1949–1959).

What is evident from the trace or equivalently from Table 10.7 is that the estimated values of the coefficients θ_1 and θ_3 are quite unstable for small values of k. The estimate of θ_1 changes rapidly from an implausible negative value of -0.339 to a stable value of about 0.43. The estimate of θ_3 goes from 1.303 to stabilize at about 0.50. The coefficient of $\tilde{X}_2$ (STOCK), θ_2 is unaffected by the multicollinearity and remains stable throughout at about 0.21.

Table 10.7 Ridge Estimates $\hat{\theta}_j(k)$, as Functions of the Ridge Parameter k, for the IMPORT Data (1949–1959).

k	$\hat{\theta}_1(k)$	$\hat{\theta}_2(k)$	$\hat{\theta}_3(k)$
0.000	−0.339	0.213	1.303
0.001	−0.117	0.215	1.080
0.003	0.092	0.217	0.870
0.005	0.192	0.217	0.768
0.007	0.251	0.217	0.709
0.009	0.290	0.217	0.669
0.010	0.304	0.217	0.654
0.012	0.328	0.217	0.630
0.014	0.345	0.217	0.611
0.016	0.359	0.217	0.597
0.018	0.370	0.216	0.585
0.020	0.379	0.216	0.575
0.022	0.386	0.216	0.567
0.024	0.392	0.215	0.560
0.026	0.398	0.215	0.553
0.028	0.402	0.215	0.548
0.030	0.406	0.214	0.543
0.040	0.420	0.213	0.525
0.050	0.427	0.211	0.513
0.060	0.432	0.209	0.504
0.070	0.434	0.207	0.497
0.080	0.436	0.206	0.491
0.090	0.436	0.204	0.486
0.100	0.436	0.202	0.481
0.200	0.426	0.186	0.450
0.300	0.411	0.173	0.427
0.400	0.396	0.161	0.408
0.500	0.381	0.151	0.391
0.600	0.367	0.142	0.376
0.700	0.354	0.135	0.361
0.800	0.342	0.128	0.348
0.900	0.330	0.121	0.336
1.000	0.319	0.115	0.325

The next step in the ridge analysis is to select a value of k and to obtain the corresponding estimates of the regression coefficients. If multicollinearity is a serious problem, the ridge estimators will vary dramatically as k is slowly increased from zero. As k increases, the coefficients will eventually stabilize. Since k is a bias parameter, it is desirable to select the smallest value of k for which stability occurs since the size of k is directly related to the amount of bias introduced. Several methods have been suggested for the choice of k. These methods include:

1. *Fixed Point.* Hoerl, Kennard, and Baldwin (1975) suggest estimating k by

$$k = \frac{p\hat{\sigma}^2(0)}{\sum\limits_{j=1}^{p} [\hat{\theta}_j(0)]^2} \,, \tag{10.19}$$

   where $\hat{\theta}_1(0), \ldots, \hat{\theta}_p(0)$ are the least squares estimates of $\theta_1, \ldots, \theta_p$ when the model in (10.15) is fitted to the data (i.e., when $k = 0$), and $\hat{\sigma}^2(0)$ is the corresponding residual mean square.

2. *Iterative Method.* Hoerl and Kennard (1976) propose the following iterative procedure for selecting k: Start with the initial estimate of k in (10.19). Denote this value by k_0. Then, calculate

$$k_1 = \frac{p\hat{\sigma}^2(0)}{\sum\limits_{j=1}^{p} [\hat{\theta}_j(k_0)]^2} \,. \tag{10.20}$$

   Then use k_1 to calculate k_2 as

$$k_2 = \frac{p\hat{\sigma}^2(0)}{\sum\limits_{j=1}^{p} [\hat{\theta}_j(k_1)]^2} \,. \tag{10.21}$$

   Repeat this process until the difference between two successive estimates of k is negligible.

3. *Ridge Trace.* The behavior of $\hat{\theta}_j(k)$ as a function of k is easily observed from the ridge trace. The value of k selected is the smallest value for which all the coefficients $\hat{\theta}_j(k)$ are stable. In addition, at the selected value of k, the residual sum of squares should remain close to its minimum value. The variance inflation factors,[5] $\text{VIF}_j(k)$, should also get down to less than 10. (Recall that a value of 1 is a characteristic of an orthogonal system and a value less than 10 would indicate a non-collinear or stable system.)

---

[5]The formula for $\text{VIF}_j(k)$ is given in the Appendix to this chapter.

4. *Other Methods.* Many other methods for estimating k have been suggested in the literature. See, for example, Marquardt (1970), Mallows (1973), Goldstein and Smith (1974), McDonald and Galarneau (1975), Dempster et al. (1977), and Wahba, Golub, and Health (1979). The appeal of the ridge trace, however, lies in its graphical representation of the effects that multicollinearity has on the estimated coefficients.

For the IMPORT data, the fixed point formula in (10.19) gives

$$k = \frac{3 \times 0.0101}{(-0.339)^2 + (0.213)^2 + (1.303)^2} = 0.0164. \tag{10.22}$$

The iterative method gives the following sequence: $k_0 = 0.0164$, $k_1 = 0.0161$, and $k_2 = 0.0161$. So, it converges after two iterations to $k = 0.0161$. The ridge trace in Figure 10.2 (see also Table 10.7) appears to stabilize for k around 0.04. We therefore have three estimates of k (0.0164, 0.0161, and 0.04).

From Table 10.7, we see that at any of these values the improper negative sign on the estimate of θ_1 has disappeared and the coefficient has stabilized (at 0.359 for $k = 0.016$ and at 0.42 for $k = 0.04$). From Table 10.8, we see that the sum of squared residuals $(SSE(k))$ has only increased from 0.081 at $k = 0$ to 0.108 at $k = 0.016$, and to 0.117 at $k = 0.04$. Also, the variance inflation factors, $\text{VIF}_1(k)$ and $\text{VIF}_3(k)$, decreased from about 185 to values between 1 and 4. It is clear that values of k in the interval (0.016 to 0.04) appear to be satisfactory.

The estimated coefficients from the model stated in standardized and original variables units are summarized in Table 10.9. The original coefficient $\hat{\beta}_j$ is obtained from the standardized coefficient $\hat{\theta}_j$ using (10.3). For example, $\hat{\beta}_1$ is calculated by

$$\hat{\beta}_{1j} = (s_y/s_1)\hat{\theta}_1 = (4.5437/29.9995)(0.4196) = 0.0635.$$

Thus, the resulting model in terms for the original variables fitted by ridge method using $k = 0.04$ is

$$\begin{aligned} \text{IMPORT} = \quad &{-8.5537} + 0.0635 \cdot \text{DOPROD} \\ &+ 0.5859 \cdot \text{STOCK} + 0.1156 \cdot \text{CONSUM}. \end{aligned}$$

The equation gives a plausible representation of the relationship. Note that the final equation for these data is not particularly different from the result obtained by using the first two principal components (see Table 10.3), although the two computational methods appear to be very different.

10.8   RIDGE REGRESSION: SOME REMARKS

Ridge regression provides a tool for judging the stability of a given body of data for analysis by least squares. In highly collinear situations, as has

Table 10.8 Residual Sum of Squares, $SSE(k)$, and Variance Inflation Factors, $VIF_j(k)$, as Functions of the Ridge Parameter k, for the IMPORT Data (1949–1959).

k	$SSE(k)$	$VIF_1(k)$	$VIF_2(k)$	$VIF_3(k)$
0.000	0.0810	186.11	1.02	186.00
0.001	0.0837	99.04	1.01	98.98
0.003	0.0911	41.80	1.00	41.78
0.005	0.0964	23.00	0.99	22.99
0.007	0.1001	14.58	0.99	14.57
0.009	0.1027	10.09	0.98	10.09
0.010	0.1038	8.60	0.98	8.60
0.012	0.1056	6.48	0.98	6.48
0.014	0.1070	5.08	0.97	5.08
0.016	0.1082	4.10	0.97	4.10
0.018	0.1093	3.39	0.97	3.39
0.020	0.1102	2.86	0.96	2.86
0.022	0.1111	2.45	0.96	2.45
0.024	0.1118	2.13	0.95	2.13
0.026	0.1126	1.88	0.95	1.88
0.028	0.1132	1.67	0.95	1.67
0.030	0.1139	1.50	0.94	1.50
0.040	0.1170	0.98	0.93	0.98
0.050	0.1201	0.72	0.91	0.72
0.060	0.1234	0.58	0.89	0.58
0.070	0.1271	0.49	0.87	0.49
0.080	0.1310	0.43	0.86	0.43
0.090	0.1353	0.39	0.84	0.39
0.100	0.1400	0.35	0.83	0.35
0.200	0.2052	0.24	0.69	0.24
0.300	0.2981	0.20	0.59	0.20
0.400	0.4112	0.18	0.51	0.18
0.500	0.5385	0.17	0.44	0.17
0.600	0.6756	0.15	0.39	0.15
0.700	0.8191	0.14	0.35	0.14
0.800	0.9667	0.13	0.31	0.13
0.900	1.1163	0.12	0.28	0.12
1.000	1.2666	0.11	0.25	0.11

Table 10.9 OLS and Ridge Estimates of the Regression Coefficients for IMPORT Data (1949–1959).

Variable	OLS ($k = 0$)		Ridge ($k = 0.04$)	
	Standardized Coefficients	Original Coefficients	Standardized Coefficients	Original Coefficients
Constant	0	−10.1300	0	−8.5537
DOPROD	−0.3393	−0.0514	0.4196	0.0635
STOCK	0.2130	0.5869	0.2127	0.5859
CONSUM	1.3027	0.2868	0.5249	0.1156
	$R^2 = 0.992$		$R^2 = 0.988$	

been pointed out, small changes (perturbations) in the data cause very large changes in the estimated regression coefficients. Ridge regression will reveal this condition. Least squares regression should be used with caution in these situations. Ridge regression provides estimates that are more robust than least squares estimates for small perturbations in the data. The method will indicate the sensitivity (or the stability) of the least squares coefficients to small changes in the data.

The ridge estimators are stable in the sense that they are not affected by slight variations in the estimation data. Because of the smaller mean square error property, values of the ridge estimated coefficients are expected to be closer than the OLS estimates to the true values of the regression coefficients. Also, forecasts of the response variable corresponding to values of the predictor variables not included in the estimation set tend to be more accurate.

The estimation of the bias parameter k is rather subjective. There are many methods for estimating k but there is no consensus as to which method is preferable. Regardless of the method of choice for estimating the ridge parameter k, the estimated parameter can be affected by the presence of outliers in the data. Therefore a careful checking for outliers should accompany any method for estimating k to ensure that the obtained estimate is not unduly influenced by outliers in the data.

As with the principal components method, the criteria for deciding when the ridge estimators are superior to the OLS estimators depend on the values of the true regression coefficients in the model. Although these values cannot be known, we still suggest that ridge analysis is useful in cases where extreme multicollinearity is suspected. The ridge coefficients can suggest an alternative interpretation of the data that may lead to a better understanding of the process under study.

Another practical problem with ridge regression is that it has not been implemented in some statistical packages. If a statistical package does not have a routine for ridge regression, ridge regression estimates can be obtained from the standard least squares package by using a slightly altered data set. Specifically, the ridge estimates of the regression coefficients can be obtained from the regression of Y^* on $X_1^*, \ldots, X_p^*$. The new response variable Y^* is obtained by augmenting $\tilde{Y}$ by p new fictitious observations, each of which is equal to zero. Similarly, the new predictor variable X_j^* is obtained by augmenting $\tilde{X}_j$ by p new fictitious observations, each of which is equal to zero except the one in the jth position which is equal to $\sqrt{k}$, where k is the chosen value of the ridge parameter. It can be shown that the ridge estimates $\hat{\theta}_1(k)$, $\ldots, \hat{\theta}_p(k)$ are obtained by the least squares regression of Y^* on $X_1^*, \ldots, X_p^*$ without having a constant term in the model.

10.9   SUMMARY

Both alternative estimation methods, ridge regression and principal components regression, provide additional information about the data being analyzed. We have seen that the eigenvalues of the correlation matrix of predictor variables play an important role in detecting multicollinearity and in analyzing its effects. The regression estimates produced by these methods are biased but may be more accurate than OLS estimates in terms of mean square error. It is impossible to evaluate the gain in accuracy for a specific problem since a comparison of the two methods to OLS requires knowledge of the true values of the coefficients. Nevertheless, when severe multicollinearity is suspected, we recommend that at least one set of estimates in addition to the OLS estimates be calculated. The estimates may suggest an interpretation of the data that were not previously considered.

There is no strong theoretical justification for using principal components or ridge regression methods. We recommend that the methods be used in the presence of severe multicollinearity as a visual diagnostic tool for judging the suitability of the data for least squares analysis. When principal components or ridge regression analysis reveal the instability of a particular data set, the analyst should first consider using least squares regression on a reduced set of variables (as indicated in Chapter 9). If least squares regression is still unsatisfactory (high VIFs, coefficients with wrong signs, large condition number), only then should principal components or ridge regression be used.

EXERCISES

10.1   Longley's (1967) data set is a classic example of multicollinear data. The data (Table 10.10) consist of a response variable S and six predictor variables $X_1, \ldots, X_6$. The data can be found in the book's Web site. The initial model

$$S = \beta_0 + \beta_1 X_1 + \ldots + \beta_6 X_6 + \varepsilon, \qquad (10.23)$$

in terms of the original variables, can be written in terms of the standardized variables as

$$\tilde{S} = \theta_1 \tilde{X}_1 + \ldots + \theta_6 \tilde{X}_6 + \varepsilon'. \qquad (10.24)$$

(a) Fit the model (10.24) to the data using least squares. What conclusion can you draw from the data?

(b) From the results you obtained from the model in (10.24), obtain the least squares estimated regression coefficients in model (10.23).

(c) Now fit the model in (10.23) to the data using least squares and verify that the obtained results are consistent with those obtained above.

Table 10.10   Longley (1967) Data.

Y	X_1	X_2	X_3	X_4	X_5	X_6
60323	830	234289	2356	1590	107608	1947
61122	885	259426	2325	1456	108632	1948
60171	882	258054	3682	1616	109773	1949
61187	895	284599	3351	1650	110929	1950
63221	962	328975	2099	3099	112075	1951
63639	981	346999	1932	3594	113270	1952
64989	990	365385	1870	3547	115094	1953
63761	1000	363112	3578	3350	116219	1954
66019	1012	397469	2904	3048	117388	1955
67857	1046	419180	2822	2857	118734	1956
68169	1084	442769	2936	2798	120445	1957
66513	1108	444546	4681	2637	121950	1958
68655	1126	482704	3813	2552	123366	1959
69564	1142	502601	3931	2514	125368	1960
69331	1157	518173	4806	2572	127852	1961
70551	1169	554894	4007	2827	130081	1962

(d) Compute the correlation matrix of the six predictor variables and the corresponding scatter plot matrix. Do you see any evidence of collinearity?

(e) Compute the corresponding PCs, their sample variances, and the condition number. How many different sets of multicollinearity exist in the data? What are the variables involved in each set?

(f) Based on the number of PCs you choose to retain, obtain the PC estimates of the coefficients in (10.23) and (10.24).

(g) Using the ridge method, construct the ridge trace. What value of k do you recommend to be used in the estimation of the parameters in (10.23) and (10.24)? Use the chosen value of k and compute the ridge estimates of the regression coefficients in (10.23) and (10.24).

(h) Compare the estimates you obtained by the three methods. Which one would you recommend? Explain.

10.2 Repeat Exercise 10.1 using the Hald's data discussed in Section 10.5 but using the original response variable Y and the four predictors X_1, ..., X_4. The data appear in Table 10.11.

10.3 From your analysis of the Longley and Hald data sets, do you observe the sort of problems pointed out in Section 10.5?

Table 10.11   Hald's Data.

Y	X_1	X_2	X_3	X_4
78.5	7	26	6	60
74.3	1	29	15	52
104.3	11	56	8	20
87.6	11	31	8	47
95.9	7	52	6	33
109.2	11	55	9	22
102.7	3	71	17	6
72.5	1	31	22	44
93.1	2	54	18	22
115.9	21	47	4	26
83.8	1	40	23	34
113.3	11	66	9	12
109.4	10	68	8	12

Source: Draper and Smith (1998), p. 348.

Appendix: Ridge Regression

In this appendix we present ridge regression method in matrix notation.

A. The Model

The regression model can be expressed as

$$\mathbf{Y} = \mathbf{Z}\boldsymbol{\theta} + \boldsymbol{\varepsilon}, \tag{A.1}$$

where $\mathbf{Y}$ is an $n \times 1$ vector of observations on the response variable, $\mathbf{Z} = (\mathbf{Z}_1, \ldots, \mathbf{Z}_p)$ is an $n \times p$ matrix of n observations on p predictor variables, $\boldsymbol{\theta}$ is a $p \times 1$ vector of regression coefficients, and $\boldsymbol{\varepsilon}$ is an $n \times 1$ vector of random errors. It is assumed that $E(\boldsymbol{\varepsilon}) = \mathbf{0}$, $E(\boldsymbol{\varepsilon}\boldsymbol{\varepsilon}^T) = \sigma^2 \mathbf{I}$, where $\mathbf{I}$ is the identity matrix of order n. It is also assumed, without loss of generality, that $\mathbf{Y}$ and $\mathbf{Z}$ have been centered and scaled so that $\mathbf{Z}^T\mathbf{Z}$ and $\mathbf{Z}^T\mathbf{Y}$ are matrices of correlation coefficients.[6]

The least squares estimator for $\boldsymbol{\theta}$ is $\hat{\boldsymbol{\theta}} = (\mathbf{Z}^T\mathbf{Z})^{-1}\mathbf{Z}^T\mathbf{Y}$. It can be shown that

$$E[(\hat{\boldsymbol{\theta}} - \boldsymbol{\theta})^T(\hat{\boldsymbol{\theta}} - \boldsymbol{\theta})] = \sigma^2 \sum_{j=1}^{p} \lambda_j^{-1}, \tag{A.2}$$

---

[6]Note that Z_j is obtained by transforming the original predictor variable X_j by $z_{ij} = (x_{ij} - \bar{x}_j)/\sqrt{\sum(x_{ij} - \bar{x}_j)^2}$. Thus, Z_j is centered and scaled to have unit length, that is, $\sum z_{ij}^2 = 1$.

where $\lambda_1 \geq \lambda_2 \geq \cdots \geq \lambda_p$ are the eigenvalues of $\mathbf{Z}^T\mathbf{Z}$. The left-hand side of (A.2) is called the *total mean square error*. It serves as a composite measure of the squared distance of the estimated regression coefficients from their true values.

B. Effect of Multicollinearity

It was argued in Chapter 9 and in the Appendix to Chapter 9 that multicollinearity is synonymous with small eigenvalues. It follows from Equation (A.2) that when one or more of the λ's are small, the total mean square error of $\hat{\theta}$ is large, suggesting imprecision in the least squares estimation method. The ridge regression approach is an attempt to construct an alternative estimator that has a smaller total mean square error value.

C. Ridge Regression Estimators

Hoerl and Kennard (1970) suggest a class of estimators indexed by a parameter $k > 0$. The estimator is (for a given value of k)

$$\hat{\theta}(k) = (\mathbf{Z}^T\mathbf{Z} + k\mathbf{I})^{-1}\mathbf{Z}^T\mathbf{Y} = (\mathbf{Z}^T\mathbf{Z} + k\mathbf{I})^{-1}\mathbf{Z}^T\mathbf{Z}\hat{\theta}. \tag{A.3}$$

The expected value of $\hat{\theta}(k)$ is

$$E[\hat{\theta}(k)] = (\mathbf{Z}^T\mathbf{Z} + k\mathbf{I})^{-1}\mathbf{Z}^T\mathbf{Z}\theta \tag{A.4}$$

and the variance-covariance matrix is

$$Var[\hat{\theta}(k)] = (\mathbf{Z}^T\mathbf{Z} + k\mathbf{I})^{-1}\mathbf{Z}^T\mathbf{Z}(\mathbf{Z}^T\mathbf{Z} + k\mathbf{I})^{-1}\sigma^2. \tag{A.5}$$

The variance inflation factor, $\mathrm{VIF}_j(k)$, as a function of k is the jth diagonal element of the matrix $(\mathbf{Z}^T\mathbf{Z} + k\mathbf{I})^{-1}\mathbf{Z}^T\mathbf{Z}(\mathbf{Z}^T\mathbf{Z} + k\mathbf{I})^{-1}$.

The residual sum of squares can be written as

$$\begin{aligned} SSE(k) &= (\mathbf{Y} - \mathbf{Z}\hat{\theta}(k))^T(\mathbf{Y} - \mathbf{Z}\hat{\theta}(k)) \\ &= (\mathbf{Y} - \mathbf{Z}\hat{\theta})^T(\mathbf{Y} - \mathbf{Z}\hat{\theta}) + (\hat{\theta}(k) - \hat{\theta})^T\mathbf{Z}^T\mathbf{Z}(\hat{\theta}(k) - \hat{\theta}). \end{aligned} \tag{A.6}$$

The total mean square error is

$$\begin{aligned} \mathrm{TMSE}(k) &= E[(\hat{\theta}(k) - \theta)^T(\hat{\theta}(k) - \theta)] \\ &= \sigma^2 \, \mathrm{trace}[(\mathbf{Z}^T\mathbf{Z} + k\mathbf{I})^{-1}\mathbf{Z}^T\mathbf{Z}(\mathbf{Z}^T\mathbf{Z} + k\mathbf{I})^{-1}] \\ &\quad + k^2\theta^T(\mathbf{Z}^T\mathbf{Z} + k\mathbf{I})^{-2}\theta \\ &= \sigma^2 \sum_{j=1}^{p} \lambda_j(\lambda_j + k)^{-2} + k^2\theta^T(\mathbf{Z}^T\mathbf{Z} + k\mathbf{I})^{-2}\theta. \end{aligned} \tag{A.7}$$

Note that the first term on the right-hand side of Equation (A.7) is the sum of the variances of the components of $\hat{\theta}(k)$ (total variance) and the second term is the square of the bias. Hoerl and Kennard (1970) prove that there exists a value of $k > 0$ such that

$$E[(\hat{\boldsymbol{\theta}}(k) - \boldsymbol{\theta})^T(\hat{\boldsymbol{\theta}}(k) - \boldsymbol{\theta})] < E[(\hat{\boldsymbol{\theta}} - \boldsymbol{\theta})^T(\hat{\boldsymbol{\theta}} - \boldsymbol{\theta})],$$

that is, the mean square error of the ridge estimator, $\hat{\boldsymbol{\theta}}(k)$, is less than the mean square error of the OLS estimator, $\hat{\boldsymbol{\theta}}$. Hoerl and Kennard (1970) suggest that an appropriate value of k may be selected by observing the ridge trace and some complementary summary statistics for $\hat{\boldsymbol{\theta}}(k)$ such as $SSE(k)$ and $\text{VIF}_j(k)$. The value of k selected is the smallest value for which $\hat{\boldsymbol{\theta}}(k)$ is stable. In addition, at the selected value of k, the residual sum of squares should remain close to its minimum value, and the variance inflation factors are less than 10, as discussed in Chapter 9.

Ridge estimators have been generalized in several ways. They are sometimes generically referred to as *shrinkage estimators*, because these procedures tend to shrink the estimates of the regression coefficients toward zero. To see one possible generalization, consider the regression model restated in terms of the principal components, $\mathbf{C} = (\mathbf{C}_1, \ldots, \mathbf{C}_p)$, discussed in the Appendix to Chapter 9. The general model takes the form

$$\mathbf{Y} = \mathbf{C}\boldsymbol{\alpha} + \boldsymbol{\varepsilon}, \tag{A.8}$$

where

$$\mathbf{C} = \mathbf{ZV}, \quad \boldsymbol{\alpha} = \mathbf{V}^T\boldsymbol{\theta}, \tag{A.9}$$

$$\mathbf{V}^T\mathbf{Z}^T\mathbf{ZV} = \boldsymbol{\Lambda}, \quad \mathbf{V}^T\mathbf{V} = \mathbf{VV}^T = \mathbf{I},$$

and

$$\boldsymbol{\Lambda} = \begin{pmatrix} \lambda_1 & 0 & 0 & \cdots & 0 & 0 \\ 0 & \lambda_2 & 0 & \cdots & 0 & 0 \\ \vdots & \vdots & \vdots & \ddots & \vdots & \vdots \\ 0 & 0 & 0 & \cdots & \lambda_{p-1} & 0 \\ 0 & 0 & 0 & \cdots & 0 & \lambda_p \end{pmatrix}, \quad \lambda_1 \geq \lambda_2 \geq \cdots \geq \lambda_p,$$

is a diagonal matrix consisting of the ordered eigenvalues of $\mathbf{Z}^T\mathbf{Z}$. The total mean square error in (A.7) becomes

$$\begin{aligned} \text{TMSE}(k) &= E[(\hat{\boldsymbol{\theta}}(k) - \boldsymbol{\theta})^T(\hat{\boldsymbol{\theta}}(k) - \boldsymbol{\theta})] \\ &= \sigma^2 \sum_{j=1}^{p} \frac{\lambda_j}{(\lambda_j + k)^2} + \sum_{j=1}^{p} \frac{k^2\alpha_j^2}{(\lambda_j + k)^2}, \end{aligned} \tag{A.10}$$

where $\alpha^T = (\alpha_1, \alpha_2, \ldots, \alpha_p)$. Instead of taking a single value for k, we can consider several different values k, say $k_1, k_2, \ldots, k_p$. We consider separate ridge parameters (i.e., shrinkage factors) for each of the regression coefficients.

The quantity k, instead of being a scalar, is now a vector and denoted by $\mathbf{k}$. The total mean square error given in (A.10) now becomes

$$
\begin{aligned}
\mathrm{TMSE}(\mathbf{k}) \;=\;\; & E[(\hat{\boldsymbol{\theta}}(\mathbf{k}) - \boldsymbol{\theta})^T (\hat{\boldsymbol{\theta}}(\mathbf{k}) - \boldsymbol{\theta})] \\
=\;\; & \sigma^2 \sum_{j=1}^{p} \frac{\lambda_j}{(\lambda_j + k_j)^2} + \sum_{j=1}^{p} \frac{k_j^2 \alpha_j^2}{(\lambda_j + k_j)^2} \, .
\end{aligned}
\qquad \text{(A.11)}
$$

The total mean square error given in (A.11) is minimized by taking $k_j = \sigma^2/\alpha_j^2$. An iterative estimation procedure is suggested. At Step 1, k_j is computed by using ordinary least squares estimates for σ^2 and α_j. Then a new value of $\hat{\boldsymbol{\alpha}}(\mathbf{k})$ is computed,

$$
\hat{\boldsymbol{\alpha}}(\mathbf{k}) = (\mathbf{C}^T \mathbf{C} + \mathbf{K})^{-1} \mathbf{C}^T \mathbf{Y},
$$

where $\mathbf{K}$ is a diagonal matrix with diagonal elements $k_1, \ldots, k_p$ from Step 1. The process is repeated until successive changes in the components of $\hat{\alpha}(k)$ are negligible. Then, using Equation (A.9), the estimate of $\boldsymbol{\theta}$ is

$$
\hat{\boldsymbol{\theta}}(\mathbf{k}) = \mathbf{V}\hat{\boldsymbol{\alpha}}(\mathbf{k}).
\qquad \text{(A.12)}
$$

The two ridge-type estimators (one value of k, several values of k) defined previously, as well as other related alternatives to ordinary least-squares estimation, are discussed by Dempster et al. (1977). The different estimators are compared and evaluated by Monte Carlo techniques. In general, the choice of the best estimation method for a particular problem depends on the specific model and data. Dempster et al. (1977) hint at an analysis that could be used to identify the best estimation method for a given set of data. At the present time, our preference is for the simplest version of the ridge method, a single ridge parameter k, chosen after an examination of the ridge trace.

11

Variable Selection
Procedures

11.1 INTRODUCTION

In our discussion of regression problems so far we have assumed that the variables that go into the equation were chosen in advance. Our analysis involved examining the equation to see whether the functional specification was correct, and whether the assumptions about the error term were valid. The analysis presupposed that the set of variables to be included in the equation had already been decided. In many applications of regression analysis, however, the set of variables to be included in the regression model is not predetermined, and it is often the first part of the analysis to select these variables. There are some occasions when theoretical or other considerations determine the variables to be included in the equation. In those situations the problem of variable selection does not arise. But in situations where there is no clear-cut theory, the problem of selecting variables for a regression equation becomes an important one.

The problems of variable selection and the functional specification of the equation are linked to each other. The questions to be answered while formulating a regression model are: Which variables should be included, and in what form should they be included; that is, should they enter the equation as an original variable X, or as some transformed variable such as X^2, $\log X$, or a combination of both? Although ideally the two problems should be solved simultaneously, we shall for simplicity propose that they be treated sequentially. We first determine the variables that will be included in the equation, and after that investigate the exact form in which the variables enter it. This approach is a simplification, but it makes the problem of variable selection more tractable. Once the variables that are to be included in the equation have been selected, we can apply the methods described in the earlier chapters to arrive at the actual form of the equation.

11.2   FORMULATION OF THE PROBLEM

We have a response variable Y and q predictor variables $X_1, X_2, \ldots, X_q$. A linear model that represents Y in terms of q variables is

$$y_i = \beta_0 + \sum_{j=1}^{q} \beta_j x_{ij} + \varepsilon_i, \tag{11.1}$$

where β_j are parameters and ε_i represents random disturbances. Instead of dealing with the full set of variables (particularly when q is large), we might delete a number of variables and construct an equation with a subset of variables. This chapter is concerned with determining which variables are to be retained in the equation. Let us denote the set of variables retained by $X_1, X_2, \ldots, X_p$ and those deleted by $X_{p+1}, X_{p+2}, \ldots, X_q$. Let us examine the effect of variable deletion under two general conditions:

1. The model that connects Y to the X's has all β's $(\beta_0, \beta_1, \ldots, \beta_q)$ nonzero.

2. The model has $\beta_0, \beta_1, \ldots, \beta_p$ nonzero, but $\beta_{p+1}, \beta_{p+2}, \ldots, \beta_q$ zero.

Suppose that instead of fitting (11.1) we fit the subset model

$$y_i = \beta_0 + \sum_{j=1}^{p} \beta_j x_{ij} + \varepsilon_i. \tag{11.2}$$

We shall describe the effect of fitting the model to the full and partial set of X's under the two alternative situations described previously. In short, what are the effects of including variables in an equation when they should be properly left out (because the population regression coefficients are zero) and the effect of leaving out variables when they should be included (because the population regression coefficients are not zero)? We will examine the effect of deletion of variables on the estimates of parameters and the predicted values of Y. The solution to the problem of variable selection becomes a little clearer once the effects of retaining unessential variables or the deletion of essential variables in an equation are known.

11.3   CONSEQUENCES OF VARIABLES DELETION

Denote the estimates of the regression parameters by $\hat{\beta}_0^*, \hat{\beta}_1^*, \ldots, \hat{\beta}_q^*$ when the model (11.1) is fitted to the full set of variables $X_1, X_2, \ldots, X_q$. Denote the estimates of the regression parameters by $\hat{\beta}_0, \hat{\beta}_1, \ldots, \hat{\beta}_p$ when the model (11.2) is fitted. Let $\hat{y}_i^*$ and $\hat{y}_i$ be the predicted values from the full and partial set of variables corresponding to an observation $(x_{i1}, x_{i2}, \ldots, x_{iq})$. The results can now be summarized as follows (a summary using matrix notation is given in the Appendix to this chapter): $\hat{\beta}_0, \hat{\beta}_1, \ldots, \hat{\beta}_p$ are biased estimates of $\beta_0, \beta_1, \ldots, \beta_p$ unless the remaining β's in the model $(\beta_{p+1}, \beta_{p+2}, \ldots, \beta_q)$ are zero or the variables $X_1, X_2, \ldots, X_p$ are orthogonal to the variable set $(X_{p+1}, X_{p+2}, \ldots, X_q)$. The estimates $\hat{\beta}_0^*, \hat{\beta}_1^*, \ldots, \hat{\beta}_p^*$ have less precision than $\hat{\beta}_0, \hat{\beta}_1, \ldots, \hat{\beta}_p$; that is,

$$Var(\hat{\beta}_j^*) \geq Var(\hat{\beta}_j), \quad j = 0, 1, \ldots, p.$$

The variance of the estimates of regression coefficients for variables in the reduced equation are not greater than the variances of the corresponding estimates for the full model. Deletion of variables decreases or, more correctly, never increases, the variances of estimates of the retained regression coefficients. Since $\hat{\beta}_j$ are biased and $\hat{\beta}_j^*$ are not, a better comparison of the precision of estimates would be obtained by comparing the mean square errors of $\hat{\beta}_j$ with the variances of $\hat{\beta}_j^*$. The mean squared errors (MSE) of $\hat{\beta}_j$ will be smaller than the variances of $\hat{\beta}_j^*$, only if the deleted variables have regression coefficients smaller in magnitude than the standard deviations of the estimates of the corresponding coefficients. The estimate of σ^2, based on the subset model, is generally biased upward.

Let us now look at the effect of deletion of variables on prediction. The prediction $\hat{y}_i$ is biased unless the deleted variables have zero regression coefficients, or the set of retained variables are orthogonal to the set of deleted variables. The variance of a predicted value from the subset model is smaller than or equal to the variance of the predicted value from the full model; that is,

$$Var(\hat{y}_i) \leq Var(\hat{y}_i^*).$$

The conditions for $MSE(\hat{y}_i)$ to be smaller than $Var(\hat{y}_i^*)$ are identical to the conditions for $MSE(\hat{\beta}_j)$ to be smaller than $Var(\hat{\beta}_j^*)$, which we have already stated. For further details, refer to Chatterjee and Hadi (1988).

The rationale for variable selection can be outlined as follows: Even though the variables deleted have nonzero regression coefficients, the regression coefficients of the retained variables may be estimated with smaller variance from the subset model than from the full model. The same result also holds for the variance of a predicted response. The price paid for deleting variables is in the introduction of bias in the estimates. However, there are conditions (as we have described above), when the MSE of the biased estimates will be smaller than the variance of their unbiased estimates; that is, the gain in precision is not offset by the square of the bias. On the other hand, if some of the retained variables are extraneous or unessential, that is, have zero coefficients or coefficients whose magnitudes are smaller than the standard deviation of the estimates, the inclusion of these variables in the equation leads to a loss of precision in estimation and prediction.

The reader is referred to Sections 3.5, 4.12, and 4.13 for further elaboration on the interpretation of regression coefficients and the role of variables in regression modeling.

11.4   USES OF REGRESSION EQUATIONS

A regression equation has many uses. These are broadly summarized below.

11.4.1   Description and Model Building

A regression equation may be used to describe a given process or as a model for a complex interacting system. The purpose of the equation may be purely descriptive, to clarify the nature of this complex interaction. For this use there are two conflicting requirements: (1) to account for as much of the variation as possible, which points in the direction for inclusion of a large number of variables; and (2) to adhere to the principle of parsimony, which suggests that we try, for ease of understanding and interpretation, to describe the process with as few variables as possible. In situations where description is the prime goal, we try to choose the smallest number of predictor variables that accounts for the most substantial part of the variation in the response variable.

11.4.2   Estimation and Prediction

A regression equation is sometimes constructed for prediction. From the regression equation we want to predict the value of a future observation or estimate the mean response corresponding to a given observation. When a regression equation is used for this purpose, the variables are selected with an eye toward minimizing the MSE of prediction.

11.4.3   Control

A regression equation may be used as a tool for control. The purpose for constructing the equation may be to determine the magnitude by which the value of a predictor variable must be altered to obtain a specified value of the response (target) variable. Here the regression equation is viewed as a response function, with Y as the response variable. For control purposes it is desired that the coefficients of the variables in the equation be measured accurately; that is, the standard errors of the regression coefficients are small.

These are the broad uses of a regression equation. Occasionally, these functions overlap and an equation is constructed for some or all of these purposes. The main point to be noted is that the purpose for which the regression equation is constructed determines the criterion that is to be optimized in its formulation. It follows that a subset of variables that may be best for one purpose may not be best for another. The concept of the "best" subset of variables to be included in an equation always requires additional qualification.

Before discussing actual selection procedures we make two preliminary remarks. First, it is not usually meaningful to speak of the "best set" of variables to be included in a multiple regression equation. There is no unique "best set" of variables. A regression equation can be used for several purposes. The set of variables that may be best for one purpose may not be best for another. The purpose for which a regression equation is constructed should be kept in mind in the variable selection process. We shall show later that the purpose for which an equation is constructed determines the criteria for selecting and evaluating the contributions of different variables.

Second, since there is no best set of variables, there may be several subsets that are adequate and could be used in forming an equation. A good variable selection procedure should point out these several sets rather than generate a so-called single "best" set. The various sets of adequate variables throw light on the structure of data and help us in understanding the underlying process. In fact, the process of variable selection should be viewed as an intensive analysis of the correlational structure of the predictor variables and how they individually and jointly affect the response variable under study. These two points influence the methodology that we present in connection with variable selection.

11.5   CRITERIA FOR EVALUATING EQUATIONS

To judge the adequacy of various fitted equations we need a criterion. Several have been proposed in the statistical literature. We describe the two that we consider most useful. An exhaustive list of criteria is found in Hocking (1976).

11.5.1   Residual Mean Square

One measure that is used to judge the adequacy of a fitted equation is the residual mean square (RMS). With a p-term equation, the RMS is defined as

$$\text{RMS}_p = \frac{\text{SSE}_p}{n - p}. \tag{11.3}$$

where $(\text{SSE})_p$ is the residual sum of squares for a p-term equation. Between two equations, the one with the smaller RMS is usually preferred, especially if the objective is forecasting.

It is clear that RMS_p is related to the square of the multiple correlation coefficient R_p^2 and the square of the adjusted multiple correlation coefficient R_{ap}^2 which have already been described (Chapter 3) as measures for judging the adequacy of fit of an equation. Here we have added a subscript to R^2 and R_a^2 to denote their dependence on the number of terms in an equation. The relationship between these quantities are given by

$$R_p^2 = 1 - (n - p)\frac{\text{RMS}_p}{(\text{SST})} \tag{11.4}$$

and

$$R_{ap}^2 = 1 - (n - 1)\frac{\text{RMS}_p}{(\text{SST})}, \tag{11.5}$$

where

$$\text{SST} = \sum (y_i - \bar{y})^2.$$

Note that R_{ap}^2 is more appropriate than R_p^2 when comparing models with different number of predictors because R_{ap}^2 adjusts (penalizes) for the number of predictor variables in the model.

11.5.2   Mallows C_p

We pointed out earlier that predicted values obtained from a regression equation based on a subset of variables are generally biased. To judge the performance of an equation we should consider the mean square error of the predicted value rather than the variance. The standardized total mean squared error of prediction for the observed data is measured by

$$J_p = \frac{1}{\sigma^2} \sum_{i=1}^{n} \text{MSE}(\hat{y}_i), \tag{11.6}$$

where $\text{MSE}(\hat{y}_i)$ is the mean squared error of the ith predicted value from a p-term equation, and σ^2 is the variance of the random errors. The $\text{MSE}(\hat{y}_i)$ has two components, the variance of prediction arising from estimation, and a bias component arising from the deletion of variables.

To estimate J_p, Mallows (1973) uses the statistic

$$C_p = \frac{\text{SSE}_p}{\hat{\sigma}^2} + (2p - n), \tag{11.7}$$

where $\hat{\sigma}^2$ is an estimate of σ^2 and is usually obtained from the linear model with the full set of q variables. It can be shown that the expected value of C_p is p when there is no bias in the fitted equation containing p terms. Consequently, the deviation of C_p from p can be used as a measure of bias. The C_p statistic therefore measures the performance of the variables in terms of the standardized total mean square error of prediction for the observed data points irrespective of the unknown true model. It takes into account both the bias and the variance. Subsets of variables that produce values of C_p that are close to p are the desirable subsets. The selection of "good" subsets is done graphically. For the various subsets a graph of C_p is plotted against p. The line $C_p = p$ is also drawn on the graph. Sets of variables corresponding to points close to the line $C_p = p$ are the good or desirable subsets of variables to form an equation. The use of C_p plots is illustrated and discussed in more detail in the example that is given in Section 11.10. A very thorough treatment of the C_p statistic is given in Daniel and Wood (1980).

11.6   MULTICOLLINEARITY AND VARIABLE SELECTION

In discussing variable selection procedures we distinguish between two broad situations:

1. The predictor variables are not collinear; that is, there is no strong evidence of multicollinearity.

2. The predictor variables are collinear; that is, the data are highly multicollinear.

Depending on the correlation structure of the predictor variables, we propose different approaches to the variable selection procedure. If the data analyzed are not collinear, we proceed in one manner, and if collinear, we proceed in another.

As a first step in variable selection procedure we recommend calculating the variance inflation factors (VIFs) or the eigenvalues of the correlation matrix of the predictor variables. If none of the VIFs are greater than 10, collinearity is not a problem. Further, as we explained in Chapter 9, the presence of small eigenvalues indicates collinearity. If the condition number[1] is larger than 15, the variables are collinear. We may also look at the sum of the reciprocals of the eigenvalues. If any of the individual eigenvalues are less than 0.01, or the sum of the reciprocals of the eigenvalues is greater than, say, five times the number of predictor variables in the problem, we say that the variables are collinear. If the conditions above do not hold, the variables are regarded as noncollinear.

11.7   EVALUATING ALL POSSIBLE EQUATIONS

The first procedure described is very direct and applies equally well to both collinear and noncollinear data. The procedure involves fitting all possible

---

[1] Recall from Chapter 9 that the condition number is defined by $\kappa = \sqrt{\lambda_{max}/\lambda_{min}}$ where λ_{max} and λ_{min} are the maximum and minimum eigenvalues of the matrix of correlation coefficients.

subset equations to a given body of data. With q variables the total number of equations fitted is 2^q (including an equation that contains all the variables and another that contains no variables). The latter is simply $\hat{y}_i = \bar{y}$, which is obtained from fitting the model $Y = \beta_0 + \varepsilon$. This method clearly gives an analyst the maximum amount of information available concerning the nature of relationships between Y and the set of X's. However, the number of equations and supplementary information that must be looked at may be prohibitively large. Even with only six predictor variables, there are 64 (2^6) equations to consider; with seven variables the number grows to 128 (2^7), neither feasible nor practical. An efficient way of using the results from fitting all possible equations is to pick out the three "best" (on the basis of R^2, C_p, or RMS) equations containing a specified number of variables. This smaller subset of equations is then analyzed to arrive at the final model. When using the method of all subset regressions, the most promising ones are identified using either C_p or RMS. These regressions are then carefully analyzed by examining the residuals for outliers, autocorrelation, or the need for transformations before deciding on the final model. The various subsets that are investigated may suggest interpretations of the data that might have been overlooked in a more restricted variable selection approach.

When the number of variables is large, the evaluation of all possible equations may not be practically feasible. Certain shortcuts have been suggested (Furnival and Wilson, 1974; La Motte and Hocking, 1970) which do not involve computing the entire set of equations while searching for the desirable subsets. But with a large number of variables these methods still involve a considerable amount of computation. There are variable selection procedures that do not require the evaluation of all possible equations. Employing these procedures will not provide the analyst with as much information as the fitting of all possible equations, but it will entail considerably less computation and may be the only available practical solution. These are discussed in Section 11.8. These procedures are quite efficient with noncollinear data. We do not, however, recommend them for collinear data.

11.8   VARIABLE SELECTION PROCEDURES

For cases when there are a large number of potential predictor variables, a set of procedures that does not involve computing of all possible equations has been proposed. These procedures have the feature that the variables are introduced or deleted from the equation one at a time, and involve examining only a subset of all possible equations. With q variables these procedures will involve evaluation of at most $(q + 1)$ equations, as contrasted with the evaluation of 2^q equations necessary for examining all possible equations. The procedures can be classified into two broad categories: (1) the *forward selection* procedure (FS), and (2) the *backward elimination* procedure (BE). There is also a very popular modification of the FS procedure called the *stepwise* method. The three procedures are described and compared below.

11.8.1 Forward Selection Procedure

The forward selection procedure starts with an equation containing no predictor variables, only a constant term. The first variable included in the equation is the one which has the highest simple correlation with the response variable Y. If the regression coefficient of this variable is significantly different from zero it is retained in the equation, and a search for a second variable is made. The variable that enters the equation as the second variable is one which has the highest correlation with Y, after Y has been adjusted for the effect of the first variable, that is, the variable with the highest simple correlation coefficient with the residuals from Step 1. The significance of the regression coefficient of the second variable is then tested. If the regression coefficient is significant, a search for a third variable is made in the same way. The procedure is terminated when the last variable entering the equation has an insignificant regression coefficient or all the variables are included in the equation. The significance of the regression coefficient of the last variable introduced in the equation is judged by the standard t-test computed from the latest equation. Most forward selection algorithms use a low t cutoff value for testing the coefficient of the newly entered variable; consequently, the forward selection procedure goes through the full set of variables and provides us with $q + 1$ possible equations.

11.8.2 Backward Elimination Procedure

The backward elimination procedure starts with the full equation and successively drops one variable at a time. The variables are dropped on the basis of their contribution to the reduction of error sum of squares. The first variable deleted is the one with the smallest contribution to the reduction of error sum of squares. This is equivalent to deleting the variable which has the smallest t-test in the equation. If all the t-tests are significant, the full set of variables is retained in the equation. Assuming that there are one or more variables that have insignificant t-tests, the procedure operates by dropping the variable with the smallest insignificant t-test. The equation with the remaining $(q-1)$ variables is then fitted and the t-tests for the new regression coefficients are examined. The procedure is terminated when all the t-tests are significant or all variables have been deleted. In most backward elimination algorithms the cutoff value for the t-test is set high so that the procedure runs through the whole set of variables, that is, starting with the q-variable equation and ending up with an equation containing only the constant term. The backward elimination procedure involves fitting at most $q + 1$ regression equations

11.8.3 Stepwise Method

The stepwise method is essentially a forward selection procedure but with the added proviso that at each stage the possibility of deleting a variable, as in backward elimination, is considered. In this procedure a variable that entered in the earlier stages of selection may be eliminated at later stages. The calculations made for inclusion and deletion of variables are the same as FS and BE procedures. Often, different levels of significance are assumed for inclusion and exclusion of variables from the equation.

11.9 GENERAL REMARKS ON VARIABLE SELECTION METHODS

The variable selection procedures discussed above should be used with caution. These procedures should not be used mechanically to determine the "best" variables. The order in which the variables enter or leave the equation in variable selection procedures should not be interpreted as reflecting the relative importance of the variables. If these caveats are kept in mind, the variable selection procedures are useful tools for variable selection in noncollinear situations. All three procedures will give nearly the same selection of variables with noncollinear data. They entail much less computing than that in the analysis of all possible equations.

Several stopping rules have been proposed for the variable selection procedures. A stopping rule that has been reported to be quite effective is as follows:

- In FS: Stop if minimum t-test is less than 1.

- In BE: Stop if minimum t-test is greater than 1.

In the following example we illustrate the effect of different stopping rules in variable selection.

We recommend the BE procedure over FS procedure for variable selection. One obvious reason is that in BE procedure the equation with the full variable set is calculated and available for inspection even though it may not be used as the final equation. Although we do not recommend the use of variable selection procedures in a collinear situation, the BE procedure is better able to handle multicollinearity than the FS procedure (Mantel, 1970).

In an application of variable selection procedures several equations are generated, each equation containing a different number of variables. The various equations generated can then be evaluated using a statistic such as C_p or RMS. The residuals for the various equations should also be examined. Equations with unsatisfactory residual plots are rejected. Only a total and comprehensive analysis will provide an adequate selection of variables and a useful regression equation. This approach to variable selection is illustrated by the following example.

Table 11.1   Correlation Matrix for the Supervisor Performance Data in Table 3.3.

	X_1	X_2	X_3	X_4	X_5	X_6
X_1	1.000					
X_2	0.558	1.000				
X_3	0.597	0.493	1.000			
X_4	0.669	0.445	0.640	1.000		
X_5	0.188	0.147	0.116	0.377	1.000	
X_6	0.225	0.343	0.532	0.574	0.283	1.000

11.10   A STUDY OF SUPERVISOR PERFORMANCE

To illustrate variable selection procedures in a noncollinear situation, consider the Supervisor Performance data discussed in Section 3.3. A regression equation was needed to study the qualities that led to the characterization of good supervisors by the people being supervised. The equation is to be constructed in an attempt to understand the supervising process and the relative importance of the different variables. In terms of the use for the regression equation, this would imply that we want accurate estimates of the regression coefficients, in contrast to an equation that is to be used only for prediction. The variables in the problem are given in Table 3.2. The data are shown in Table 3.3 and can also be obtained from the book's Web site.[2]

The VIFs resulting from regressing Y on $X_1, X_2, \ldots, X_6$ are

$$\text{VIF}_1 = 2.7, \quad \text{VIF}_2 = 1.6, \quad \text{VIF}_3 = 2.3,$$

$$\text{VIF}_4 = 3.1, \quad \text{VIF}_5 = 1.2, \quad \text{VIF}_6 = 2.0.$$

The range of the VIFs (1.2 to 3.1) shows that collinearity is not a problem for these data. The same picture emerges if we examine the eigenvalues of the correlation matrix of the data (Table 11.1). The eigenvalues of the correlation matrix are:

$$\lambda_1 = 3.169, \quad \lambda_2 = 1.006, \quad \lambda_3 = 0.763,$$

$$\lambda_4 = 0.553, \quad \lambda_5 = 0.317, \quad \lambda_6 = 0.192.$$

The sum of the reciprocals of the eigenvalues is 12.8. Since none of the eigenvalues are small (the condition number is 4.1) and the sum of the reciprocals of the eigenvalues is only about twice the number of variables, we conclude that the data in the present example are not seriously collinear and we can apply the variable selection procedures just described.

[2]http://www.ilr.cornell.edu/~hadi/RABE

Table 11.2   Variables Selected by the Forward Selection Method.

Variables in Equation	min($|t|$)	RMS	C_p	p	Rank
X_1	7.74	6.993	1.41	2	1
$X_1 X_3$	1.57	6.817	1.11	3	1
$X_1 X_3 X_6$	1.29	6.734	1.60	4	1
$X_1 X_3 X_6 X_2$	0.59	6.820	3.28	5	1
$X_1 X_3 X_6 X_2 X_4$	0.47	6.928	5.07	6	1
$X_1 X_3 X_6 X_2 X_4 X_5$	0.26	7.068	7.00	7	—

The result of forward selection procedure is given in Table 11.2. For successive equations we show the variables present, the RMS, and the value of the C_p statistic. The last column shows the rank of the subset obtained by FS relative to best subset (on the basis of RMS) of same size. The value of p is the number of predictor variables in the equation, including a constant term. Two stopping rules are used:

1. Stop if minimum absolute t-test is less than $t_{0.05}(n - p)$.

2. Stop if minimum absolute t-test is less than 1.

The first rule is more stringent and terminates with variables X_1 and X_3. The second rule is less stringent and terminates with variables X_1, X_3, and X_6.

The results of applying the BE procedure are presented in Table 11.3. They are identical in structure to Table 11.2. For the BE we will use the stopping rules:

1. Stop if minimum absolute t-test is greater than $t_{0.05}(n - p)$.

2. Stop if minimum absolute t-test is greater than 1.

With the first stopping rule the variables selected are X_1 and X_3. With the second stopping rule the variables selected are X_1, X_3, and X_6. The FS and BE give identical equations for this problem, but this is not always the case (an example is given in Section 11.12). An application of the variable selection procedure yielded the same results. To describe the supervisor performance, the equation

$$Y = 13.58 + 0.62X_1 + 0.31X_3 - 0.19X_6$$

is chosen. The residual plots (not shown) for this equation are satisfactory. Since the present problem has only six variables, the total number of equations that can be fitted which contain at least one variable is 63. The C_p values for all 63 equations are shown in Table 11.4. The C_p values are plotted against p in Figure 11.1. The best subsets of variables based on C_p values are given in Table 11.5.

It is seen that the subsets selected by C_p are different from those arrived at by the variable selection procedures as well as those selected on the basis of

Table 11.3 Variables Selected by Backward Elimination Method.

Variables in Equation	min($|t|$)	RMS	C_p	p	Rank
$X_1X_2X_3X_4X_5X_6$	0.26	7.068	7.00	7	–
$X_1X_2X_3X_4X_6$	0.47	6.928	5.07	6	1
$X_1X_2X_3X_6$	0.59	6.820	3.28	5	1
$X_1X_3X_6$	1.29	6.734	1.60	4	1
X_1X_3	1.57	6.817	1.11	3	1
X_1	7.74	6.993	1.41	2	1

Table 11.4 Values of C_p Statistic (All Possible Equations).

Variables	C_p	Variables	C_p	Variables	C_p	Variables	C_p
1	1.41	1 5	3.41	1 6	3.33	1 5 6	5.32
2	44.40	2 5	45.62	2 6	46.39	2 5 6	47.91
1 2	3.26	1 2 5	5.26	1 2 6	5.22	1 2 5 6	7.22
3	26.56	3 5	27.94	3 6	24.82	3 5 6	25.02
1 3	1.11	1 3 5	3.11	1 3 6	1.60	1 3 5 6	3.46
2 3	26.96	2 3 5	28.53	2 3 6	24.62	2 3 5 6	25.11
1 2 3	2.51	1 2 3 5	4.51	1 2 3 6	3.28	1 2 3 5 6	5.14
4	30.06	4 5	31.62	4 6	27.73	4 5	29.50
1 4	3.19	1 4 5	5.16	1 4 6	4.70	1 4 5 6	6.69
2 4	29.20	2 4 5	30.82	2 4 6	25.91	2 4 5 6	27.74
1 2 4	4.99	1 2 4 5	6.97	1 2 4 6	6.63	1 2 4 5 6	8.61
3 4	23.25	3 4 5	25.23	3 4 6	16.50	3 4 5 6	18.42
1 3 4	3.09	1 3 4 5	5.09	1 3 4 6	3.35	1 3 4 5 6	5.29
2 3 4	24.56	2 3 4 5	26.53	2 3 4 6	17.57	2 3 4 5 6	19.51
1 2 3 4	4.49	1 2 3 4 5	6.48	1 2 3 4 6	5.07	1 2 3 4 5 6	19.51
5	57.91	6	57.95	5 6	58.76		

Table 11.5 Variables Selected on the Basis of C_p Statistic.

Variables in Equation	min($|t|$)	RMS	C_p	p	Rank
X_1	7.74	6.993	1.41	2	1
X_1X_4	0.47	7.093	3.19	3	2
$X_1X_4X_6$	0.69	7.163	4.70	4	5
$X_1X_3X_4X_5$	0.07	7.080	5.09	5	6
$X_1X_2X_3X_4X_5$	0.11	7.139	6.48	6	4
$X_1X_2X_3X_4X_5X_6$	0.26	7.068	7.00	7	–

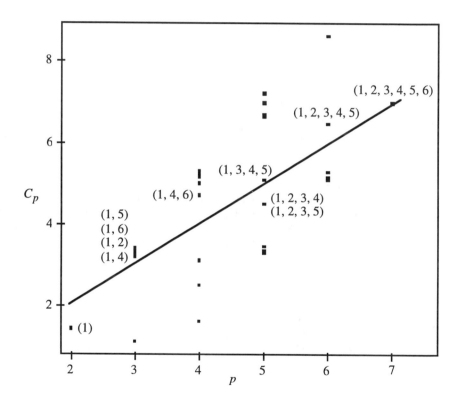

Fig. 11.1 Supervisor's Performance Data: Scatter plot of C_p versus p for subsets with $C_p < 10$.

residual mean square. This anomaly suggests an important point concerning the C_p statistic that the reader should bear in mind. For applications of the C_p statistic, an estimate of σ^2 is required. Usually, the estimate of σ^2 is obtained from the residual sum of squares from the full model. If the full model has a large number of variables with no explanatory power (i.e., population regression coefficients are zero), the estimate of σ^2 from the residual sum of squares for the full model would be large. The loss in degrees of freedom for the divisor would not be balanced by a reduction in the error sum of squares. If $\hat{\sigma}^2$ is large, then the value of C_p is small. For C_p to work properly, a good estimate of σ^2 must be available. When a good estimate of σ^2 is not available, C_p is of only limited usefulness. In our present example, the RMS for the full model with six variables is larger than the RMS for the model with three variables X_1, X_3, X_6. Consequently, the C_p values are distorted and not very useful in variable selection in the present case. The type of situation we have described can be spotted by looking at the RMS for different values of p. RMS will at first tend to decrease with p, but increase at later stages. This behavior indicates that the latter variables are not contributing significantly to the reduction of error sum of squares. Useful application of C_p requires a parallel monitoring of RMS to avoid distortions.

11.11 VARIABLE SELECTION WITH COLLINEAR DATA

In Chapter 9 it was pointed out that serious distortions are introduced in standard analysis with collinear data. Consequently, we recommend a different set of procedures for selecting variables in these situations. Collinearity is indicated when the correlation matrix has one or more small eigenvalues. With a small number of collinear variables we can evaluate all possible equations and select an equation by methods that have already been described. But with a larger number of variables this method is not feasible.

Two different approaches to the problem have been proposed. The first approach tries to break down the collinearity of the data by deleting variables. The collinear structure present in the variables is revealed by the eigenvectors corresponding to the very small eigenvalues (see Chapters 9 and 10). Once the collinearities are identified, a set of variables can then be deleted to produce a reduced noncollinear data set. We can then apply the methods described earlier. The second approach uses ridge regression as the main tool. We assume that the reader is familiar with the basic terms and concepts of ridge regression (Chapter 10). The first approach (by judicious dropping of correlated variables) is the one that is almost always used in practice

Table 11.6   Homicide Data: Description of Variables.

Variable	Symbol	Description
1	FTP	Number of full-time police per 100,000 population
2	UEMP	Percent of the population unemployed
3	M	Number of manufacturing workers (in thousands)
4	LIC	Number of handgun licenses issued per 100,000 population
5	GR	No. of handgun registration issued per 100,000 population
6	CLEAR	Percent of homicides cleared by arrest
7	W	Number of white males in the population
8	NMAN	Number of nonmanufacturing workers (in thousands)
9	G	Number of government workers (in thousands)
10	HE	Average hourly earnings
11	WE	Average weekly earnings
12	H	Number of homicides per 100,000 population

11.12   THE HOMICIDE DATA

In a study investigating the role of firearms in accounting for the rising homicide rate in Detroit, data were collected for the years 1961–1973. The data are reported in Gunst and Mason (1980), p. 360. The response variable (the homicide rate) and the predictor variables believed to influence or be related to the rise in the homicide rate are defined in Table 11.6 and given in Tables 11.7 and 11.8. The data can also be found in the book's Web site.

We use these data to illustrate the danger of mechanical variable selection procedures, such as the FS and BE, in collinear situations. We are interested in fitting the model

$$H = \beta_0 + \beta_1\, G + \beta_2\, M + \beta_3\, W + \varepsilon.$$

In terms of the centered and scaled version of the variables, the model becomes

$$\tilde{H} = \theta_1\, \tilde{G} + \theta_2\, \tilde{M} + \theta_3\, \tilde{W} + \varepsilon'. \tag{11.8}$$

The OLS results are shown in Table 11.9. Can the number of predictor variables in this model be reduced? If the standard assumptions hold, the small t-test for the variable G (0.68) would indicate that the corresponding regression coefficient is insignificant and G can be omitted from the model. Let us now apply the forward selection and the backward elimination procedures to see which variables are selected. The regression output that we need to implement the two methods on the standardized versions of the variables are summarized in Table 11.10. In this table we give the estimated coefficients, their t-tests, and the adjusted squared multiple correlation coefficient, R_a^2 for each model for comparison purposes.

Table 11.7   First Part of the Homicides Data.

Year	FTP	UNEMP	M	LIC	GR	CLEAR
1961	260.35	11.0	455.5	178.15	215.98	93.4
1962	269.80	7.0	480.2	156.41	180.48	88.5
1963	272.04	5.2	506.1	198.02	209.57	94.4
1964	272.96	4.3	535.8	222.10	231.67	92.0
1965	272.51	3.5	576.0	301.92	297.65	91.0
1966	261.34	3.2	601.7	391.22	367.62	87.4
1967	268.89	4.1	577.3	665.56	616.54	88.3
1968	295.99	3.9	596.9	1131.21	1029.75	86.1
1969	319.87	3.6	613.5	837.80	786.23	79.0
1970	341.43	7.1	569.3	794.90	713.77	73.9
1971	356.59	8.4	548.8	817.74	750.43	63.4
1972	376.69	7.7	563.4	583.17	1027.38	62.5
1973	390.19	6.3	609.3	709.59	666.50	58.9

Source: Gunst and Mason (1980), p. 360.

Table 11.8   Second Part of the Homicide Data.

Year	W	NMAN	G	HE	WE	H
1961	558724	538.1	133.9	2.98	117.18	8.60
1962	538584	547.6	137.6	3.09	134.02	8.90
1963	519171	562.8	143.6	3.23	141.68	8.52
1964	500457	591.0	150.3	3.33	147.98	8.89
1965	482418	626.1	164.3	3.46	159.85	13.07
1966	465029	659.8	179.5	3.60	157.19	14.57
1967	448267	686.2	187.5	3.73	155.29	21.36
1968	432109	699.6	195.4	2.91	131.75	28.03
1969	416533	729.9	210.3	4.25	178.74	31.49
1970	401518	757.8	223.8	4.47	178.30	37.39
1971	398046	755.3	227.7	5.04	209.54	46.26
1972	373095	787.0	230.9	5.47	240.05	47.24
1973	359647	819.8	230.2	5.76	258.05	52.33

Source: Gunst and Mason (1980), p. 360.

Table 11.9   Homicide Data: The OLS Results From Fitting Model (11.8).

Variable	Coefficient	s.e.	t-test	VIF
G	0.235	0.345	0.68	42
M	−0.405	0.090	−4.47	3
W	−1.025	0.378	−2.71	51
$n = 13$	$R^2 = 0.975$	$R_a^2 = 0.966$	$\hat{\sigma} = 0.0531$	$d.f. = 9$

Table 11.10 Homicide Data: The Estimated Coefficients, Their t-tests, and the Adjusted Squared Multiple Correlation Coefficient, R_a^2.

Variable	Model (a)	Model (b)	Model (c)	Model (d)	Model (e)	Model (f)	Model (g)
G: Coeff.	0.96			1.15	0.87	0.24	
t-test	11.10			11.90	1.62	0.68	
M: Coeff.		0.55		−0.27		−0.40	−0.43
t-test		2.16		−2.79		−4.47	−5.35
W: Coeff.			−0.95		−0.09	−1.02	−1.28
t-test			−9.77		−0.17	−2.71	−15.90
R_a^2	0.91	0.24	0.89	0.95	0.90	0.97	0.97

The first variable to be selected by the FS is G because it has the largest t-test among the three models that contain a single variable (Models (a) to (c) in Table 11.10). Between the two candidates for the two-variable models (Models (d) and (e)), Model (d) is better than Model (e). Therefore, the second variable to enter the equation is M. The third variable to enter the equation is W (Model (f)) because it has a significant t-test. Note, however, the dramatic change of the significance of G in Models (a), (d), and (f). It was highly significant coefficient in Models (a) and (d), but became insignificant in Model (f). Collinearity is a suspect!

The BE methods starts with the three-variable Model (f). The first variable to leave is G (because it has the lowest t-test), which leads to Model (g). Both M and W in Model (g) have significant t-tests and the BE procedure terminates.

Observe that the first variable eliminated by the BE (G) is the same as the first variable selected by the FS. That is, the variable G, which was selected by the FS as the most important of the three variables, was regarded by the BE as the least important! Among other things, the reason for this anomalous result is collinearity. The eigenvalues of the correlation matrix, $\lambda_1 = 2.65$, $\lambda_2 = 0.343$, and $\lambda_3 = 0.011$, give a large condition number ($\kappa = 15.6$). Two of the three variables (G and W) have large VIF (42 and 51). The sum of the reciprocals of the eigenvalues is also very large (96). In addition to collinearity, since the observations were taken over time (for the years 1961–1973), we are dealing with time series data here. Consequently, the error terms can be autocorrelated (see Chapter 8). Examining the pairwise scatter plots of the data will reveal other problems with the data.

This example shows clearly that automatic applications of variable selection procedure in multicollinear data can lead to the selection of a wrong model. In Sections 11.13 and 11.14 we make use of ridge regression for the process of variable selection in multicollinear situations.

11.13 VARIABLE SELECTION USING RIDGE REGRESSION

One of the goals of ridge regression is to produce a regression equation with stable coefficients. The coefficients are stable in the sense that they are not affected by slight variations in the estimation data. The objectives of a good variable selection procedure are (1) to select a set of variables that provides a clear understanding of the process under study, and (2) to formulate an equation that provides accurate forecasts of the response variable corresponding to values of the predictor variables not included in the study. It is seen that the objectives of a good variable selection procedure and ridge regression are very similar and, consequently, one (ridge regression) can be employed to accomplish the other (variable selection).

The variable selection is done by examining the ridge trace, a plot of the ridge regression coefficients against the ridge parameter k. For a collinear system, the characteristic pattern of ridge trace has been described in Chapter 10. The ridge trace is used to eliminate variables from the equation. The guidelines for elimination are:

1. Eliminate variables whose coefficients are stable but small. Since ridge regression is applied to standardized data, the magnitude of the various coefficients are directly comparable.

2. Eliminate variables with unstable coefficients that do not hold their predicting power, that is, unstable coefficients that tend to zero.

3. Eliminate one or more variables with unstable coefficients. The variables remaining from the original set, say p in number, are used to form the regression equation.

At the end of each of the above steps, we refit the model that includes the remaining variables before we proceed to the next step.

The subset of variables remaining after elimination should be examined to see if collinearity is no longer present in the subset. We illustrate this procedure by an example.

11.14 SELECTION OF VARIABLES IN AN AIR POLLUTION STUDY

McDonald and Schwing (1973) present a study that relates total mortality to climate, socioeconomic, and pollution variables. Fifteen predictor variables selected for the study are listed in Table 11.11. The response variable is the total age-adjusted mortality from all causes. We will not comment on the epidemiological aspects of the study, but merely use the data as an illustrative example for variable selection. A very detailed discussion of the problem is presented by McDonald and Schwing in their paper and we refer the interested reader to it for more information.

Table 11.11 Description of Variables, Means, and Standard Deviations, SD ($n = 60$).

Variable	Description	Mean	SD
X_1	Mean annual precipitation (inches)	37.37	9.98
X_2	Mean January temperature (degrees Fahrenheit)	33.98	10.17
X_3	Mean July temperature (degrees Fahrenheit)	74.58	4.76
X_4	Percent of population over 65 years of age	8.80	1.46
X_5	Population per household	3.26	0.14
X_6	Median school years completed	10.97	0.85
X_7	Percent of housing units that are sound	80.92	5.15
X_8	Population per square mile	3876.05	1454.10
X_9	Percent of nonwhite population	11.87	8.92
X_{10}	Percent employment in white-collar jobs	46.08	4.61
X_{11}	Percent of families with income under \$3000	14.37	4.16
X_{12}	Relative pollution potential of hydrocarbons	37.85	91.98
X_{13}	Relative pollution potential of oxides of nitrogen	22.65	46.33
X_{14}	Relative pollution potential of sulfur dioxide	53.77	63.39
X_{15}	Percent relative humidity	57.67	5.37
Y	Total age-adjusted mortality from all causes.	940.36	62.21

The original data are not available to us, but the correlation matrix of the response and the 15 predictor variables is given in Table 11.12. It is not a good practice to perform the analysis based only on the correlation matrix because without the original data we will not be able to perform diagnostics checking which is necessary in any thorough data analysis. To start the analysis we shall assume that the standard assumptions of the linear regression model hold. As can be expected from the nature of the variables, some of them are highly correlated with each other. The evidence of collinearity is clearly seen if we examine the eigenvalues of the correlation matrix. The eigenvalues are

$$\lambda_1 = 4.5272, \quad \lambda_6 = 0.9605, \quad \lambda_{11} = 0.1665,$$

$$\lambda_2 = 2.7547, \quad \lambda_7 = 0.6124, \quad \lambda_{12} = 0.1275,$$

$$\lambda_3 = 2.0545, \quad \lambda_8 = 0.4729, \quad \lambda_{13} = 0.1142,$$

$$\lambda_4 = 1.3487, \quad \lambda_9 = 0.3708, \quad \lambda_{14} = 0.0460,$$

$$\lambda_5 = 1.2227, \quad \lambda_{10} = 0.2163, \quad \lambda_{15} = 0.0049.$$

There are two very small eigenvalues; the largest eigenvalue is nearly 1000 times larger than the smallest eigenvalue. The sum of the reciprocals of the eigenvalues is 263, which is nearly 17 times the number of variables. The data show strong evidence of collinearity.

Table 11.12 Correlation Matrix for the Variables in Table 11.11.

	X_1	X_2	X_3	X_4	X_5	X_6	X_7	X_8
X_1	1.0000	.0922	.5033	.1011	.2634	−.4904	−.4903	−.0035
X_2		1.0000	.3463	−.3981	−.2092	.1163	.0139	−.1001
X_3			1.0000	−.4340	.2623	−.2385	−.4155	−.0610
X_4				1.0000	−.5091	−.1389	.0649	.1620
X_5					1.0000	−.3951	−.4095	−.1843
X_6						1.0000	.5515	−.2439
X_7							1.0000	.1806
X_8								1.0000

	X_9	X_{10}	X_{11}	X_{12}	X_{13}	X_{14}	X_{15}	Y
X_1	.4132	−.2973	.5066	−.5318	−.4873	−.1069	−.0773	.5095
X_2	.4538	.2380	.5653	.3508	.3210	−.1078	.0679	−.0300
X_3	.5753	−.0214	.6193	−.3565	−.3377	−.0993	−.4528	.2770
X_4	−.6378	−.1177	−.3098	−.0205	−.0021	.0172	.1124	−.1746
X_5	.4194	−.4257	.2599	−.3882	−.3584	−.0041	−.1357	.3573
X_6	−.2088	.7032	−.4033	.2868	.2244	−.2343	.1765	−.5110
X_7	−.4091	.3376	−.6806	.3859	.3476	.1180	.1224	−.4248
X_8	−.0057	−.0318	−.1629	.1203	.1653	.4321	−.1250	.2655
X_9	1.0000	−.0044	.7049	−.0259	.0184	.1593	−.1180	.6437
X_{10}		1.0000	−.1852	.2037	.1600	−.0685	.0607	−.2848
X_{11}			1.0000	−.1298	−.1025	−.0965	−.1522	.4105
X_{12}				1.0000	.9838	.2823	−.0202	−.1772
X_{13}					1.0000	.4094	−.0459	−.0774
X_{14}						1.0000	−.1026	.4259
X_{15}							1.0000	−.0885
Y								1.0000

Table 11.13   OLS Regression Output for the Air Pollution Data (Fifteen Predictor Variables).

Variable	Coefficient	s.e.	t-test	VIF
X_1	0.306	0.148	2.063	4.11
X_2	−0.318	0.181	−1.755	6.13
X_3	−0.237	0.146	−1.627	3.97
X_4	−0.213	0.200	−1.064	7.46
X_5	−0.232	0.152	−1.527	4.31
X_6	−0.233	0.161	−1.448	4.85
X_7	−0.052	0.146	−0.356	3.97
X_8	0.084	0.094	0.890	1.66
X_9	0.640	0.190	3.359	6.78
X_{10}	−0.014	0.123	−0.112	2.84
X_{11}	−0.010	0.216	−0.042	8.72
X_{12}	−0.979	0.724	−1.353	97.92
X_{13}	0.983	0.747	1.316	104.22
X_{14}	0.090	0.150	0.599	4.21
X_{15}	0.009	0.101	0.093	1.91
$n = 60$	$R^2 = 0.764$	$R_a^2 = 0.648$	$\hat{\sigma} = 0.073$	$d.f. = 44$

The initial OLS results from fitting a linear model to the centered and scaled data are given in Table 11.13. Although the model has a high R^2, some of the estimated coefficients have small t-tests. In the presence of multicollinearity, a small t-test does not necessarily mean that the corresponding variable is not important. The small t-test might be due of variance inflation because of the presence of multicollinearity. As can be seen in Table 11.13, VIF_{12} and VIF_{13} are very large.

The ridge trace for the 15 regression coefficients are shown in Figures 11.2 to 11.4. Each Figure shows five curves. If we put all 15 curves, the graph would be quite cluttered and the curves would be difficult to trace. To make the three graphs comparable, the scale is kept the same for all graphs. From the ridge trace, we see that some of the coefficients are quite unstable and some are small regardless of the value of the ridge parameter k.

We now follow the guidelines suggested for the selection of variables in multicollinear data. Following the first criterion we eliminate variables 7, 8, 10, 11, and 15. These variables all have fairly stable coefficients, as shown by the flatness of their ridge traces, but are very small. Although variable 14 has a small coefficient at $k = 0$ (see Table 11.13), its value increases sharply as k increases from zero. So, it should not be eliminated at this point.

We now repeat the analysis using the ten remaining variables: 1, 2, 3, 4, 5, 6, 9, 12, 13, and 14. The corresponding OLS results are given in Table 11.14. There is still an evidence of multicollinearity. The largest eigenvalue,

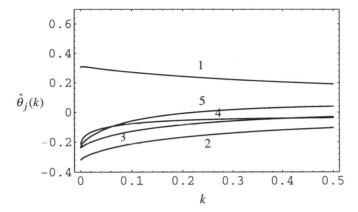

Fig. 11.2 Air Pollution Data: Ridge traces for $\hat{\theta}_1, \ldots, \hat{\theta}_5$ (the 15-variable-model).

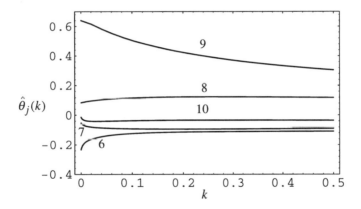

Fig. 11.3 Air Pollution Data: Ridge traces for $\hat{\theta}_6, \ldots, \hat{\theta}_{10}$ (the 15-variable-model).

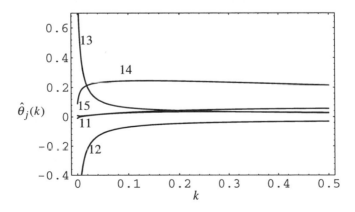

Fig. 11.4 Air Pollution Data: Ridge traces for $\hat{\theta}_{11}, \ldots, \hat{\theta}_{15}$ (the 15-variable-model).

Table 11.14   OLS Regression Output for the Air Pollution Data (Ten Predictor Variables).

Variable	Coefficient	s.e.	t-test	VIF
X_1	0.306	0.135	2.260	3.75
X_2	−0.345	0.119	−2.907	2.88
X_3	−0.244	0.108	−2.256	2.39
X_4	−0.222	0.175	−1.274	6.22
X_5	−0.268	0.137	−1.959	3.81
X_6	−0.292	0.103	−2.842	2.15
X_9	0.664	0.140	4.748	3.99
X_{12}	−1.001	0.658	−1.522	88.30
X_{13}	1.001	0.673	1.488	92.40
X_{14}	0.098	0.127	0.775	3.29
$n = 60$	$R^2 = 0.760$	$R_a^2 = 0.711$	$\hat{\sigma} = 0.070$	$d.f. = 49$

$\lambda_1 = 3.377$, is about 600 times the smallest value $\lambda_{10} = 0.005$. The two VIFs for variable 12 and 13 are still high. The corresponding ridge traces are shown in Figures 11.5 and 11.6. Variable 14 continues to have a small coefficient at $k = 0$ but it increases as k increases from zero. So, it should be kept in the model at this stage. None of the other nine variables satisfy the first criterion.

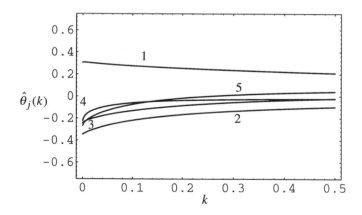

Fig. 11.5   Air Pollution Data: Ridge traces for $\hat{\theta}_1, \ldots, \hat{\theta}_5$ (the ten-variable-model).

The second criterion suggests eliminating variables with unstable coefficients that tend to zero. Examination of the ridge traces in Figures 11.5 and 11.6 shows that variables 12 and 13 fall in this category.

The OLS results for the remaining 8 variables are shown in Table 11.15. Collinearity has disappeared. Now, the largest and smallest eigenvalues are 2.886 and 0.094, which give a small condition number ($\kappa = 5.5$). The sum

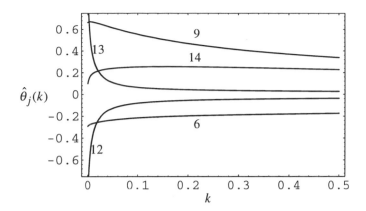

Fig. 11.6 Air Pollution Data: Ridge traces for $\hat{\theta}_6$, $\hat{\theta}_9$, $\hat{\theta}_{12}$, $\hat{\theta}_{13}$ and $\hat{\theta}_{14}$ (the ten-variable-model).

Table 11.15 OLS Regression Output for the Air Pollution Data (Eight Predictor Variables).

Variable	Coefficient	s.e.	t-test	VIF
X_1	0.331	0.120	2.765	2.911
X_2	−0.351	0.106	−3.313	2.279
X_3	−0.217	0.104	−2.087	2.191
X_4	−0.155	0.163	−0.946	5.419
X_5	−0.221	0.134	−1.656	3.621
X_6	−0.270	0.102	−2.654	2.097
X_9	0.692	0.133	5.219	3.567
X_{14}	0.230	0.083	2.767	1.405
$n = 60$	$R^2 = 0.749$	$R_a^2 = 0.709$	$\hat{\sigma} = 0.070$	$d.f. = 51$

of the reciprocals of the eigenvalues is 23.5, about twice the number of variables. All values of VIF are less than 10. Since the retained variables are not collinear, we can now apply the variables selection methods for non-collinear data discussed in Sections 11.7 and 11.8. This is left as an exercise for the reader.

An alternative way of analyzing these Air Pollution data is as follows: The collinearity in the original 15 variables is actually a simple case of multi-collinearity; it involves only two variables (12 and 13). So, the analysis can proceed by eliminating any one of the two variables. The reader can verify that the remaining 14 variables are not collinear. The standard variables selection procedures for non-collinear data can now be utilized. We leave this as an exercise for the reader.

In our analysis of the Air Pollution data, we did not use the third criterion, but there are situations where this criterion is needed. We should note that ridge regression was used successfully in this example as a tool for variable selection. Because the variables selected at an intermediate stage were found to be non-collinear, the standard OLS was utilized.

An analysis of these data not using ridge regression has been given by Henderson and Velleman (1981). They present a thorough analysis of the data and the reader is referred to their paper for details.

We hope it is clear from our discussion that variable selection is a mixture of art and science, and should be performed with care and caution. We have outlined a set of approaches and guidelines rather than prescribing a formal procedure. In conclusion, we must emphasize the point made earlier that variable selection should not be performed mechanically as an end in itself but rather as an exploration into the structure of the data analyzed, and as in all true explorations, the explorer is guided by theory, intuition, and common sense.

11.15   A POSSIBLE STRATEGY FOR FITTING REGRESSION MODELS

In the concluding section of the chapter we outline a possible sequence of steps that may be used to fit a regression model satisfactorily. Let us emphasize at the beginning that there is no single correct approach. The reader may be more comfortable with a different sequence of steps and should feel free to follow such a sequence. In almost all cases the analysis described here will lead to meaningful interpretable models useful in real-life applications.

We assume that we have a response variable Y which we want to relate to some or all of a set of variables $X_1, X_2, \ldots, X_p$. The set, $X_1, X_2, \ldots, X_p$, is often generated from external subject matter considerations. The set of variables is often large and we want to come to an acceptable reduced set. Our objective is to construct a valid and viable regression model. A possible sequence of steps are:

1. Examine the variables $(Y, X_1, X_2, \ldots, X_p)$ one at a time. This can be done by calculating the summary statistics, and also graphically by looking at histograms, dot plots, or box plots (see Chapter 4). The distributions of the values should not be too skewed, nor the range of the variables very large. Look for outliers (check for transcription errors). Make transformations to induce symmetry and reduce skewness. Logarithmic transformations are useful in this situation (see Chapter 6).

2. Construct pairwise scatter plots for each variable. When p, the number of predictor variables, is large, this may not be feasible. Pairwise scatter plots are quite informative on the relationship between two variables. A

look at the correlation matrix will point out obvious collinearity problems. Delete redundant variables. Calculate the condition number of the correlation matrix to get an idea of the severity of the collinearity (Chapters 9 and 10).

3. Fit the full linear regression model. Delete variables with no significant explanatory power (insignificant t-tests). For the reduced model, examine the residuals:

   (a) Check linearity. If none, make a transformation on the variable (Chapter 6).

   (b) Check for heteroscedasticity and autocorrelation (for time series data). If present, take appropriate action (Chapters 7 and 8).

   (c) Look for outliers, high leverage points, and influential points. If present, take appropriate action (Chapter 4).

4. Examine if additional variables can be dropped without compromising the integrity of the model. Examine if new variables are to be brought into the model (added variable plots, residual plus component plots) (Chapters 4 and 11). Repeat Step 3.

5. For the final fitted model, check variance inflation factors. Ensure satisfactory residual plots and no negative diagnostic messages (Chapters 3, 5, 6, and 9). If need be, repeat Step 4.

6. Attempt should then be made to validate the fitted model. When the amount of data is large, the model may be fitted by part of the data and validated by the remainder of the data. Resampling methods such as bootstrap, jackknife, and cross-validation are also possibilities, particularly when the amount of data available is not large [see Efron (1982) and Diaconis and Efron (1983)].

The steps we have described are, in practice, often not done sequentially but implemented synchronously. The process described is an iterative process and it may be necessary to recycle through the outlined steps several times to arrive at a satisfactory model. They enumerate the factors that must be considered for constructing a satisfactory model.

One important component that we have not included in our outlined steps is the subject matter knowledge of the analyst in the area in which the model is constructed. This knowledge should always be incorporated in the model-building process. Incorporation of this knowledge will often accelerate the process of arriving at a satisfactory model because it will help considerably in the appropriate choice of variables and corresponding transformations. After all is said and done, statistical model building is an art. The techniques that we have described are the tools by which this task can be attempted methodically.

11.16   BIBLIOGRAPHIC NOTES

There is a vast amount of literature on variable selection scattered in statistical journals. A very comprehensive review with an extensive bibliography may be found in Hocking (1976). A detailed treatment on variable selection with special emphasis on C_p statistic is given in the book by Daniel and Wood (1980). Refinements on the application of C_p statistic are given by Mallows (1973). The variable selection procedures are discussed in the book by Draper and Smith (1998). Use of ridge regression in connection with variable selection is discussed by Hoerl and Kennard (1970) and by McDonald and Schwing (1973).

EXERCISES

11.1   As we have seen in Section 11.14, the three noncollinear subsets of predictor variables below have emerged. Apply one or more variable selection methods to each subset and compare the resulting final models:

(a)   The subset of eight variables: 1, 2, 3, 4, 5, 6, 9, and 14.

(b)   The subset of 14 variables obtained after omitting variable 12.

(c)   The subset of 14 variables obtained after omitting variable 13.

11.2   The estimated regression coefficients in Table 11.13 correspond to the standardized versions of the variables because they are computed using the correlation matrix of the response and predictor variables. Using the means and standard deviations of the variables in Table 11.11, write the estimated regresion equation in terms of the original variables (before centering and scaling).

11.3   In the Homicide data discussed in Section 11.12, we observed that when fitting the model in (11.8), the FS and BE methods give contradictory results. In fact, there are several other subsets in the data (not necessarily with three predictor variables) for which the FS and BE methods give contradictory results. Find one or more of these subsets.

11.4   Use the variable selection methods, as appropriate, to find one or more subsets of the predictor variables in Tables 11.7 and 11.8 that best account for the variability in the response variable H.

11.5   Property Valuation: Scientific mass appraisal is a technique in which linear regression methods applied to the problem of property valuation. The objective in scientific mass appraisal is to predict the sale price of a home from selected physical characteristics of the building and taxes (local, school, county) paid on the building. Twenty-four observations were obtained from *Multiple Listing* (Vol. 87) for Erie, PA, which is designated as Area 12 in the directory. These data (Table 11.17) were originally presented by Narula and Wellington (1977). The list of variables are given in Table 11.16.

Table 11.16   List of Variables for Data in Table 11.17.

Variable	Definition
Y	Sale price of the house in thousands of dollars
X_1	Taxes (local, county, school) in thousands of dollars
X_2	Number of bathrooms
X_3	Lot size (in thousands of square feet)
X_4	Living space (in thousands of square feet)
X_5	Number of garage stalls
X_6	Number of rooms
X_7	Number of bedrooms
X_8	Age of of the home (years)
X_9	Number of fireplaces

Table 11.17   Building Characteristics and Sales Price.

Row	X_1	X_2	X_3	X_4	X_5	X_6	X_7	X_8	X_9	Y
1	4.918	1.000	3.472	0.998	1.0	7	4	42	0	25.90
2	5.021	1.000	3.531	1.500	2.0	7	4	62	0	29.50
3	4.543	1.000	2.275	1.175	1.0	6	3	40	0	27.90
4	4.557	1.000	4.050	1.232	1.0	6	3	54	0	25.90
5	5.060	1.000	4.455	1.121	1.0	6	3	42	0	29.90
6	3.891	1.000	4.455	0.988	1.0	6	3	56	0	29.90
7	5.898	1.000	5.850	1.240	1.0	7	3	51	1	30.90
8	5.604	1.000	9.520	1.501	0.0	6	3	32	0	28.90
9	5.828	1.000	6.435	1.225	2.0	6	3	32	0	35.90
10	5.300	1.000	4.988	1.552	1.0	6	3	30	0	31.50
11	6.271	1.000	5.520	0.975	1.0	5	2	30	0	31.00
12	5.959	1.000	6.666	1.121	2.0	6	3	32	0	30.90
13	5.050	1.000	5.000	1.020	0.0	5	2	46	1	30.00
14	8.246	1.500	5.150	1.664	2.0	8	4	50	0	36.90
15	6.697	1.500	6.902	1.488	1.5	7	3	22	1	41.90
16	7.784	1.500	7.102	1.376	1.0	6	3	17	0	40.50
17	9.038	1.000	7.800	1.500	1.5	7	3	23	0	43.90
18	5.989	1.000	5.520	1.256	2.0	6	3	40	1	37.90
19	7.542	1.500	5.000	1.690	1.0	6	3	22	0	37.90
20	8.795	1.500	9.890	1.820	2.0	8	4	50	1	44.50
21	6.083	1.500	6.727	1.652	1.0	6	3	44	0	37.90
22	8.361	1.500	9.150	1.777	2.0	8	4	48	1	38.90
23	8.140	1.000	8.000	1.504	2.0	7	3	3	0	36.90
24	9.142	1.500	7.326	1.831	1.5	8	4	31	0	45.80

Answer the following questions, in each case justifying your answer by appropriate analyses.

(a) In a fitted regression model that relates the sale price to taxes and building characteristics, would you include all the variables?

(b) A veteran real estate agent has suggested that local taxes, number of rooms, and age of the house would adequately describe the sale price. Do you agree?

(c) A real estate expert who was brought into the project reasoned as follows: The selling price of a home is determined by its desirability and this is certainly a function of the physical characteristic of the building. This overall assessment is reflected in the local taxes paid by the homeowner; consequently, the best predictor of sale price is the local taxes. The building characteristics are therefore redundant in a regression equation which includes local taxes. An equation that relates sale price solely to local taxes would be adequate. Examine this assertion by examining several models. Do you agree? Present what you consider to be the most adequate model or models for predicting sale price of homes in Erie, PA.

11.6   Refer to the Gasoline Consumption data in Tables 9.18 and 9.19.

(a) Would you include all the variables to predict the gasoline consumption of the cars? Explain, giving reasons.

(b) Six alternative models have been suggested:
  (a)   Regress Y on X_1.
  (b)   Regress Y on X_{10}.
  (c)   Regress Y on X_1 and X_{10}.
  (d)   Regress Y on X_2 and X_{10}.
  (e)   Regress Y on X_8 and X_{10}.
  (f)   Regress Y on X_8 and X_5, and X_{10}.

  Among these regression models, which would you choose to predict the gasoline consumption of automobiles? Can you suggest a better model?

(c) Plot Y against X_1, X_2, X_8, and X_{10} (one at a time). Do the plots suggest that the relationship between Y and the 11 predictor variables may not be linear?

(d) The gasoline consumption was determined by driving each car with the same load over the same track (a road length of about 123 miles). Instead of using Y (miles per gallon), it was suggested that we consider a new variable, $W = 100/Y$ (gallons per hundred miles). Plot W against X_1, X_2, X_8, and X_{10} and examine if the relationship between W and the 11 predictor variables is more linear than that between Y and the 11 predictor variables.

(e) Repeat Exercise 11.6b using W in place of Y. What are your conclusions?

(f) Regress Y on X_{13}, where $X_{13} = X_8/X_{10}$.

(g) Write a brief report describing your findings. Make a recommendation on the model to be used for predicting gasoline consumption of cars.

11.7 Refer to the Presidential Election Data in Table 5.17 and, as in Exercise 9.3, consider fitting a model relating V to all the variables (including a time trend representing year of election) plus as many interaction terms involving two or three variables as you possibly can.

(a) Starting with the model in Exercise 9.3a. Apply two or more variable selection methods to choose the best model or models that might be expected to perform best in predicting future presidential elections.

(b) Repeat the above exercise starting with the model in Exercise 9.3d.

(c) Which one of the models obtained above would you prefer?

(d) Use your chosen model to predict the proportion of votes expected to be obtained by a presidential candidate in United States presidential elections in the years 2000, 2004, and 2008.

(e) Which one of the above three predictions would you expect to be more accurate than the other two? Explain.

(f) The result of the 2000 presidential election was not known at the time this edition went to press. If you happen to be reading this book after the election of the year 2000 and beyond, were your predictions in Exercise correct?

11.8 Cigarette Consumption Data: Consider the Cigarette Consumption data described in Exercise 3.14 and given in Table 3.17. The organization wanted to construct a regression equation that relates statewide cigarette consumption (per capita basis) to various socioeconomic and demographic variables, and to determine whether these variables were useful in predicting the consumption of cigarettes.

(a) Construct a linear regression model that explains the per capita sale of cigarettes in a given state. In your analysis, pay particular attention to outliers. See if the deletion of an outlier affects your findings. Look at residual plots before deciding on a final model. You need not include all the variables in the model if your analysis indicates otherwise. Your objective should be to find the smallest number of variables that describes the state sale of cigarettes meaningfully and adequately.

(b) Write a report describing your findings.

Appendix: Effects of Incorrect Model Specifications

In this Appendix we discuss the effects of an incorrect model specification on the estimates of the regression coefficients and predicted values using matrix notation. Define the following matrix and vectors:

$$\mathbf{X} = \begin{bmatrix} x_{10} & x_{11} & \cdots & x_{1p} & x_{1(p+1)} & \cdots & x_{1q} \\ x_{20} & x_{21} & \cdots & x_{2p} & x_{2(p+1)} & \cdots & x_{2q} \\ \vdots & \vdots & \ddots & \vdots & \vdots & \ddots & \vdots \\ x_{n0} & x_{n1} & \cdots & x_{np} & x_{n(p+1)} & \cdots & x_{nq} \end{bmatrix}, \quad \mathbf{Y} = \begin{bmatrix} y_1 \\ \vdots \\ y_n \end{bmatrix},$$

$$\beta = \begin{bmatrix} \beta_0 \\ \beta_1 \\ \vdots \\ \beta_p \\ \hline \beta_{p+1} \\ \vdots \\ \beta_q \end{bmatrix}, \quad \varepsilon = \begin{bmatrix} \varepsilon_1 \\ \varepsilon_2 \\ \vdots \\ \varepsilon_n \end{bmatrix},$$

where $x_{i0} = 1$ for $i = 1, \ldots, n$. The matrix $\mathbf{X}$, which has n rows and $(q+1)$ columns, is partitioned into two submatrices $\mathbf{X}_p$ and $\mathbf{X}_r$, of dimensions $(n \times (p+1))$ and $(n \times r)$, where $r = q - p$. The vector β is similarly partitioned into β_p and β_r, which have $(p+1)$ and r components, respectively.

The full linear model containing all q variables is given by

$$\mathbf{Y} = \mathbf{X}\beta + \varepsilon = \mathbf{X}_p\beta_p + \mathbf{X}_r\beta_r + \varepsilon, \tag{A.1}$$

where ε_i's are independently normally distributed errors with zero means and unit variance.

The linear model containing only p variables (i.e., an equation with $(p+1)$ terms) is

$$\mathbf{Y} = \mathbf{X}_p\beta_p + \varepsilon. \tag{A.2}$$

Let us denote the least squares estimate of β obtained from the full model (A.1) by $\hat{\beta}^*$, where

$$\hat{\beta}^* = \begin{pmatrix} \hat{\beta}_p^* \\ \hat{\beta}_r^* \end{pmatrix} = (\mathbf{X}^T\mathbf{X})^{-1}\mathbf{X}^T\mathbf{Y}.$$

The estimate $\hat{\beta}_p$ of β_p obtained from the subset model (A.2) is given by

$$\hat{\beta}_p = (\mathbf{X}_p^T\mathbf{X}_p)^{-1}\mathbf{X}_p^T\mathbf{Y}.$$

Let $\hat{\sigma}_q^2$ and $\hat{\sigma}_p^2$ denote the estimates of σ^2 obtained from (A.1) and (A.2), respectively. Then it follows that

$$\hat{\sigma}_q^2 = \frac{\mathbf{Y}^T\mathbf{Y} - \hat{\boldsymbol{\beta}}^{*T}\mathbf{X}^T\mathbf{Y}}{n - q - 1}$$

and

$$\hat{\sigma}_p^2 = \frac{\mathbf{Y}^T\mathbf{Y} - \hat{\boldsymbol{\beta}}_p^T\mathbf{X}_p^T\mathbf{Y}}{n - p - 1}.$$

It is known from standard theory that $\hat{\boldsymbol{\beta}}^*$ and $\hat{\sigma}_q^2$ are unbiased estimates of $\boldsymbol{\beta}$ and σ^2. It can be shown that

$$E(\hat{\boldsymbol{\beta}}_p) = \boldsymbol{\beta}_p + \mathbf{A}\boldsymbol{\beta}_r,$$

where

$$\mathbf{A} = (\mathbf{X}_p^T\mathbf{X}_p)^{-1}\mathbf{X}_p^T\mathbf{X}_r.$$

Further,

$$\begin{aligned}
Var(\hat{\boldsymbol{\beta}}_p) &= (\mathbf{X}_p^T\mathbf{X}_p)^{-1}\sigma^2, \\
Var(\hat{\boldsymbol{\beta}}^*) &= (\mathbf{X}^T\mathbf{X})^{-1}\sigma^2,
\end{aligned}$$

and

$$MSE(\hat{\boldsymbol{\beta}}_p) = (\mathbf{X}_p^T\mathbf{X}_p)^{-1}\sigma^2 + \mathbf{A}\boldsymbol{\beta}_r\boldsymbol{\beta}_r^T\mathbf{A}^T.$$

We can summarize the properties of $\hat{\boldsymbol{\beta}}_p$ and $\hat{\boldsymbol{\beta}}_p^*$ as follows:

1. $\hat{\boldsymbol{\beta}}_p$ is a biased estimate of $\boldsymbol{\beta}_p$ unless (1) $\boldsymbol{\beta}_r = 0$ or (2) $\mathbf{X}_p^T\mathbf{X}_r = 0$.

2. The matrix $Var(\hat{\boldsymbol{\beta}}^*) - Var(\hat{\boldsymbol{\beta}}_p)$ is positive semidefinite; that is, variances of the least squares estimates of regression coefficients obtained from the full model are larger than the corresponding variances of the estimates obtained from the subset model. In other words, the deletion of variables always results in smaller variances for the estimates of the regression coefficients of the remaining variables.

3. If the matrix $Var(\hat{\boldsymbol{\beta}}_r^*) - \boldsymbol{\beta}_r\boldsymbol{\beta}_r^T$ is positive semidefinite, then the matrix $Var(\hat{\boldsymbol{\beta}}_p^*) - MSE(\hat{\boldsymbol{\beta}}_p)$ is positive semidefinite. This means that the least squares estimates of regression coefficients obtained from the subset model have smaller mean square error than estimates obtained from the full model when the variables deleted have regression coefficients that are smaller than the standard deviation of the estimates of the coefficients.

4. $\hat{\sigma}_p^2$ is generally biased upward as an estimate of σ^2.

To see the effect of model misspecification on prediction, let us examine the prediction corresponding to an observation, say $\mathbf{x}^T = (\mathbf{x}_p^T : \mathbf{x}_r^T)$. Let $\ddot{y}^*$

denote the predicted value corresponding to $\mathbf{x}^T$ when the full set of variables are used. Then $\hat{y}^* = \mathbf{x}^T \hat{\beta}^*$ with mean $\mathbf{x}^T \beta$ and prediction variance $Var(\hat{y}^*)$:

$$Var(\hat{y}^*) = \sigma^2(1 + \mathbf{x}^T(\mathbf{X}^T\mathbf{X})^{-1}\mathbf{x}).$$

On the other hand, if the subset model (A.2) is used, the estimated predicted value $\hat{y} = \mathbf{x}_p^T \hat{\beta}_p$ with mean

$$E(\hat{y}) = \mathbf{x}_p^T \beta_p + \mathbf{x}_p^T \mathbf{A}\beta_r$$

and prediction variance

$$Var(\hat{y}) = \sigma^2(1 + \mathbf{x}_p^T(\mathbf{X}_p^T\mathbf{X}_p)^{-1}\mathbf{x}_p).$$

The prediction mean square error is given by

$$MSE(\hat{y}) = \sigma^2(1 + \mathbf{x}_p^T(\mathbf{X}_p^T\mathbf{X}_p)^{-1}\mathbf{x}_p) + (\mathbf{x}_p^T \mathbf{A}\beta_r - \mathbf{x}_r^T \beta_r)^2.$$

The properties of $\hat{y}^*$ and $\hat{y}$ can be summarized as follows:

1. $\hat{y}$ is biased unless $\mathbf{X}_p^T \mathbf{X}_r \beta_r = 0$.

2. $Var(\hat{y}^*) \geq Var(\hat{y})$.

3. If the matrix $Var(\hat{\beta}_r^*) - \beta_r \beta_r^T$ is positive semidefinite, then $Var(\hat{y}^*) \geq MSE(\hat{y})$.

The significance and interpretation of these results in the context of variable selection are given in the main body of the chapter.

$$12$$

Logistic Regression

12.1 INTRODUCTION

In our discussion of regression analysis so far the response variable Y has been regarded as a continuous quantitative variable. The predictor variables, however, have been both quantitative, as well as qualitative. Indicator variables, which we have described earlier, fall into the second category. There are situations, however, where the response variable is qualitative. In this chapter we present methods for dealing with this situation. The methods presented in this chapter are very different from the method of least squares considered in earlier chapters.

Consider a procedure in which individuals are selected on the basis of their scores in a battery of tests. After five years the candidates are classified as "good" or "poor". We are interested in examining the ability of the tests to predict the job performance of the candidates. Here the response variable, performance, is dichotomous. We can code "good" as 1 and "poor" as 0, for example. The predictor variables are the scores in the tests.

In a study to determine the risk factors for cancer, health records of several people were studied. Data were collected on several variables, such as age, sex, smoking, diet, and the family's medical history. The response variable was, the person had cancer ($Y = 1$), or did not have cancer ($Y = 0$).

In the financial community the "health" of a business is of primary concern. The response variable is solvency of the firm (bankrupt = 0, solvent =1), and the predictor variables are the various financial characteristics associated with the firm. Situations where the response variable is a dichotomous variable are quite common and occur extensively in statistical applications.

12.2   MODELING QUALITATIVE DATA

The qualitative data with which are dealing, the binary response variable, can always be coded as having two values, 0 or 1. Rather than predicting these two values we try to model the probabilities that the response takes one of these two values. The limitation of the previously considered standard linear regression model is obvious.

We illustrate this point by considering a simple regression problem, in which we have only one predictor. The same considerations hold for the multiple regression case. Let π denote the probability that $Y = 1$ when $X = x$. If we use the standard linear model to describe π, then our model for the probability would be

$$\pi = Pr(Y = 1 | X = x) = \beta_0 + \beta_1 x + \varepsilon. \tag{12.1}$$

Since π is a probability it must lie between 0 and 1. The linear function given in (12.1) is unbounded, and hence cannot be used to model probability. There is another reason why ordinary least squares method is unsuitable. The response variable Y is a binomial random variable, consequently its variance will be a function of π, and depends on X. The assumption of equal variance (homoscedasticity) does not hold. We could use the weighted least squares, but there are problems with that approach. The values of π are not known. In order to use weighted least squares approach, we will have to start with an initial guess for the value of π, and then iterate. Instead of this complex method we will describe an alternative method for modeling probabilities.

12.3   THE LOGIT MODEL

The relationship between the probability π and X can often be represented by a *logistic response function*. It resembles a S-shaped curve, a sketch of which is given in Figure 12.1. The probability π initially increases slowly with increase in X, then the increase accelerates, finally stabilizes, but does not increase beyond 1. Intuitively this makes sense. Consider the probability of a questionnaire being returned as a function of cash reward, or the probability of passing a test as a function of the time put in studying for it.

The shape of the S-curve given in Figure 12.1 can be reproduced if we model the probabilities as follows:

$$\pi = \Pr(Y = 1 | X = x) = \frac{e^{\beta_0 + \beta_1 x}}{1 + e^{\beta_0 + \beta_1 x}}, \tag{12.2}$$

where e is the base of the natural logarithm. The probabilities here are modeled by the distribution function (cumulative probability function) of the lo-

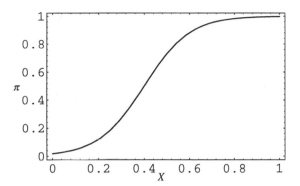

Fig. 12.1   Logistic response function.

gistic distribution. There are other ways of modeling the probabilities that would also produce the S-curve. The cumulative distribution of the normal curve has also been used. This gives rise to the *probit* model. We will not discuss the probit model here, as we consider the logistic model simpler and superior to the probit model.

The logistic model can be generalized directly to the situation where we have several predictor variables. The probability π is modeled as

$$
\begin{aligned}
\pi &= \Pr(Y = 1 | X_1 = x_1, \ldots, X_p = x_p) \\
&= \frac{e^{\beta_0 + \beta_1 x_1 + \beta_2 x_2 + \ldots + \beta_p x_p}}{1 + e^{\beta_0 + \beta_1 x_1 + \ldots + \beta_p x_p}} .
\end{aligned} \tag{12.3}
$$

The equation in (12.3) is called the *logistic regression function*. It is nonlinear in the parameters β_0, β_1, ..., β_p. However, it can be linearized by the *logit transformation*.[1] Instead of working directly with π we work with a transformed value of π. If π is the probability of an event happening, the ratio $\pi/(1 - \pi)$ is called the *odds ratio* for the event. Since

$$
1 - \pi = \Pr(Y = 0 | X_1 = x_1, \ldots, X_p = x_p) = \frac{1}{1 + e^{\beta_0 + \beta_1 x_1 + \ldots + \beta_p x_p}} ,
$$

then

$$
\frac{\pi}{1 - \pi} = e^{\beta_0 + \beta_1 x_1 + \ldots + \beta_p x_p}. \tag{12.4}
$$

Taking the natural logarithm of both sides of (12.4), we obtain

$$
\begin{aligned}
g(x_1, \ldots, x_p) &= log\left(\frac{\pi}{1 - \pi}\right) \\
&= \beta_0 + \beta_1 x_1 + \ldots + \beta_p x_p.
\end{aligned} \tag{12.5}
$$

[1]See Chapter 6 for transformation of variables.

The logarithm of the odds ratio is called the *logit*. It can be seen from (12.5) that the logit transformation produces a linear function of the parameters $\beta_0, \beta_1, \ldots, \beta_p$. Note also that while the range of values of π in (12.3) is between 0 and 1, the range of values of $log(\pi/(1 - \pi))$ is between $-\infty$ and $+\infty$, which makes the logits (the logarithm of the odds ratio) more appropriate for linear regression fitting.

Modeling the response probabilities by the logistic distribution and estimating the parameters of the model given in (12.3) constitutes fitting a logistic regression. In logistic regression the fitting is carried out by working with the logits. The logit transformation produces a model that is linear in the parameters. The method of estimation used is the *maximum likelihood* method. The maximum likelihood estimates are obtained numerically, using an iterative procedure. Unlike least squares fitting, no closed-form expression exists for the estimates of the parameters. We will not go into the computational aspects of the problem but refer the reader to McCullagh and Nelder (1983), Seber (1984), and Hosmer and Lemeshow (1989).

To fit a logistic regression in practice a computer program is essential. Most regression packages have a logistic regression option. After the fitting one looks at the same set of questions that are usually considered in linear regression. Questions about the suitability of the model, the variables to be retained, and goodness of fit are all considered. Tools used are not the usual R^2, t, and F tests, the ones employed in least squares regression, but others which provide answers to these same questions. Hypothesis testing is done by different methods, since the method of estimation is maximum likelihood as opposed to least squares.

12.4   EXAMPLE: ESTIMATING PROBABILITY OF BANKRUPTCIES

Detecting ailing financial and business establishments is an important function of audit and control. Systematic failure to do audit and control can lead to grave consequences, such as the savings-and-loan fiasco of the 1980s in the United States. Table 12.1 gives some of the operating financial ratios of 33 firms that went bankrupt after 2 years and 33 that remained solvent during the same period. The data can also be found in the book's Web site.[2] A multiple logistic regression model is fitted using variables X_1, X_2, and X_3. The output from fitting the model is given in Table 12.2.

Three financial ratios were available for each firm:

$$X_1 = \frac{\text{Retained Earnings}}{\text{Total Assets}},$$

$$X_2 = \frac{\text{Earnings Before Interest and Taxes}}{\text{Total Assets}},$$

[2]http://www.ilr.cornell.edu/~hadi/RABE

Table 12.1   Financial Ratios of Solvent and Bankrupt Firms.

Row	Y	X_1	X_2	X_3	Row	Y	X_1	X_2	X_3
1	0	−62.8	−89.5	1.7	34	1	43.0	16.4	1.3
2	0	3.3	−3.5	1.1	35	1	47.0	16.0	1.9
3	0	−120.8	−103.2	2.5	36	1	−3.3	4.0	2.7
4	0	−18.1	−28.8	1.1	37	1	35.0	20.8	1.9
5	0	−3.8	−50.6	0.9	38	1	46.7	12.6	0.9
6	0	−61.2	−56.2	1.7	39	1	20.8	12.5	2.4
7	0	−20.3	−17.4	1.0	40	1	33.0	23.6	1.5
8	0	−194.5	−25.8	0.5	41	1	26.1	10.4	2.1
9	0	20.8	−4.3	1.0	42	1	68.6	13.8	1.6
10	0	−106.1	−22.9	1.5	43	1	37.3	33.4	3.5
11	0	−39.4	−35.7	1.2	44	1	59.0	23.1	5.5
12	0	−164.1	−17.7	1.3	45	1	49.6	23.8	1.9
13	0	−308.9	−65.8	0.8	46	1	12.5	7.0	1.8
14	0	7.2	−22.6	2.0	47	1	37.3	34.1	1.5
15	0	−118.3	−34.2	1.5	48	1	35.3	4.2	0.9
16	0	−185.9	−280.0	6.7	49	1	49.5	25.1	2.6
17	0	−34.6	−19.4	3.4	50	1	18.1	13.5	4.0
18	0	−27.9	6.3	1.3	51	1	31.4	15.7	1.9
19	0	−48.2	6.8	1.6	52	1	21.5	−14.4	1.0
20	0	−49.2	−17.2	0.3	53	1	8.5	5.8	1.5
21	0	−19.2	−36.7	0.8	54	1	40.6	5.8	1.8
22	0	−18.1	−6.5	0.9	55	1	34.6	26.4	1.8
23	0	−98.0	−20.8	1.7	56	1	19.9	26.7	2.3
24	0	−129.0	−14.2	1.3	57	1	17.4	12.6	1.3
25	0	−4.0	−15.8	2.1	58	1	54.7	14.6	1.7
26	0	−8.7	−36.3	2.8	59	1	53.5	20.6	1.1
27	0	−59.2	−12.8	2.1	60	1	35.9	26.4	2.0
28	0	−13.1	−17.6	0.9	61	1	39.4	30.5	1.9
29	0	−38.0	1.6	1.2	62	1	53.1	7.1	1.9
30	0	−57.9	0.7	0.8	63	1	39.8	13.8	1.2
31	0	−8.8	−9.1	0.9	64	1	59.5	7.0	2.0
32	0	−64.7	−4.0	0.1	65	1	16.3	20.4	1.0
33	0	−11.4	4.8	0.9	66	1	21.7	−7.8	1.6

$$X_3 = \frac{\text{Sales}}{\text{Total Assets}} .$$

The response variable is defined as

$$Y = \begin{cases} 0, & \text{if bankrupt after 2 years,} \\ 1, & \text{if solvent after 2 years.} \end{cases}$$

Table 12.2 has a certain resemblance to the standard regression output. Some of the output serve similar functions. We now describe and interpret the output obtained from fitting a logistic regression. If π denotes the probability

Table 12.2   Output from the Logistic Regression Using X_1, X_2, and X_3.

Variable	Coefficient	s.e.	Z-test	p-value	Odds Ratio	95% C.I. Lower	Upper
Constant	-10.15	10.84	-0.94	0.349			
X_1	0.33	0.30	1.10	0.27	1.39	0.77	2.51
X_2	0.18	0.11	1.69	0.09	1.20	0.97	1.48
X_3	5.09	5.08	1.00	0.32	161.98	0.01	3.43×10^6

Log-Likelihood $= -2.906$ $G = 85.683$ $d.f. = 3$ p-value < 0.000

of a firm remaining solvent after 2 years, the fitted logit is given by:

$$\hat{g}(x_1, \ldots, x_p) = -10.15 + 0.33\, x_1 + 0.18\, x_2 + 5.09\, x_3. \tag{12.6}$$

This corresponds to the fitted regression equation in standard analysis. Here instead of predicting Y we obtain a model to predict the logits, $log(\pi/(1-\pi))$. From the logits, after transformation, we can get the predicted probabilities. The constant and the coefficients are read directly from the second column in the table. The standard errors (s.e.) of the coefficients are given in the third column. The fourth column headed by Z is the ratio of the coefficient and the standard deviation. The Z corresponding to the coefficient of X_2 is obtained from dividing 0.181 by 0.107. In the standard regression this would be the t-test. This ratio for the logistic regression has a normal distribution as opposed to a t-distribution that we get in linear regression. The fifth column gives the p-value corresponding to the observed Z value, and should be interpreted like any p-value (see Chapters 2 and 3). These p-values are used to judge the significance of the coefficient. Values smaller than 0.05 would lead us to conclude that the coefficient is significantly different from 0 at the 5% significance level. From the p-values in Table 12.2, we see that none of the variables individually are significant for predicting the logits of the observations.

In the standard regression output the regression coefficients have a simple interpretation. The regression coefficient of the jth predictor variable X_j is the expected change in Y for unit change in X_j when other variables are held fixed. The coefficient of X_2 in (12.6) is the expected change in the logit for unit change in X_2 when the other variables are held fixed. The coefficients of a logistic regression fit have another interpretation that is of major practical importance. Keeping X_1 and X_3 fixed, for unit increase in X_2 the relative odds of

$$\frac{\text{Pr(Firm solvent after 2 years)}}{\text{Pr(Firm bankrupt)}}$$

is multiplied by $e^{\hat{\beta}_2} = e^{0.181} = 1.198$, that is there is an increase of 20%. These values for each of the variables is given in the sixth column headed by Odds

Ratio. They represent the change in odds ratio for unit change of a particular variable while the others are held constant. The change in odds ratio for unit change in variable X_j, while the other variables are held fixed, is $e^{\hat{\beta}_j}$. If X_j was a binary variable, taking values 1 or 0, then $e^{\hat{\beta}_j}$ would be the actual value of the odds ratio rather than the change in the value of the odds ratio.

The 95% confidence intervals of the odds ratios are given in the last two columns of the table. If the confidence interval does not contain the value 1 the variable has a significant effect on the odds ratio. If the interval is below 1 the variable lowers significantly the relative odds. On the other hand, if the interval lies above 1 the relative odds is significantly increased by the variable.

To see whether the variables collectively contribute in explaining the logits a test that examines whether the coefficients $\beta_1, \ldots, \beta_p$ are all zero is performed. This corresponds to the case in multiple regression analysis where we test whether all the regression coefficients can be taken to be zero. The statistic G given at the bottom of Table 12.2 performs that task. The statistic G has a chi-square distribution. The p-value is considerably smaller than .05, and indicates that the variables collectively influence the logits.

12.5   LOGISTIC REGRESSION DIAGNOSTICS

After fitting a logistic regression model certain diagnostic measures can be examined for the detection of outliers, high leverage points, influential observations, and other model deficiencies. The diagnostic measures developed in Chapter 4 for the standard linear regression model can be adapted to the logistic regression model. Regression packages with a logistic regression option usually give various diagnostic measures. These include:

1. The estimated probabilities $\hat{\pi}_i$, $i = 1, \ldots, n$.

2. One or more types of residuals, for example, the *standardized deviance residuals*, DR_i, and the *standardized Personian residuals*, PR_i, $i = 1, \ldots, n$.

3. The *weighted leverages*, p_{ii}^*, which measure the potential effects of the observations in the predictor variables on the obtained logistic regression results.

4. The scaled difference in the regression coefficients when the ith observation is deleted: $DBETA_i$, $i = 1, \ldots, n$.

5. The change in the chi-squared statistics G when the ith observation is deleted: DFG_i, $i = 1, \ldots, n$.

The formulas and derivations of these measures are beyond the scope of this book. The interested reader is referred to Pregibon (1981), Landwehr, Pregibon, and Shoemaker (1984), Hosmer and Lemeshow (1989) and the references therein. The above measures, however, can be used in the same way as the corresponding measures obtained from a linear fit (Chapter 4). For example, the following graphical displays can be examined:

1. The scatter plot of DR_i versus $\hat{\pi}_i$.

2. The scatter plot of PR_i versus $\hat{\pi}_i$.

3. The index plots of DR_i, $DBETA_i$, DG_i, and p_{ii}^*.

As an illustrative example using the Bankruptcy data, the index plots of DR_i, $DBETA_i$, and DG_i obtained from the fitted logistic regression model in (12.6), are shown in Figures 12.2, 12.3, and 12.4, respectively. It can easily be seen from these graphs that observations 9, 52, and 36 are unusual and that they may have undue influence on the logistic regression results. We leave it as an exercise for the reader to determine if their deletion would make a significant difference in the results and the conclusion drawn from the analysis.

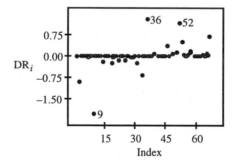

Fig. 12.2 Bankruptcy data: Index plot of DR_i, the standardized deviance residuals.

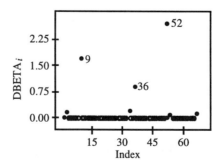

Fig. 12.3 Bankruptcy data: Index plot of $DBETA_i$, the scaled difference in the regression coefficients when the ith observation is deleted.

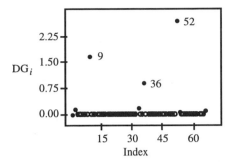

Fig. 12.4 Bankruptcy data: Index plot of DG_i, the change in the chi-squared statistics G when the ith observation is deleted.

12.6 DETERMINATION OF VARIABLES TO RETAIN

In the analysis of the Bankruptcy data we have determined so far that the variables X_1, X_2, and X_3, collectively have explanatory power. Do we need all three variables? This is analogous to the problem of variable selection in multiple regression that was discussed in Chapter 11. Instead of looking at the reduction in the error sum of squares we look at the change in the likelihood (more precisely, the logarithm of the likelihood) for the two fitted models. The reason for this is that in logistic regression the fitting criterion is the likelihood, whereas in least squares it is the sum of squares. Let $L(p)$ denote the logarithm of the likelihood when we have a model with p variables and a constant. Similarly, let $L(p+q)$ be the logarithm of the likelihood for a model in which we have $p+q$ variables and a constant. To see whether the q additional variables contribute significantly we look at $2(L(p+q) - L(p))$. This quantity is twice the difference between the log-likelihood for the two models. This difference is distributed as a chi-square variable with q degrees of freedom (see Table A.3).

The magnitude of this quantity determines the significance of the test. A small value of chi-square would lead to the conclusion that the q variables do not add significantly to the improvement in prediction of the logits, and is therefore not necessary in the model. A large value of chi-square would call for the retention of the q variables in the model. The critical value is determined by the significance level of the test. This test procedure is valid when n, the number of observations available for fitting the model, is large.

In the Bankruptcy data we are analyzing, let us see if the variable X_3 can be deleted without degrading the model. We want to answer the question: Should the variable X_3 be retained in the model? We fit a logistic regression using X_1 and X_2. The results are given in Table 12.3. The log-likelihood for the model with X_1, X_2, and X_3 is -2.906, whereas with only X_1 and X_2 it is -4.736. Here $p = 2$ and $q = 1$, and $2(L(3) - L(2)) = 3.66$. This is a chi-square variable with 1 degree of freedom. From Table A.3, we find that

Table 12.3 Output From the Logistic Regression Using X_1 and X_2.

Variable	Coefficient	s.e.	Z-test	p-value	Odds Ratio	95% C.I. Lower	95% C.I. Upper
Constant	−0.550	0.951	−0.58	0.563			
X_1	0.157	0.075	2.10	0.036	1.17	1.01	1.36
X_2	0.195	0.122	1.59	0.112	1.21	0.96	1.54

Log-Likelihood $= -4.736$ $G = 82.024$ $d.f. = 2$ p-value < 0.000

Table 12.4 Output from the Logistic Regression Using X_1.

Variable	Coefficient	s.e.	Z-test	p-value	Odds Ratio	95% C.I. Lower	95% C.I. Upper
Constant	−1.167	0.816	−1.43	0.153			
X_1	0.177	0.057	3.09	0.002	1.19	1.07	1.33

Log-Likelihood $= -7.902$ $G = 75.692$ $d.f. = 1$ p-value < 0.000

the 5% critical value of the chi-square distribution with 1 degree of freedom is 3.84. At the 5% level we can conclude that the variable X_3 can be deleted without affecting the effectiveness of the model.

Let us now see if we can delete X_2. The result of regressing Y on X_1 is given in Table 12.4. The resulting log-likelihood is -7.902. The test statistic, which we have described earlier, has a value of 6.332. This is distributed as a chi-square random variable with 1 degree of freedom. The 5% value, as we saw earlier, was 3.84. The analysis indicates that we should not delete X_2 from our model. The p-value for this test, as can be verified, is 0.019. To predict probabilities of bankruptcies of firms in our data we should include both X_1 and X_2 in our model.

The procedure that we have outlined above enables us to test any *nested model*. A set of models are said to be nested if they can be obtained from a larger model as special cases. The methodology is similar to that used in analyzing nested models in multiple regression. The only difference is that here our test statistic is based on log of the likelihood instead of sum of squares.

12.7   JUDGING THE FIT OF A LOGISTIC REGRESSION

The overall fit of a multiple regression model is judged, for example, by the value of R^2 from the fitted model. No such simple satisfactory measure exists for logistic regression. Some ad hoc measures have been proposed which are based on the ratio of likelihoods. Most of these are functions of the ratio of the likelihood for the model and the likelihood of the data under a binomial model. These measures are not particularly informative and we will consider a different approach.

The logistic regression equation attempts to model probabilities for the two values of Y (0 or 1). To judge how well the model is doing we will determine the number of observations in the sample that the model is classifying correctly. Our approach will be to fit the logistic model to the data, and calculate the fitted logits. From the fitted logits we will calculate the fitted probabilities for each observation. If the fitted probability for an observation is greater than 0.5 we will assign it to Group 1 ($Y = 1$), and if less than 0.5 we will classify it in Group 0 ($Y = 0$). We will then determine what proportion of the data is classified correctly. A high proportion of correct classification will indicate to us that the logistic model is working well. A low proportion of correct classification will indicate poor performance.

Different cutoff values, other than 0.5, have been suggested in the literature. In most practical situations, without any auxiliary information, such as the relative cost of misclassification or the relative frequency of the two categories in the population, 0.5 is recommended as a cutoff value.

A slightly more problematical question is how high the correct classification probability has to be before logistic regression is thought to be effective. Suppose that in sample of size n there are n_1 observations from Group 1, and n_2 from Group 2. If we classify all the observations into one group or the other, then we will get either n_1/n or n_2/n proportions of observations classified correctly. As a base level for correct classification we can take the $max(n_1/n, n_2/n)$. The proportion of observation classified correctly by the logistic regression should be much higher than the base level for the logistic model to be deemed useful.

For the Bankruptcy data that we have been analyzing logistic regression performs very well. Using variables X_1 and X_2, we find that the model misclassifies one observation from the solvent group (observation number 36), and one observation from the bankruptcy group (observation number 9). The overall correct classification rate $(64/66) = 0.97$. This is considerably higher than the base level rate of 0.5.

The observed correct classification rate should be treated with caution. In practice, if this logistic regression was applied to a new set of observations from this population, it would be very unlikely to do as well. The classification probability has an upward bias. The bias arises due to the fact that the same data that were used to fit the model, was used to judge the performance of the model. The model fitted to a given body of data is expected to perform well on the same body of data. The true measure of the performance of the logistic regression model for classification is the probability of classifying a future observation correctly and not a sample observation. This upward bias in the estimate of correct classification probability can be reduced by using resampling methods, such as jack-knife or bootstrap. These will not be discussed here. The reader is referred to Efron (1982) and Diaconis and Efron (1983).

12.8   CLASSIFICATION PROBLEM: ANOTHER APPROACH

The method of logistic regression has been used to model the probability that an observation belongs to one group given the measurements on several characteristics. We have described how the fitted logits could then be used for classifying an observation into one of two categories. A different statistical methodology is available if our primary interest is *classification*. When the sole interest is to predict the group membership of each observation a statistical method called *discriminant analysis* is commonly used. Without discussing discriminant analysis here, we indicate a simple regression method that will accomplish the same task. The reader can find a discussion of discriminant analysis in McLachlan (1992), Rencher (1995), and Johnson (1998).

The essential idea in discriminant analysis is to find a linear combination of the predictor variables X_1, ..., X_p, such that the scores given by this linear combination separates the observations from the two groups as far as possible. One way that this separation can be accomplished is by fitting a multiple regression model to the data. The response variable is Y, taking values 0 and 1, and the predictors are X_1, ..., X_p. As has been pointed out earlier, some of the fitted values will be outside the range of 0 and 1. This does not matter here, as we are not trying to model probabilities, but only to predict group membership. We calculate the average of the predicted values of all the observations. If the predicted value for a given observation is greater than the average predicted value we assign that observation to the group which has $Y = 1$; if the predicted value is smaller than the average predicted value we assign it to the group with $Y = 0$. From this assignment we determine the number of observations classified correctly in the sample. The variables used in this classification procedure are determined exactly by the same methods as those used for variable selection in multiple regression.

We illustrate this method by applying it to the Bankruptcy data that we have used earlier to illustrate least squares regression. Table 12.5 gives the OLS regression results using the three predictor variables X_1, X_2, and X_3. All three variables have significant regression coefficients and should be retained for classification equation.

Table 12.6 displays the observed Y, the predicted Y, and the assigned group for the Bankruptcy data. The average value of the predicted Y is 0.5. All observations with predicted value less than 0.5 is assigned to $Y = 0$, and those with predicted value greater than 0.5 is assigned to the group with $Y = 1$. The wrongly classified observations are marked by *. It is seen that 5 bankrupt firms are classified as solvent, and one solvent firm is classified as bankrupt. The logistic regression, it should be noted, classified only two observations

Table 12.5 Results from the OLS Regression of Y on X_1, X_2, X_3.

Variable	Coefficient	s.e.	t-test	p-value
Constant	0.322	0.087	3.68	0.0005
X_1	0.003	0.001	3.76	0.0004
x_2	0.004	0.001	2.96	0.0044
x_3	0.149	0.045	3.28	0.0017
$n = 66$	$R^2 = 0.57$	$R_a^2 = 0.55$	$\hat{\sigma} = 0.3383$	$d.f. = 62$

wrongly. One solvent firm and one bankrupt firm were misclassified. For the Bankruptcy data presented in Table 12.2, the logistic regression performs better than the multiple regression in classifying the sample data. In general this is true. The logistic regression does not have to make the restrictive assumption of multivariate normality for the predictor variables. For classification problems we recommend the use of logistic regression. If a logistic regression package is not available, then the multiple regression approach may be tried.

EXERCISES

12.1 The diagnostic plots in Figures 12.2, 12.3, and 12.4 show three unusual observations in the Bankruptcy data. Fit a logistic regression model to the 63 observations without these three observations and compare your results with the results obtained in Section 12.5. Does the deletion of the three points cause a substantial change in the logistic regression results?

12.2 Examine the various logistic regression diagnostics obtained from fitting the logistic regression Y on X_1 and X_2 (Table 12.3) and determine if the data contain unusual observations.

12.3 The *O-rings* in the booster rockets used in space launching play an important part in preventing rockets from exploding. Probabilities of O-ring failures are thought to be related to temperature. A detailed discussion of the background of the problem is found in The Flight of the Space Shuttle Challenger (pp. 33–35) in Chatterjee, Handcock, and Simonoff (1995). Each flight has six O-rings that could be potentially damaged in a particular flight. The data from 23 flights are given in Table 12.7 and can also be found in the the book's Web site.[3] For each flight we have the number of O-rings damaged and the temperature of the launch.

(a) Fit a logistic regression connecting the probability of an O-ring failure with temperature. Interpret the coefficients.

[3]http://www.ilr.cornell.edu/~hadi/RABE

Table 12.6   Classification of Observations by Fitted Values.

Row	Y	Fitted	Assigned	Row	Y	Fitted	Assigned
1	0	−0.00	0	34	1	0.72	1
2	0	0.48	0	35	1	0.82	1
3	0	−0.12	0	36	1	0.73	1
4	0	0.31	0	37	1	0.80	1
5	0	0.23	0	38	1	0.65	1
6	0	0.14	0	39	1	0.80	1
7	0	0.33	0	40	1	0.75	1
8	0	−0.32	0	41	1	0.76	1
9	0	0.52	1*	42	1	0.83	1
10	0	0.12	0	43	1	1.10	1
11	0	0.23	0	44	1	1.42	1
12	0	−0.07	0	45	1	0.86	1
13	0	−0.80	0	46	1	0.66	1
14	0	0.55	1*	47	1	0.81	1
15	0	0.03	0	48	1	0.58	1
16	0	−0.45	0	49	1	0.97	1
17	0	0.64	1*	50	1	1.03	1
18	0	0.45	0	51	1	0.77	1
19	0	0.44	0	52	1	0.48	0*
20	0	0.14	0	53	1	0.60	1
21	0	0.22	0	54	1	0.74	1
22	0	0.37	0	55	1	0.81	1
23	0	0.18	0	56	1	0.84	1
24	0	0.05	0	57	1	0.62	1
25	0	0.55	1*	58	1	0.81	1
26	0	0.56	1*	59	1	0.74	1
27	0	0.39	0	60	1	0.84	1
28	0	0.34	0	61	1	0.86	1
29	0	0.39	0	62	1	0.80	1
30	0	0.26	0	63	1	0.68	1
31	0	0.39	0	64	1	0.83	1
32	0	0.12	0	65	1	0.61	1
33	0	0.44	0	66	1	0.59	1

Table 12.7 Number of O-rings Damaged and the Temperature (Degrees Fahrenheit) at the Time of Launch for 23 Flights of the Space Shuttle *Challenger*.

Flight	Damaged	Temperature	Flight	Damaged	Temperature
1	2	53	13	1	70
2	1	57	14	1	70
3	1	58	15	0	72
4	1	63	16	0	73
5	0	66	17	0	75
6	0	67	18	2	75
7	0	67	19	0	76
8	0	67	20	0	78
9	0	68	21	0	79
10	0	69	22	0	81
11	0	70	23	0	76
12	0	70			

(b) The data for Flight 18 that was launched when the launch temperature was 75 was thought to be problematic, and was deleted. Fit a logistic regression to the reduced data set. Interpret the coefficients.

(c) From the fitted model, find the probability of an O-ring failure when the temperature at launch was 31 degrees. This was the temperature forecast for the day of the launching of the fatal *Challenger* flight on January 20, 1986.

(d) Would you have advised the launching on that particular day?

12.4 Field-goal-kicking data for the entire American Football League (AFL) and National Football League (NFL) for the 1969 season are given in Table 12.8 and can also be found in the the book's Web site. Let $\pi(X)$ denote the probability of kicking a field goal from a distance of X yards.

(a) For each of the leagues, fit the model

$$\pi(X) = \frac{e^{\beta_0 + \beta_1 X + \beta_2 X^2}}{1 + e^{\beta_0 + \beta_1 X + \beta_2 X^2}}.$$

(b) Let Z be an indicator variable representing the league, that is,

$$Z = \begin{cases} 1, & \text{for the AFL,} \\ 0, & \text{for the NFL.} \end{cases}$$

Fit a single model combining the data from both leagues by extending the model to include the indicator variable Z; that is, fit

$$\pi(X, Z) = \frac{e^{\beta_0 + \beta_1 X + \beta_2 X^2 + \beta_3 Z}}{1 + e^{\beta_0 + \beta_1 X + \beta_2 X^2 + \beta_3 Z}}.$$

Table 12.8 Field-Goal-Kicking Performances of the American Football League (AFL) and National Football League (NFL) for the 1969 Season. The Variable Z Is an Indicator Variable Representing League.

League	Distance	Success	Attempts	Z
NFL	14.5	68	77	0
NFL	24.5	74	95	0
NFL	34.5	61	113	0
NFL	44.5	38	138	0
NFL	52.0	2	38	0
AFL	14.5	62	67	1
AFL	24.5	49	70	1
AFL	34.5	43	79	1
AFL	44.5	25	82	1
AFL	52.0	7	24	1

Source: Morris and Rolph (1981), p. 200.

(c) Does the quadratic term contribute significantly to the model?

(d) Are the probabilities of scoring field goals from a given distance the same for each league?

Appendix:
Statistical Tables

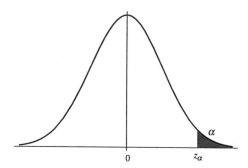

Fig. A.1 The probability density function of the standard normal distribution.

Table A.1 Critical Values z_α, Where $Pr(Z \geq z_\alpha) = \alpha$ and Z Is the Standard Normal Distribution.

α	z_α	α	z_α	α	z_α	α	z_α	α	z_α
.50	0.00	.050	1.64	.030	1.88	.020	2.05	.010	2.33
.45	0.13	.048	1.66	.029	1.90	.019	2.07	.009	2.37
.40	0.25	.046	1.68	.028	1.91	.018	2.10	.008	2.41
.35	0.39	.044	1.71	.027	1.93	.017	2.12	.007	2.46
.30	0.52	.042	1.73	.026	1.94	.016	2.14	.006	2.51
.25	0.67	.040	1.75	.025	1.96	.015	2.17	.005	2.58
.20	0.84	.038	1.77	.024	1.98	.014	2.20	.004	2.65
.15	1.04	.036	1.80	.023	2.00	.013	2.23	.003	2.75
.10	1.28	.034	1.83	.022	2.01	.012	2.26	.002	2.88
.05	1.64	.032	1.85	.021	2.03	.011	2.29	.001	3.09

Source: Adapted from Table 2 of Lindley and Miller (1958), *Cambridge Elementary Statistical Tables*, published by Cambridge University Press, with kind permission of the authors and publishers.

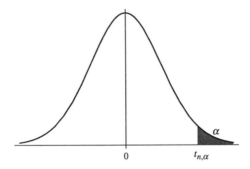

Fig. A.2 The probability density function of the Student's t-distribution with n degrees of freedom $(d.f.)$.

Table A.2 Critical Values $t_{n,\alpha}$, Where $Pr(T_n \geq t_{n,\alpha}) = \alpha$ and T_n Is the Student's t-Distribution With n Degrees of Freedom $(d.f.)$.

n $(d.f.)$	α				
	0.10	0.05	0.025	0.010	0.005
1	3.08	6.31	12.71	31.82	63.66
2	1.89	2.92	4.30	6.97	9.92
3	1.64	2.35	3.18	4.54	5.84
4	1.53	2.13	2.78	3.75	4.60
5	1.48	2.02	2.57	3.36	4.03
6	1.44	1.94	2.45	3.14	3.71
7	1.42	1.89	2.36	3.00	3.50
8	1.40	1.86	2.31	2.90	3.36
9	1.38	1.83	2.26	2.82	3.25
10	1.37	1.81	2.23	2.76	3.17
12	1.36	1.78	2.18	2.68	3.06
14	1.34	1.76	2.14	2.62	2.98
16	1.34	1.75	2.12	2.58	2.92
18	1.33	1.73	2.10	2.55	2.88
20	1.32	1.72	2.09	2.53	2.84
30	1.31	1.70	2.04	2.46	2.75
40	1.30	1.68	2.02	2.42	2.70
60	1.30	1.67	2.00	2.39	2.66
120	1.29	1.66	1.98	2.36	2.62
∞	1.28	1.64	1.96	2.33	2.58

Source: Adapted from Table III of Fisher and Yates (1963), *Statistical Tables for Biological, Agricultural and Medical Research*, 6th Ed., published by Oliver and Boyd, Edinburgh, with kind permission of the authors and publishers.

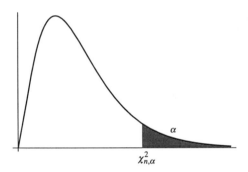

Fig. A.3 The probability density function of the χ^2 distribution with n degrees of freedom $(d.f.)$.

Table A.3   Critical Values $\chi^2_{n,\alpha}$, Where $Pr(\chi^2_n \geq \chi^2_{n,\alpha}) = \alpha$ and χ^2_n Is the χ^2 Distribution With n Degrees of Freedom ($d.f.$).

n ($d.f.$)	α				
	0.10	0.05	0.025	0.010	0.005
1	2.71	3.84	5.02	6.63	7.88
2	4.61	5.99	7.38	9.21	10.60
3	6.25	7.81	9.35	11.34	12.84
4	7.78	9.49	11.14	13.28	14.86
5	9.24	11.07	12.83	15.09	16.75
6	10.65	12.59	14.45	16.81	18.55
7	12.02	14.07	16.01	18.48	20.28
8	13.36	15.51	17.53	20.09	21.96
9	14.68	16.92	19.02	21.67	23.59
10	15.99	18.31	20.48	23.21	25.19
11	17.28	19.68	21.92	24.72	26.76
12	18.55	21.03	23.34	26.22	28.30
13	19.81	22.36	24.74	27.69	29.82
14	21.06	23.68	26.12	29.14	31.32
15	22.31	25.00	27.49	30.58	32.80
16	23.54	26.30	28.85	32.00	34.27
17	24.77	27.59	30.19	33.41	35.72
18	25.99	28.87	31.53	34.81	37.16
19	27.20	30.14	32.85	36.19	38.58
20	28.41	31.41	34.17	37.57	40.00
21	29.62	32.67	35.48	38.93	41.40
22	30.81	33.92	36.78	40.29	42.80
23	32.01	35.17	38.08	41.64	44.18
24	33.20	36.42	39.36	42.98	45.56
25	34.28	37.65	40.65	44.31	46.93
26	35.56	38.89	41.92	45.64	48.29
27	36.74	40.11	43.19	46.96	49.65
28	37.92	41.34	44.46	48.28	50.99
29	39.09	42.56	45.72	49.59	52.34
30	40.26	43.77	46.98	50.89	53.67
40	51.81	55.76	59.34	63.69	66.77
50	63.17	67.50	71.42	76.15	79.49
60	74.40	79.08	83.30	88.38	91.95
70	85.53	90.53	95.02	100.42	104.22
80	96.58	101.88	106.63	112.33	116.32
90	107.57	113.14	118.14	124.12	128.30
100	118.50	124.34	129.56	135.81	140.17

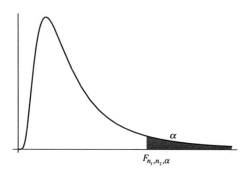

Fig. A.4 The probability density function of the F-distribution with n_1 (numerator) and n_2 (denominator) $(d.f.)$.

Table A.4   The 5% Critical Values $f_{n_1,n_2;0.05}$, Where $Pr(F_{n_1,n_2} \geq f_{n_1,n_2;0.05}) = 0.05$ and F_{n_1,n_2} Is the F-Distribution With n_1 (numerator) and n_2 (denominator) (d.f.).

n_2	n_1								
	1	2	4	6	8	10	12	24	∞
1	161.4	199.5	224.6	234.0	238.9	241.9	243.9	249.1	254.30
2	18.51	19.00	19.25	19.33	19.37	19.40	19.41	19.45	19.50
3	10.13	9.55	9.12	8.94	8.85	8.79	8.74	8.64	8.53
4	7.71	6.94	6.39	6.16	6.04	5.96	5.91	5.77	5.63
5	6.61	5.79	5.19	4.95	4.82	4.74	4.68	4.53	4.36
6	5.99	5.14	4.53	4.28	4.15	4.06	4.00	3.84	3.67
7	5.59	4.74	4.12	3.87	3.73	3.64	3.57	3.41	3.23
8	5.32	4.46	3.84	3.58	3.44	3.35	3.28	3.12	2.93
9	5.12	4.26	3.63	3.37	3.23	3.14	3.07	2.90	2.71
10	4.96	4.10	3.48	3.22	3.07	2.98	2.91	2.74	2.54
11	4.84	3.98	3.36	3.09	2.95	2.85	2.79	2.61	2.40
12	4.75	3.89	3.26	3.00	2.85	2.75	2.69	2.51	2.30
13	4.67	3.81	3.18	2.92	2.77	2.67	2.60	2.42	2.21
14	4.60	3.74	3.11	2.85	2.70	2.60	2.53	2.35	2.13
15	4.54	3.68	3.06	2.79	2.64	2.54	2.48	2.29	2.07
20	4.35	3.49	2.87	2.60	2.45	2.35	2.28	2.08	1.84
25	4.24	3.39	2.76	2.49	2.34	2.24	2.16	1.96	1.71
30	4.17	3.32	2.69	2.42	2.27	2.16	2.09	1.89	1.62
40	4.08	3.23	2.61	2.34	2.18	2.08	2.00	1.79	1.51
60	4.00	3.15	2.53	2.25	2.10	1.99	1.92	1.70	1.39
120	3.92	3.07	2.45	2.17	2.02	1.91	1.83	1.61	1.25
∞	3.84	3.00	2.37	2.10	1.94	1.83	1.75	1.52	1.00

Table A.5   The 1% Critical Values $f_{n_1,n_2;0.01}$, Where $Pr(F_{n_1,n_2} \geq f_{n_1,n_2;0.01}) = 0.01$ and F_{n_1,n_2} Is the F-Distribution With n_1 (numerator) and n_2 $(d.f.)$.

					n_1				
n_2	1	2	4	6	8	10	12	24	∞
1	4052	5000	5625	5859	5982	6056	6106	6235	6366
2	98.50	99.00	99.25	99.33	99.37	99.40	99.42	99.46	99.50
3	34.12	30.82	28.71	27.91	27.49	27.23	27.05	26.60	26.13
4	21.20	18.00	15.98	15.21	14.80	14.55	14.37	13.93	13.46
5	16.26	13.27	11.39	10.67	10.29	10.05	9.89	9.47	9.02
6	13.75	10.92	9.15	8.47	8.10	7.87	7.72	7.31	6.88
7	12.25	9.55	7.85	7.19	6.84	6.62	6.47	6.07	5.65
8	11.26	8.65	7.01	6.37	6.03	5.81	5.67	5.28	4.86
9	10.56	8.02	6.42	5.80	5.47	5.26	5.11	4.73	4.31
10	10.04	7.56	5.99	5.39	5.06	4.85	4.71	4.33	3.91
11	9.65	7.21	5.67	5.07	4.74	4.54	4.40	4.02	3.60
12	9.33	6.93	5.41	4.82	4.50	4.30	4.16	3.78	3.36
13	9.07	6.70	5.21	4.62	4.30	4.10	3.96	3.59	3.17
14	8.86	6.51	5.04	4.46	4.14	3.94	3.80	3.43	3.00
15	8.68	6.36	4.89	4.32	4.00	3.80	3.67	3.29	2.87
20	8.10	5.85	4.43	3.87	3.56	3.37	3.23	2.86	2.42
25	7.77	5.57	4.18	3.63	3.32	3.13	2.99	2.62	2.17
30	7.56	5.39	4.02	3.47	3.17	2.98	2.84	2.47	2.01
40	7.31	5.18	3.83	3.29	2.99	2.80	2.66	2.29	1.80
60	7.08	4.98	3.65	3.12	2.82	2.63	2.50	2.12	1.60
120	6.85	4.79	3.48	2.96	2.66	2.47	2.34	1.95	1.38
∞	6.63	4.61	3.32	2.80	2.51	2.32	2.18	1.79	1.00

Source: Abridged from Table 18 of Pearson and Hartley (1954), *Biometrika Tables for Statisticians, Volume 1*, published at the Cambridge University Press for the *Biometrika* Trustees, with kind permission of the authors and publishers.

Table A.6 Distribution of Durbin-Watson Statistic d: The 5% Significance Points of d_L and d_U (p Is the Number of Predictor Variables).

n	$p = 1$		$p = 2$		$p = 3$		$p = 4$		$p = 5$	
	d_L	d_U	d_L	d_U	d_L	d_U	d_L	d_U	d_L	d_U
15	1.08	1.36	0.95	1.54	0.82	1.75	0.69	1.97	0.56	2.21
16	1.10	1.37	0.98	1.54	0.86	1.73	0.74	1.93	0.62	2.15
17	1.13	1.38	1.02	1.54	0.90	1.71	0.78	1.90	0.67	2.10
18	1.16	1.39	1.05	1.53	0.93	1.69	0.82	1.87	0.71	2.06
19	1.18	1.40	1.08	1.53	0.97	1.68	0.86	1.85	0.75	2.02
20	1.20	1.41	1.10	1.54	1.00	1.68	0.90	1.83	0.79	1.99
21	1.22	1.42	1.13	1.54	1.03	1.67	0.93	1.81	0.83	1.96
22	1.24	1.43	1.15	1.54	1.05	1.66	0.96	1.80	0.86	1.94
23	1.26	1.44	1.17	1.54	1.08	1.66	0.99	1.79	0.90	1.92
24	1.27	1.45	1.19	1.55	1.10	1.66	1.01	1.78	0.93	1.90
25	1.29	1.45	1.21	1.55	1.12	1.66	1.04	1.77	0.95	1.89
26	1.30	1.46	1.22	1.55	1.14	1.65	1.06	1.76	0.98	1.88
27	1.32	1.47	1.24	1.56	1.16	1.65	1.08	1.76	1.01	1.86
28	1.33	1.48	1.26	1.56	1.18	1.65	1.10	1.75	1.03	1.85
29	1.34	1.48	1.27	1.56	1.20	1.65	1.12	1.74	1.05	1.84
30	1.35	1.49	1.28	1.57	1.21	1.65	1.14	1.74	1.07	1.83
31	1.36	1.50	1.30	1.57	1.23	1.65	1.16	1.74	1.09	1.83
32	1.37	1.50	1.31	1.57	1.24	1.65	1.18	1.73	1.11	1.82
33	1.38	1.51	1.32	1.58	1.26	1.65	1.19	1.73	1.13	1.81
34	1.39	1.51	1.33	1.58	1.27	1.65	1.21	1.73	1.15	1.81
35	1.40	1.52	1.34	1.58	1.28	1.65	1.22	1.73	1.16	1.80
36	1.41	1.52	1.35	1.59	1.29	1.65	1.24	1.73	1.18	1.80
37	1.42	1.53	1.36	1.59	1.31	1.66	1.25	1.72	1.19	1.80
38	1.43	1.54	1.37	1.59	1.32	1.66	1.26	1.72	1.21	1.79
39	1.43	1.54	1.38	1.60	1.33	1.66	1.27	1.72	1.22	1.79
40	1.44	1.54	1.39	1.60	1.34	1.66	1.29	1.72	1.23	1.79
45	1.48	1.57	1.43	1.62	1.38	1.67	1.34	1.72	1.29	1.78
50	1.50	1.59	1.46	1.63	1.42	1.67	1.38	1.72	1.34	1.77
55	1.53	1.60	1.49	1.64	1.45	1.68	1.41	1.72	1.38	1.77
60	1.55	1.62	1.51	1.65	1.48	1.69	1.44	1.73	1.41	1.77
65	1.57	1.63	1.54	1.66	1.50	1.70	1.47	1.73	1.44	1.77
70	1.58	1.64	1.55	1.67	1.52	1.70	1.49	1.74	1.46	1.77
75	1.60	1.65	1.57	1.68	1.54	1.71	1.51	1.74	1.49	1.77
80	1.61	1.66	1.59	1.69	1.56	1.72	1.53	1.74	1.51	1.77
85	1.62	1.67	1.60	1.70	1.57	1.72	1.55	1.75	1.52	1.77
90	1.63	1.68	1.61	1.70	1.59	1.73	1.57	1.75	1.54	1.78
95	1.64	1.69	1.62	1.71	1.60	1.73	1.58	1.75	1.56	1.78
100	1.65	1.69	1.63	1.72	1.61	1.74	1.59	1.76	1.57	1.78

Source: Durbin and Watson (1951)

Table A.7 Distribution of Durbin-Watson Statistic d: The 1% Significance Points of d_L and d_U (p Is the Number of Predictor Variables).

n	$p = 1$		$p = 2$		$p = 3$		$p = 4$		$p = 5$	
	d_L	d_U	d_L	d_U	d_L	d_U	d_L	d_U	d_L	d_U
15	0.81	1.07	0.70	1.25	0.59	1.46	0.49	1.70	0.39	1.96
16	0.84	1.09	0.74	1.25	0.63	1.44	0.53	1.66	0.44	1.90
17	0.87	1.10	0.77	1.25	0.67	1.43	0.57	1.63	0.48	1.85
18	0.90	1.12	0.80	1.26	0.71	1.42	0.61	1.60	0.52	1.80
19	0.93	1.13	0.83	1.26	0.74	1.41	0.65	1.58	0.56	1.77
20	0.95	1.15	0.86	1.27	0.77	1.41	0.68	1.57	0.60	1.74
21	0.97	1.16	0.89	1.27	0.80	1.41	0.72	1.55	0.63	1.71
22	1.00	1.17	0.91	1.28	0.83	1.40	0.75	1.54	0.66	1.69
23	1.02	1.19	0.94	1.29	0.86	1.40	0.77	1.53	0.70	1.67
24	1.04	1.20	0.96	1.30	0.88	1.41	0.80	1.53	0.72	1.66
25	1.05	1.21	0.98	1.30	0.90	1.41	0.83	1.52	0.75	1.65
26	1.07	1.22	1.00	1.31	0.93	1.41	0.85	1.52	0.78	1.64
27	1.09	1.23	1.02	1.32	0.95	1.41	0.88	1.51	0.81	1.63
28	1.10	1.24	1.04	1.32	0.97	1.41	0.90	1.51	0.83	1.62
29	1.12	1.25	1.05	1.33	0.99	1.42	0.92	1.51	0.85	1.61
30	1.13	1.26	1.07	1.34	1.01	1.42	0.94	1.51	0.88	1.61
31	1.15	1.27	1.08	1.34	1.02	1.42	0.96	1.51	0.90	1.60
32	1.16	1.28	1.10	1.35	1.04	1.43	0.98	1.51	0.92	1.60
33	1.17	1.29	1.11	1.36	1.05	1.43	1.00	1.51	0.94	1.59
34	1.18	1.30	1.13	1.36	1.07	1.43	1.01	1.51	0.95	1.59
35	1.19	1.31	1.14	1.37	1.08	1.44	1.03	1.51	0.97	1.59
36	1.21	1.32	1.15	1.38	1.10	1.44	1.04	1.51	0.99	1.59
37	1.22	1.32	1.16	1.38	1.11	1.45	1.06	1.51	1.00	1.59
38	1.23	1.33	1.18	1.39	1.12	1.45	1.07	1.52	1.02	1.58
39	1.24	1.34	1.19	1.39	1.14	1.45	1.09	1.52	1.03	1.58
40	1.25	1.34	1.20	1.40	1.15	1.46	1.10	1.52	1.05	1.58
45	1.29	1.38	1.24	1.42	1.20	1.48	1.16	1.53	1.11	1.58
50	1.32	1.40	1.28	1.45	1.24	1.49	1.20	1.54	1.16	1.59
55	1.36	1.43	1.32	1.47	1.28	1.51	1.25	1.55	1.21	1.59
60	1.38	1.45	1.35	1.48	1.32	1.52	1.28	1.56	1.25	1.60
65	1.41	1.47	1.38	1.50	1.35	1.53	1.31	1.57	1.28	1.61
70	1.43	1.49	1.40	1.52	1.37	1.55	1.34	1.58	1.31	1.61
75	1.45	1.50	1.42	1.53	1.39	1.56	1.37	1.59	1.34	1.62
80	1.47	1.52	1.44	1.54	1.42	1.57	1.39	1.60	1.36	1.62
85	1.48	1.53	1.46	1.55	1.43	1.58	1.41	1.60	1.39	1.63
90	1.50	1.54	1.47	1.56	1.45	1.59	1.43	1.61	1.41	1.64
95	1.51	1.55	1.49	1.57	1.47	1.60	1.45	1.62	1.42	1.64
100	1.52	1.56	1.50	1.58	1.48	1.60	1.46	1.63	1.44	1.65

Source: Durbin and Watson (1951).

References

Anscombe, F. J. (1960), "Rejection of Outliers," *Technometrics*, 2, 123–167.

Anscombe, F. J. (1973), "Graphs in Statistical Analysis," *The American Statistician*, 27, 17–21.

Atkinson, A. C. (1985), *Plots, Transformations, and Regression: An Introduction to Graphical Methods of Diagnostic Regression Analysis*, Oxford: Clarendon Press.

Barnett, V. and Lewis, T. (1994), *Outliers in Statistical Data*, 3rd ed., New York: John Wiley & Sons.

Bartlett, G., Stewart, J., and Abrahamowicz, M. (1998), "Quantitative Sensory Testing of Peripheral Nerves," *Student: A Statistical Journal for Graduate Students*, 2, 289–301.

Bates, D. M. and Watts, D. G. (1988), *Nonlinear Regression Analysis and Its Applications*, New York: John Wiley & Sons.

Becker, R. A., Cleveland, W. S., and Wilks, A. R (1987), "Dynamic Graphics for Data Analysis," *Statistical Science*, 2, 4, 355–395.

Belsley, D. A. (1991), *Conditioning Diagnostics: Collinearity and Weak Data in Regression*, New York: John Wiley & Sons.

Belsley, D. A., Kuh, E., and Welsch, R. E. (1980), *Regression Diagnostics: Identifying Influential Data and Sources of Collinearity*, New York: John Wiley & Sons.

Billor, N., Chatterjee, S., and Hadi, A. S. (1999), "A Re-Weighted Least Squares Method for Robust Regression Estimation and Outlier Detection," *Technical Report* #99-002, Department of Social Statistics, Cornell University.

Birkes, D. and Dodge, Y. (1993), *Alternative Methods of Regression*, New York: John Wiley & Sons.

Box, G. E. P. and Pierce, D. A. (1970), "Distribution of Residual Autocorrelation in Autoregressive-Integrated Moving Average Time Series Models," *Journal of the American Statistical Association*, 64, 1509–1526.

Carroll, R. J. and Ruppert, D. (1988), *Transformation and Weighting in Regression*, London: Chapman and Hall.

Chambers, J. M., Cleveland, W. S., Kleiner, B., and Tukey, P. A. (1983), *Graphical Methods for Data Analysis*, Boston: Duxbury Press.

Chatterjee, S. and Hadi, A. S. (1988), *Sensitivity Analysis in Linear Regression*, New York: John Wiley & Sons.

Chatterjee, S., Handcock, M. S., and Simonoff, J. S. (1995), *A Casebook for a First Course in Statistics and Data Analysis*, New York: John Wiley & Sons.

Chatterjee, S. and Mächler, M. (1997), "Robust Regression: A Weighted Least Squares Approach," *Communications in Statistics, Theory and Methods*, 26, 1381–1394.

Chi-Lu, C. and Van Ness, J. W. (1999), *Statistical Regression With Measurement Error*, London: Arnold.

Coakley, C. W. and Hettmansperger, T. P. (1993), "A Bounded Influence, High Breakdown, Efficient Regression Estimator," *Journal of the American Statistical Association*, 88, 872–880.

Christensen, R. (1996), *Analysis of Variance, Design and Regression: Applied Statistical Methods*, New York: Chapman and Hall.

Cochrane, D. and Orcutt, G. H. (1949), "Application of Least Squares Regression to Relationships Containing Autocorrelated Error Terms," *Journal of the American Statistical Association*, 44, 32–61.

Coleman, J. S., Cambell, E. Q., Hobson, C. J., McPartland, J., Mood, A. M., Weinfield, F. D., and York, R. L. (1966), *Equality of Educational Opportunity*, U.S. Government Printing Office, Washington, D.C.

Conover, W. J. (1980), *Practical Nonparametric Statistics*, New York: John Wiley & Sons.

Cook, R. D. (1977), "Detection of Influential Observations in Linear Regression," *Technometrics*, 19, 15–18.

Cook, R. D. and Weisberg, S. (1982), *Residuals and Influence in Regression*, London: Chapman and Hall.

Cox, D. R. (1989), *The Analysis of Binary Data*, 2nd ed., London: Methuen.

Daniel, C. and Wood, F. S. (1980), *Fitting Equations to Data: Computer Analysis of Multifactor Data*, 2nd ed., New York: John Wiley & Sons.

Dempster, A. P., Schatzoff, M. , and Wermuth, N. (1977), "A Simulation Study of Alternatives to Ordinary Least Squares," *Journal of the American Statistical Association*, 72, 77–106.

Diaconis, P. and Efron, B. (1983), "Computer Intensive Methods in Statistics," *Scientific American*, 248, 116-130.

Dodge, Y. and Hadi, A. S. (1999), "Simple Graphs and Bounds for the Elements of the Hat Matrix" *Journal of Applied Statistics*, 26, 817–823.

Draper, N. R. and Smith, H. (1998), *Applied Regression Analysis*, 3rd ed., New York: John Wiley & Sons.

Durbin, J. and Watson, G. S. (1950), "Testing for Serial Correlation in Least Squares Regression," *Biometrika*, 37, 409–428.

Durbin, J. and Watson, G. S. (1951), "Testing for Serial Correlation in Least Squares Regression, II," *Biometrika*, 38, 159–178.

Efron, B. (1982), "The Jacknife, the Bootstrap and Other Resampling Plans," *CBMS- National Science Monograph 38*, Society of Industrial and Applied Mathematics.

Ezekiel, M. (1924), "A Method for Handling Curvilinear Correlation for Any Number of Variables," *Journal of the American Statistical Association*, 19, 431–453.

Finney, D. J. (1964), *Probit Analysis*, London: Cambridge University Press.

Fox, J. (1984), *Linear Statistical Models and Related Methods*, New York: John Wiley & Sons.

Friedman, M. and Meiselman, D. (1963), "The Relative Stability of Monetary Velocity and the Investment Multiplier in the United States, 1897–1958," in *Commission on Money and Credit, Stabilization Policies*, Englewood Cliffs, N.J.: Prentice-Hall.

Fuller, W. A. (1987), *Measurement Error Models*, New York: John Wiley & Sons.

Furnival, G. M. and Wilson, R. W., Jr. (1974), " Regression by Leaps and Bounds," *Technometrics*, 16, 499–512.

Gibbons, J. D. (1993), *Nonparametric Statistics: An Introduction*, Newbury Park, CA: Sage Publications.

Goldstein, M. and Smith, A. F. M. (1974), "Ridge-Type Estimates for Regression Analysis," *Journal of the Royal Statistical Society (B)*, 36, 284–291.

Gray, J. B. (1986), "A Simple Graphic for Assessing Influence in Regression," *Journal of Statistical Computation and Simulation*, 24, 121–134.

Gray, J. B. and Ling, R. F. (1984), "K-Clustering as a Detection Tool for Influential Subsets in Regression (with Discussion)," *Technometrics*, 26, 305–330.

Graybill, F. A. (1976), *Theory and Application of the Linear Model*, Belmont, CA: Duxbury Press.

Graybill, F. A. and Iyer, H. K. (1994), *Regression Analysis: Concepts and Applications*, Belmont, CA: Duxbury Press.

Green, W. II. (1993), *Econometric Analysis*, 2nd ed., Saddle River, N.J. Prentice-Hall.

Gunst, R. F. and Mason, R. L. (1980), *Regression Analysis and Its Application: A Data-Oriented Approach*, New York: Marcel Dekker.

Hadi, A. S. (1988), "Diagnosing Collinearity-Influential Observations," *Computational Statistics and Data Analysis*, 7, 143–159.

Hadi, A. S. (1993), "Graphical Methods for Linear Models," Chapter 23 in *Handbook of Statistics: Computational Statistics*, (C. R. Rao, Ed.), Vol. 9, New York: North-Holland Publishing Company, 775–802.

Hadi, A. S. (1996), *Matrix Algebra As a Tool*, Belmont, CA: Duxbury Press.

Hadi, A. S. and Ling, R. F. (1998), "Some Cautionary Notes on the Use of Principal Components Regression," *The American Statistician*, 52, 15–19.

Hadi, A. S. and Simonoff, J. S. (1993), "Procedures for the Identification of Multiple Outliers in Linear Models," *Journal of the American Statistical Association*, 88, 1264–1272.

Hadi, A. S. and Son, M. S. (1997), "Detection of Unusual Observations in Regression and Multivariate Data," Chapter 13 in *Handbook of Applied Economic Statistics*, (A. Ullah and D. E. A. Giles, Eds.), New York: Marcel Dekker, 441–463.

Hadi, A. S. and Velleman , P. F. (1997), "Computationally Efficient Adaptive Methods for the Identification of Outliers and Homogeneous Groups in Large Data Sets," *Proceedings of the Statistical Computing Section, American Statistical Association*, 124–129.

Haith, D. A. (1976), "Land Use and Water Quality in New York Rivers," *Journal of the Environmental Engineering Division*, ASCE 102 (No. EEI. Proc. Paper 11902, Feb. 1976), 1–15.

Hamilton, D. J. (1987), "Sometimes $R^2 > r_{y \cdot x_1}^2 + r_{y \cdot x_2}^2$, Correlated Variables Are Not Always Redundant," *The American Statistician*, 41, 2, 129–132.

Hamilton, D. J. (1994), *Time Series Analysis*, Princeton, NJ: Princeton University Press.

Hampel, F. R., Ronchetti, E. M., Rousseeuw, P. J., and Stahel, W. A. (1986), *Robust Statistics: The Approach Based on Influence Functions*, New York: John Wiley & Sons.

Hand, D. J., Daly, F., Lunn, A. D., McConway, K. J., and Ostrowski, E. (1994), *A Handbook of Small Data Sets*, New York: Chapman and Hall.

Hawkins, D. M. (1980), *Identification of Outliers*, London: Chapman and Hall.

Henderson, H. V. and Velleman, P. F. (1981), "Building Multiple Regression Models Interactively," *Biometrics*, 37, 391–411.

Hildreth, C. and Lu, J. (1960), "Demand Relations With Autocorrelated Disturbances," *Technical Bulletin No. 276*, Michigan State University, Agricultural Experiment Station.

Hoaglin, D. C. and Welsch, R. E. (1978), "The Hat Matrix in Regression and ANOVA," *The American Statistician*, 32, 17–22.

Hocking, R. R., (1976), "The Analysis and Selection of Variables in Linear Regression," *Biometrics*, 32, 1–49.

Hoerl, A. E. (1959), "Optimum Solution of Many Variables," *Chemical Engineering Quart. Progr.*, 55, 69–78.

Hoerl, A. E. and Kennard, R. W. (1970), "Ridge Regression: Biased Estimation for Nonorthogonal Problems," *Technometrics*, 12, 69–82.

Hoerl, A. E. and Kennard, R. W. (1976), "Ridge Regression: Iterative Estimation of the Biasing Parameter," *Communications in Statistics, Theory and Methods*, A5, 77–88.

Hoerl, A. E., Kennard, R. W., and Baldwin, K. F. (1975), "Ridge Regression: Some Simulations," *Communications in Statistics, Theory and Methods*, 4, 105–123.

Hollander, M. and Wollfe, D. A. (1999), *Nonparametric Statistical Methods*, New York: John Wiley & Sons.

Hosmer, D. W. and Lemeshow, S. (1989), *Applied Logistic Regression*, New York: John Wiley & Sons.

Huber, P. J. (1981), *Robust Statistics*, New York: John Wiley & Sons.

Huber, P. J. (1991), "Between Robustness and Diagnostics," in *Directions in Robust Statistics and Diagnostics*, (W. Stahel and S. Weisberg, Eds.), New York: Springer-Verlag.

Iversen, G. R. (1976), *Analysis of Variance*, Beverly Hills, CA: Sage Publications.

Iversen, G. R. and Norpoth, H. (1987), *Analysis of Variance*, Beverly Hills, CA: Sage Publications.

Jerison, H. J. (1973), *Evolution of the Brain and Intelligence*, New York: Academic Press.

Johnson, D. E. (1998), *Applied Multivariate Methods for Data Analysts*, Belmont, CA: Duxbury Press.

Johnson, R. A. and Wichern, D. W. (1992), *Applied Multivariate Statistical Analysis*, 3rd ed., Englewood Cliffs, N.J.: Prentice-Hall.

Johnston, J. (1984), *Econometric Methods, 2nd ed.*, New York: McGraw-Hill.

Kmenta, J. (1986), *Elements of Econometrics*, New York: Macmillan.

Krasker, W. S. and Welsch, R. E. (1982), "Efficient Bounded-Influence Regression Estimation," *Journal of the American Statistical Association*, 77, 595–604.

Krishnaiah, P. R. (Ed.) (1980), *Analysis of Variance*, New York: North-Holland Publishing Co.

La Motte, L. R. and Hocking, R. R. (1970), "Computational Efficiency in the Selection of Regression Variables," *Technometrics*, 12, 83–93.

Landwehr, J., Pregibon, D., and Shoemaker, A. (1984), "Graphical Methods for Assessing Logistic Regression Models," *Journal of the American Statistical Association*, 79, 61–83.

Larsen, W. A., and McCleary, S. J. (1972), "The Use of Partial Residual Plots in Regression Analysis," *Technometrics*, 14, 781–790.

Lawless, J. F. and Wang, P. (1976), "A Simulation of Ridge and Other Regression Estimators," *Communications in Statistics, Theory and Methods*, A5, 307–323.

Lehmann, E. L. (1975), *Nonparametric Statistical Methods Based on Ranks*, New York: McGraw-Hill.

Lindman, H. R. (1992), *Analysis of Variance in Experimental Design*, New York: Springer Verlag.

Malinvaud, E. (1968), *Statistical Methods of Econometrics*, Chicago: Rand McNally.

Mallows, C. L. (1973), "Some Comments on C_p," *Technometrics*, 15, 661–675.

Manly, B. F. J. (1986), *Multivariate Statistical Methods*, New York: Chapman and Hall.

Mantel, N. (1970), "Why Stepdown Procedures in in Variable Selection," *Technometrics*, 12, 591–612.

Manly, B. F. J. (1986), *Multivariate Statistical Methods*, New York: Chapman and Hall.

Marquardt, D. W. (1970), "Generalized Inverses, Ridge Regression, Biased Linear Estimation and Nonlinear Estimation," *Technometrics*, 12, 591–612.

McCallum, B. T. (1970), "Artificial Orthogonalization in Regression Analysis," *Review of Economics and Statistics*, 52, 110–113.

McCullagh, P. and Nelder, J. A. (1983), *Generalized Linear Models*, London: Chapman and Hall.

McCulloch, C. E. and Meeter, D. (1983), Discussion of "Outlier ....s," by R. J. Beckman and R. D. Cook, *Technometrics*, 25, 119–163.

McDonald, G. C. and Galarneau, D. I. (1975), "A Monte Carlo Evaluation of Some Ridge Type Estimators," *Journal of the American Statistical Association*, 70, 407–416.

McDonald, G. C. and Schwing, R. C. (1973), "Instabilities of Regression Estimates Relating Air Pollution to Mortality," *Technometrics*, 15, 463–481.

McLachlan, G. J. (1992), *Discriminant Analysis and Statistical Pattern Recognition*, New York: John Wiley & Sons.

Moore, D. S. and McCabe, G. P. (1993), *Introduction to the Practice of Statistics*, New York: W. H. Freeman and Company.

Morris, C. N. and Rolph, J. E. (1981), *Introduction to Data Analysis and Statistical Inference*, Englewood Cliffs, NJ: Prentice-Hall.

Mosteller, F. and Moynihan, D. F. (Eds.) (1972), *On Equality of Educational Opportunity*, New York: Random House.

Mosteller, F. and Tukey, J. W. (1977), *Data Analysis and Regression*, Reading, MA: Addison-Wesley.

Myers, R. H. (1990), *Classical and Modern Regression with Applications*, 2nd ed., Boston: PWS-KENT Publishing Company.

Narula, S. C. and Wellington, J. F. (1977), "Prediction, Linear Regression, and the Minimum Sum of Relative Errors," *Technometrics*, 19, 2, 185–190.

Obenchain, R. L. (1975), "Ridge Analysis Following a Preliminary Test of the Shrunken Hypothesis," *Technometrics*, 17, 431–441.

Pregibon, D. (1981), "Logistic Regression Diagnostics," *The Annals of Statistics*, 9, 705–724.

Rao, C. R. (1973), *Linear Statistical Inference and Its Applications*, New York: John Wiley & Sons.

Ratkowsky, D. A. (1983), *Nonlinear Regression Modeling: A Unified Practical Approach*, New York: Marcel Dekker.

Ratkowsky, D. A. (1990), *Handbook of Nonlinear Regression Models*, New York: Marcel Dekker.

Rencher, A. C. (1995), *Methods of Multivariate Analysts*, New York: John Wiley & Sons.

Rousseeuw, P. J. and Leroy, A. M. (1987), *Robust Regression and Outlier Detection*, New York: John Wiley & Sons.

Scheffé, H. (1959), *The Analysis of Variance*, New York: John Wiley & Sons.

Searle, S. R. (1971), *Linear Models*, New York: John Wiley & Sons.

Seber, G. A. F. (1977), *Linear Regression Analysis*, New York: John Wiley & Sons.

Seber, G. A. F. (1984), *Multivariate Observations*, New York: John Wiley & Sons.

Seber, G. A. F. and Wild, C. J. (1989), *Nonlinear Regression*, New York: John Wiley & Sons.

Sen, A. and Srivastava, M. (1990), *Regression Analysis: Theory, Methods, and Applications*, New York: Springer-Verlag.

Shumway, R. H. (1988), *Applied Statistical Time Series Analysis*, Englewood Cliffs, NJ: Prentice-Hall.

Silvey, S. D. (1969), "Multicollinearity and Imprecise Estimation," *Journal of the Royal Statistical Society*, (B), 31, 539–552.

Snedecor, G. W. and Cochran, W. G. (1980), *Statistical Methods*, 7th ed., Ames, IA: Iowa State University Press.

Staudte, R. G. and Sheather, S. J. (1990), *Robust Estimation and Testing*, New York: John Wiley & Sons.

Strang, G. (1988), *Linear Algebra and Its Applications*, 3rd ed., San Diego: Harcourt Brace Jovanovich.

Thomson, A. and Randall-Maciver, R. (1905), *Ancient Races of the Thebaid*, Oxford: Oxford University Press.

Velleman, P. F. (1999), *Data Desk*, Ithaca, NY: Data Description.

Velleman, P. F. and Welsch, R. E. (1981), "Efficient Computing of Regression Diagnostics," *The American Statistician*, 35, 234–243.

Vinod, H. D. and Ullah, A. (1981), *Recent Advances in Regression Methods*, New York: Marcel Dekker.

Wahba, G., Golub, G. H., and Health, C. G. (1979), "Generalized Cross-Validation as a Method for Choosing a Good Ridge Parameter," *Technometrics*, 21, 215–223.

Welsch, R. E. and Kuh, E. (1977), "Linear Regression Diagnostics," *Technical Report* 923-77, Sloan School of Management, Cambridge, MA.

Wildt, A. R. and Ahtola, O. (1978), *Analysis of Covariance*, Beverly Hills, CA: Sago Publications.

Wood, F. S. (1973), "The Use of Individual Effects and Residuals in Fitting Equations to Data," *Technometrics*, 15, 677–695.

Index

WILEY SERIES IN PROBABILITY AND STATISTICS
ESTABLISHED BY WALTER A. SHEWHART AND SAMUEL S. WILKS

Editors
Vic Barnett, Noel A. C. Cressie, Nicholas I. Fisher,
Iain M. Johnstone, J. B. Kadane, David G. Kendall, David W. Scott,
Bernard W. Silverman, Adrian F. M. Smith, Jozef L. Teugels;
Ralph A. Bradley, Emeritus, J. Stuart Hunter, Emeritus

Probability and Statistics Section

*ANDERSON · The Statistical Analysis of Time Series
ARNOLD, BALAKRISHNAN, and NAGARAJA · A First Course in Order Statistics
ARNOLD, BALAKRISHNAN, and NAGARAJA · Records
BACCELLI, COHEN, OLSDER, and QUADRAT · Synchronization and Linearity:
    An Algebra for Discrete Event Systems
BASILEVSKY · Statistical Factor Analysis and Related Methods: Theory and
    Applications
BERNARDO and SMITH · Bayesian Statistical Concepts and Theory
BILLINGSLEY · Convergence of Probability Measures, *Second Edition*
BOROVKOV · Asymptotic Methods in Queuing Theory
BOROVKOV · Ergodicity and Stability of Stochastic Processes
BRANDT, FRANKEN, and LISEK · Stationary Stochastic Models
CAINES · Linear Stochastic Systems
CAIROLI and DALANG · Sequential Stochastic Optimization
CONSTANTINE · Combinatorial Theory and Statistical Design
COOK · Regression Graphics
COVER and THOMAS · Elements of Information Theory
CSÖRGÖ and HORVÁTH · Weighted Approximations in Probability Statistics
CSÖRGÖ and HORVÁTH · Limit Theorems in Change Point Analysis
DETTE and STUDDEN · The Theory of Canonical Moments with Applications in
    Statistics, Probability, and Analysis
DEY and MUKERJEE · Fractional Factorial Plans
*DOOB · Stochastic Processes
DRYDEN and MARDIA · Statistical Analysis of Shape
DUPUIS and ELLIS · A Weak Convergence Approach to the Theory of Large Deviations
ETHIER and KURTZ · Markov Processes: Characterization and Convergence
FELLER · An Introduction to Probability Theory and Its Applications, Volume 1,
    Third Edition, Revised; Volume II, *Second Edition*
FULLER · Introduction to Statistical Time Series, *Second Edition*
FULLER · Measurement Error Models
GHOSH, MUKHOPADHYAY, and SEN · Sequential Estimation
GIFI · Nonlinear Multivariate Analysis
GUTTORP · Statistical Inference for Branching Processes
HALL · Introduction to the Theory of Coverage Processes
HAMPEL · Robust Statistics: The Approach Based on Influence Functions
HANNAN and DEISTLER · The Statistical Theory of Linear Systems
HUBER · Robust Statistics
IMAN and CONOVER · A Modern Approach to Statistics
JUREK and MASON · Operator-Limit Distributions in Probability Theory
KASS and VOS · Geometrical Foundations of Asymptotic Inference

*Now available in a lower priced paperback edition in the Wiley Classics Library.

Applied Probability and Statistics Section

*Now available in a lower priced paperback edition in the Wiley Classics Library.

*Now available in a lower priced paperback edition in the Wiley Classics Library.

*Now available in a lower priced paperback edition in the Wiley Classics Library.

*Now available in a lower priced paperback edition in the Wiley Classics Library.

Applied Probability and Statistics (Continued)

VIDAKOVIC · Statistical Modeling by Wavelets

WEISBERG · Applied Linear Regression, *Second Edition*

WESTFALL and YOUNG · Resampling-Based Multiple Testing: Examples and Methods for *p*-Value Adjustment

WHITTLE · Systems in Stochastic Equilibrium

WOODING · Planning Pharmaceutical Clinical Trials: Basic Statistical Principles

WOOLSON · Statistical Methods for the Analysis of Biomedical Data

*ZELLNER · An Introduction to Bayesian Inference in Econometrics

Texts and References Section

AGRESTI · An Introduction to Categorical Data Analysis

ANDERSON · An Introduction to Multivariate Statistical Analysis, *Second Edition*

ANDERSON and LOYNES · The Teaching of Practical Statistics

ARMITAGE and COLTON · Encyclopedia of Biostatistics: Volumes 1 to 6 with Index

BARTOSZYNSKI and NIEWIADOMSKA-BUGAJ · Probability and Statistical Inference

BERRY, CHALONER, and GEWEKE · Bayesian Analysis in Statistics and Econometrics: Essays in Honor of Arnold Zellner

BHATTACHARYA and JOHNSON · Statistical Concepts and Methods

BILLINGSLEY · Probability and Measure, *Second Edition*

BOX · R. A. Fisher, the Life of a Scientist

BOX, HUNTER, and HUNTER · Statistics for Experimenters: An Introduction to Design, Data Analysis, and Model Building

BOX and LUCEÑO · Statistical Control by Monitoring and Feedback Adjustment

BROWN and HOLLANDER · Statistics: A Biomedical Introduction

CHATTERJEE and PRICE · Regression Analysis by Example, *Third Edition*

COOK and WEISBERG · Applied Regression Including Computing and Graphics

COOK and WEISBERG · An Introduction to Regression Graphics

COX · A Handbook of Introductory Statistical Methods

DILLON and GOLDSTEIN · Multivariate Analysis: Methods and Applications

DODGE and ROMIG · Sampling Inspection Tables, *Second Edition*

DRAPER and SMITH · Applied Regression Analysis, *Third Edition*

DUDEWICZ and MISHRA · Modern Mathematical Statistics

DUNN · Basic Statistics: A Primer for the Biomedical Sciences, *Second Edition*

FISHER and VAN BELLE · Biostatistics: A Methodology for the Health Sciences

FREEMAN and SMITH · Aspects of Uncertainty: A Tribute to D. V. Lindley

GROSS and HARRIS · Fundamentals of Queueing Theory, *Third Edition*

HALD · A History of Probability and Statistics and their Applications Before 1750

HALD · A History of Mathematical Statistics from 1750 to 1930

HELLER · MACSYMA for Statisticians

HOEL · Introduction to Mathematical Statistics, *Fifth Edition*

HOLLANDER and WOLFE · Nonparametric Statistical Methods, *Second Edition*

HOSMER and LEMESHOW · Applied Survival Analysis: Regression Modeling of Time to Event Data

JOHNSON and BALAKRISHNAN · Advances in the Theory and Practice of Statistics: A Volume in Honor of Samuel Kotz

JOHNSON and KOTZ (editors) · Leading Personalities in Statistical Sciences: From the Seventeenth Century to the Present

JUDGE, GRIFFITHS, HILL, LÜTKEPOHL, and LEE · The Theory and Practice of Econometrics, *Second Edition*

KHURI · Advanced Calculus with Applications in Statistics

KOTZ and JOHNSON (editors) · Encyclopedia of Statistical Sciences: Volumes 1 to 9 wtih Index

*Now available in a lower priced paperback edition in the Wiley Classics Library.

WILEY SERIES IN PROBABILITY AND STATISTICS
ESTABLISHED BY WALTER A. SHEWHART AND SAMUEL S. WILKS

Editors
*Robert M. Groves, Graham Kalton, J. N. K. Rao, Norbert Schwarz,
Christopher Skinner*

Survey Methodology Section

*Now available in a lower priced paperback edition in the Wiley Classics Library.

*Now available in a lower priced paperback edition in the Wiley Classics Library.